Rotation and Momentum Transport in Magnetized Plasmas

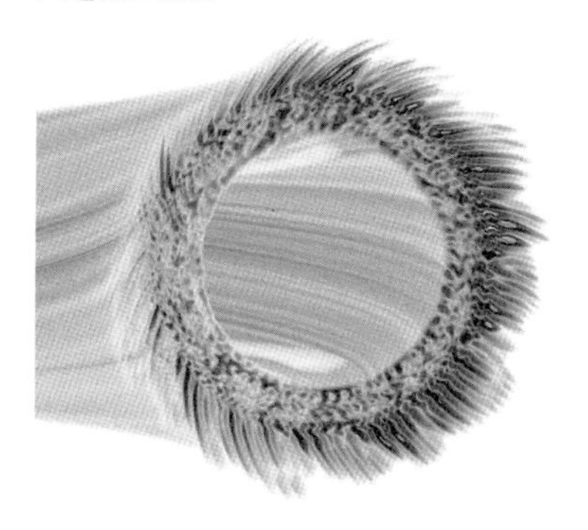

Reviews of the Theory of Magnetized Plasmas

Series Editors: Patrick H. Diamond *(Univ. of California, San Diego, USA)*
Xavier Garbet *(CEA, IRFM, France)*
Yanick Sarazin *(CEA, IRF, France)*

Vol. 1 Relaxation Dynamics in Laboratory and Astrophysical Plasmas
edited by Patrick H. Diamond, Xavier Garbet, Philippe Ghendrih and Yanick Sarazin

Vol. 2 Rotation and Momentum Transport in Magnetized Plasmas
edited by Patrick H. Diamond, Xavier Garbet, Philippe Ghendrih and Yanick Sarazin

Reviews of the Theory of Magnetized Plasmas Volume 2

Rotation and Momentum Transport in Magnetized Plasmas

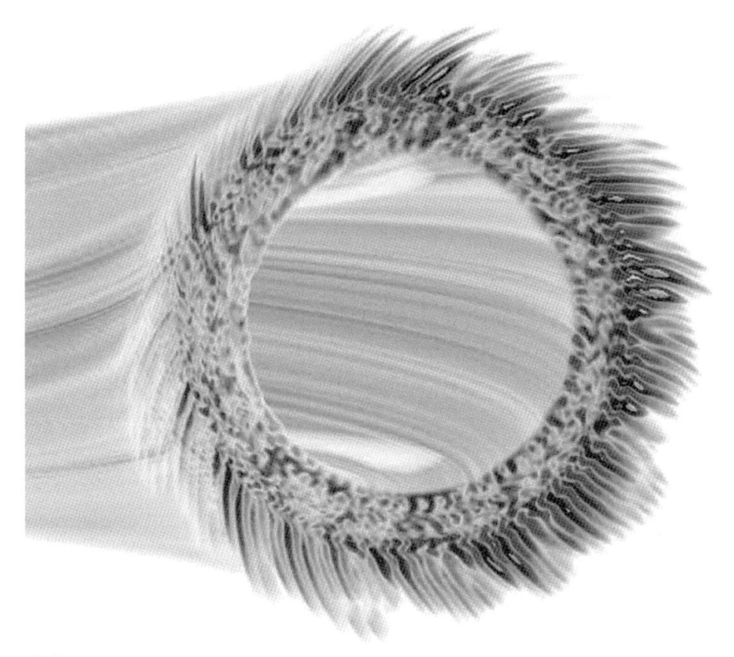

Editors

Patrick H. Diamond

University of California at San Diego, USA

Xavier Garbet, Philippe Ghendrih
& Yanick Sarazin

CEA, IRFM, France

NEW JERSEY • LONDON • SINGAPORE • BEIJING • SHANGHAI • HONG KONG • TAIPEI • CHENNAI

Published by

World Scientific Publishing Co. Pte. Ltd.
5 Toh Tuck Link, Singapore 596224
USA office: 27 Warren Street, Suite 401-402, Hackensack, NJ 07601
UK office: 57 Shelton Street, Covent Garden, London WC2H 9HE

Library of Congress Cataloging-in-Publication Data
Rotation and momentum transport in magnetized plasmas / editors, Philippe Ghendrih, CEA, IRFM, France, Xavier Garbet, CEA, IRFM, France, Yanick Sarazin, CEA, IRFM, France, Patrick H Diamond, UC San Diego.
pages cm. -- (Reviews of the theory of magnetized plasmas ; volume 2)
Includes bibliographical references and index.
ISBN 978-9814644822 (hardcover : alk. paper) -- ISBN 981464482X (hardcover : alk. paper)
1. Plasma (Ionized gases) 2. Plasma dynamics. 3. Electromagnetic waves. I. Ghendrih, Philippe, editor. II. Garbet, X. (Xavier), editor. III. Sarazin, Yanick, editor. IV. Diamond, Patrick H., editor.
QC718.R68 2015
530.13'8--dc23

2014042305

British Library Cataloguing-in-Publication Data
A catalogue record for this book is available from the British Library.

Printed in Singapore

Dedicated to our friend and colleague

Annick Pouquet

in support of all her scientific projects

"On ne fait jamais attention à ce qui a été fait ; on ne voit que ce qui reste à faire"

Marie Curie

Foreword

Annick Pouquet[1,2,3]

[1] *Visiting Research Scientist,*
Laboratory for Atmospheric and Space Science,
Colorado University, Boulder CO 80304, USA
[2] *Adjunct Professor,*
Department of Applied Mathematics,
Colorado University, Boulder CO 80304, USA
[3] *Senior Scientist, Emeritus,*
National Center for Atmospheric Research,
Boulder, CO 80307-3000, USA

Turbulence is a tough subject, and thus some may think or say that it is not even worth trying. But the progress that has been made over the years, thanks in part to the fantastic increase in the power of observational, experimental and numerical tools, together with computers and theoretical modeling, mean that we are not where we were when I started in this job in 1971. There is in fact a renewal of the topic, that gains from cross-disciplinary research, and the Festival de Théorie in Aix-en-Provence, is one place, thanks to the dedication of its organizers, that builds on such cross-learning. The mold of homogeneous isotropic turbulence is being broken, adding complexities such as anisotropy and waves, and surprises abound.

Did you know that the large-scale velocity can be substantially more intermittent in a range of parameters in stratified turbulence, the internal gravity waves enhancing the production of negative gradients? Of course, this has been known, observationally, for a while as observed in the very stable nocturnal planetary boundary layer, but writing a simple model for this phenomenon may help modeling it in weather and climate systems[1].

Did you know that the cascade of energy can be both to the large scales and the small scales, in both cases with a constant flux, in rotating stratified Boussinesq turbulence? This resolves in part a vexing paradox of geophysical fluid dynamics for the atmosphere and the ocean, dominated at large scale by geostrophic balance between pressure gradient, Coriolis and gravity, and yet also displaying enhanced transport through wave breaking and frontogenesis[2].

These are the few topics we are working on right now, having fun and at the same time discovering perhaps not a new paradigm but certainly a broadened vision of turbulent flows.

It is easy to guess what this last paragraph is about. It is in time of crisis that you find your friends, sometimes in unexpected corners. I am deeply thankful to all those who have walked with me during these wonderful years, on both sides of the Atlantic and beyond, mentors, colleagues, past and present students and post-doctoral fellows, friends, family. From chagrin to anger and then to projecting myself into a possible future, please know that you have helped through your support.

I was a student in Aix for many years before going to university in Nice, I have walked these streets extensively, smelled the air, enjoyed the platanes, the art and the music, and this town is dear to my heart. Thank you all.

[1]Cecilia Rorai, Pablo Mininni and Annick Pouquet, "Turbulence comes in bursts in stably stratified flows," submitted to Phys. Rev. E Rapid Comm., 2013; see also arXiv: 1308:6564

[2]Raffaele Marino, Pablo Mininni, Duane Rosenberg and Annick Pouquet, "Inverse cascades in rotating stratified turbulence: fast growth of large scales," EuroPhys. Lett. **102**, 44006, 2013,

Annick Pouquet and Raffaele Marino, "Geophysical turbulence and the duality of the energy flow across scales," Phys. Rev. Lett. **111**, 234501, 2013.

Preface

The material of this book was presented during the "Festival de Théorie", held in Aix-en-Provence (France) in July, 2009 focussed on **Rotation and Momentum Transport in Magnetised Plasmas**. The contributions were written shortly after this fifth venue of the "Festival de Théorie".

This Festival de Théorie was initiated in 2000 by Jean Jacquinot, and at that time René Pellat chaired the Director Committee. From the very beginning, the Festival de Théorie was also supported by Professor Marshall Rosenbluth and so quite naturally the opening lecture of the "Festival de Théorie" is the Marshall Rosenbluth Memorial Lecture.

The present book is dedicated to Annick Pouquet, our colleague and friend, who is now looking ahead to continue a demanding and challenging research activity on MHD turbulence. Annick joined CNRS, the French Centre National de la Recherche Scientifique, in 1974 and defended her PhD in Nice in 1976. With the distinction of Directeur de Recherche de Première Classe, she left CNRS in 2000 to join NCAR, National Center for Atmospheric Research, as Director of the Geophysical Turbulence Program. As she states it, the goal of her Research Projects is understanding turbulence: *Understanding turbulence remains a major stumbling block of classical physics: a large number of spatial and temporal modes are coupled leading to power-law spectra, exponential wings of pdf, long-lived coherent structures and intermittency, all of which are observed in geophysical and astrophysical fluids and plasmas. The solar and planetary environments (photosphere, corona and the Solar Wind, the magnetosphere of Jupiter or the liquid core of the Earth as the locus of the dynamo phenomenon) are specific applications of my research. The challenge is to put together a team that develops powerful new tools — harnessing the increase in power of both computer technology and algorithms — and in parallel contribute to theoretical research in turbulence and build phenomenological and numerical models, and follow this path all the way to its consequences for geophysical and astrophysical flows.* Such a research program is very closely related to that of the Festival de Théorie. We very readily express our support to Annick and her scientific projects by dedicating this book to her.

The Festival de Théorie takes place every two years in Aix-en-Provence and began in 2001. Its scientific committee is chaired by Professor Patrick Diamond and co-chaired by Doctor Xavier Garbet. This international meeting during three weeks gathers senior experts on well-focused subjects in magnetised plasma physics, both in controlled fusion and in astrophysics. The objective is threefold. First, the participants present the most recent results in the field and promote new ideas. Secondly sub-groups are formed and start new common work on the subject. The third objective is to provide educational training for PhD and post-doc students. So-called Student Lectures have been organised for them. These lectures serve as introductions to the tutorial presentations of the Festival de Théorie. The presentations selected for this book stem from the tutorial and lecture presentations. Students have been asked to work on the notes and this has led them to participate at various degrees to the present book. This foreword is the perfect occasion to thank both lecturers and students for their effort.

The backbone of this book is the various facets of "Rotation and Momentum Transport in complex systems", including magnetised plasmas of course, but also atmosphere–ocean turbulence and its disparate scale interaction as addressed in *The atmospheric wave–turbulence jigsaw*, Chapter 1, which was presented for the **Marshall Rosenbluth Memorial Lecture** by **Michael McIntyre**. The important concept of potential vorticity, or PV, is shown to provide an efficient tool for understanding jets and eddy transport barriers in the atmosphere. In Chapter 2, **Annick Pouquet** presents *A review of the possible role of constraints in MHD turbulence* and therefore elements that can contribute to the self-organization of the turbulent energy spectra. This chapter emphasises the interaction between modes with neighbouring wave vectors. This approach is somewhat in contrast to that presented by Michael McIntyre. Chapter 3, based on the tutorial and lecture given by **Patrick Diamond**, *Dynamics of Structures in Configuration Space and Phase Space: An Introductory Tutorial*, is another illustration of the concept of potential vorticity and is therefore closely related to Chapter 1 by Michael McIntyre. It addresses the reorganisation of the distribution function by phase space transport and relaxation driven by phase space structures such as clumps and holes. This approach advances beyond traditional mean field quasi-linear theory. This chapter examines the PV concept from a plasmas perspective, and so is complementary to that of Chapter 1. The topic of *Fast Dynamos*, reviewed by **David Hughes** in Chapter 4, is one of considerable astrophysical interest, addressing the idea of magnetohydrodynamic dynamo action in the asymptotic limit of infinite magnetic Reynolds number. Fast dynamo action can occur only in chaotic fluid flows and thus there are close links, brought out by so-called dynamo maps, to the mathematical ideas of chaos in nonlinear dynamical systems. In Chapter 5, *The Effect of Flow on Ideal Magnetohydrodynamic Ballooning Instabilities: a tutorial*, by **Howard Wilson** and colleagues, the symmetry breaking introduced by the large scale flows in the ballooning theory is presented, starting with a pedagogical introduction to the standard ballooning theory itself, before discussing the impact of the large scale flows. This chapter is centered on Tokamak physics and in particular to MHD relaxation events at the plasma edge, the ELMs, a region where plasma rotation is expected to play a role. Chapter 6 is used to gather all figures that require a color version, figure captions are recalled together with references to the grayscale figures in the corresponding chapters. A tutorial contribution on *Elements of Neoclassical Theory and Plasma Rotation in a Tokamak* by **Andrei Smolyakov**, is given in Chapter 7. The transport process between neighboring flux surfaces governed by collisions, usually referred to as neoclassical transport, is also an important piece of tokamak theory. The symmetry, namely the ambipolar degeneracy, introduced by the quasineutrality asymptotic limit is examined and discussed. *The general fishbone like dispersion relation*, Chapter 8 written by **Fulvio Zonca**, is dedicated to the interplay between fast-particles and MHD modes. This is a key issue in burning plasmas and an important physics to be addressed in the tokamak ITER when completed. Finally, Chapter 9, by **Dominique Escande** closes this book and presents an alternative configuration to the tokamak, the Reversed Field Pinch. The operation of existing RFP facilities in a single helicity, is presented, backed by simulations and theory, and discussed as a possible alternative to the tokamak configuration.

This book is suited for physicists and advanced students in the field, as well as researchers in other fields who would like to broaden their perspective on these problems. An Index covering all chapters should ease the use of the book. In this Index, bold font entries indicate Chapter and Section titles (including also all levels of sub-Section titles).

Acknowledgments

This book would not be without the full support of the Institut de Recherche sur la Fusion par confinement Magnétique, IRFM, that hosts the research effort on fusion plasma physics within CEA (Commissariat à l'énergie atomique). IRFM is located in Cadarache and belongs to the Direction des Sciences de la matière which is the CEA branch focussed on Physics. We also acknowledge support from the Région Provence Alpes Côte d'Azur.

Contents

Foreword vii

Preface ix

Acknowledgments xiii

1. The Atmospheric Wave–Turbulence Jigsaw 1

Michael E. McIntyre

1.1 Introduction . . . 1
1.2 On eddy-transport barriers . . . 3
1.3 The generic dynamics, also called PV dynamics . . . 5
1.4 Cyclone and jet structure . . . 9
1.5 Strong nonlinearity is ubiquitous . . . 12
1.6 Rossby waves and drift waves . . . 15
1.7 The PV Phillips effect . . . 17
1.8 Some simple inversion operators . . . 19
1.9 Pseudomomentum and the Taylor identity . . . 24
1.10 PV staircases . . . 29
1.11 Jet self-sharpening and meandering: a toy model . . . 33
1.12 Concluding remarks . . . 36
References . . . 38

2. A Review of the Possible Role of Constraints in MHD Turbulence 45

Annick Pouquet

2.1 Introduction . . . 45

2.1.1 The context 45
2.1.2 The Kolmogorov law 47
2.1.3 The equations 49
2.2 What can we learn from statistical mechanics 51
2.3 Energy spectra 57
2.3.1 Weak MHD turbulence 57
2.3.2 When the weak regime breaks down 59
2.3.3 Intermittency 61
2.4 Exact laws 62
2.5 Explicit examples of non-universal behavior in MHD . . . 64
2.6 The role of alignment between the velocity and the magnetic field 66
2.7 Can the dynamics of magnetic helicity play a role for energy scaling? 69
2.8 Can models help? 70
2.9 Conclusions 72
References 73

3. Dynamics of Structures in Configuration Space and Phase Space: A Tutorial 81

P. H. Diamond, Y. Kosuga, M. Lesur

3.1 Introduction and basic considerations 82
3.2 Rayleigh criterion, potential vorticity, and zonal flow momentum 85
3.2.1 Something old: modal derivation of Rayleigh criterion 85
3.2.2 Something newer: G. I. Taylor's stability criterion derived from PV dynamics 87
3.2.3 Something further: zonal flow evolution and wave-flow interaction 89
3.3 Dynamics of phase space structures 93
3.3.1 Basic concepts 93
3.3.2 Local structure growth 95
3.4 Phase space structure dynamics in drift turbulence 101
3.4.1 Introducing the Darmet model 101
3.4.2 Phase space structures and zonal flows 104
3.4.3 Drift-hole growth 107

3.5 Dupree-Lenard-Balescu theory for mean evolution 109
3.6 Conclusion 111
References 111

4. Fast Dynamos 115

D. W. Hughes

4.1 Introduction 115
4.2 Perfectly conducting fluids 118
4.3 Map dynamos 120
4.3.1 Fractal dimension 121
4.3.2 Cancellation 122
4.3.3 An expression for the fast dynamo growth rate . . 123
4.4 A necessary condition for fast dynamo action 125
4.5 Fast dynamo flows 126
4.5.1 An 'almost fast' dynamo 126
4.5.2 Three-dimensional flows 128
4.5.3 Unsteady two-dimensional flows 128
4.5.4 Solving the Cauchy problem 131
4.6 Nonlinear fast dynamos 131
4.7 Future challenges 133
References 134

5. The Effect of Flow on Ideal Magnetohydrodynamic Ballooning Instabilities 137

H. R. Wilson, P. B. Buxton, J. W. Connor

5.1 Introduction 137
5.2 Geometry 138
5.3 Ballooning mode physics 140
5.4 Ballooning theory 144
5.5 Ballooning modes: the effect of flow 152
5.6 Conclusions 162
References 163

6. Color Figures 165

7. Elements of Neoclassical Theory and Plasma Rotation in a Tokamak 173

A. Smolyakov

7.1 Introduction 173
7.1.1 Quasineutrality condition 174
7.1.2 Diffusion in fully ionized magnetized plasma and automatic ambipolarity 176
7.2 Toroidal geometry and neoclassical diffusion 178
7.3 Diffusion and ambipolarity in toroidal plasmas 180
7.4 Ambipolarity and equilibrium poloidal rotation 181
7.5 Ambipolarity paradox and damping of poloidal rotation . 183
7.6 Neoclassical plasma inertia 186
7.7 Oscillatory modes of poloidal plasma rotation 188
7.8 Dynamics of the toroidal momentum 191
7.8.1 Momentum diffusion in strongly collisional, short mean free path regime 191
7.8.2 Diffusion of toroidal momentum in the weak collision (banana) regime 193
7.8.3 Toroidal momentum diffusion and momentum damping from drift-kinetic theory and fluid moment equations 194
7.8.4 Comments on non-axisymmetric effects 197
7.9 Summary 198
7.10 Appendix: Trapped (banana) particles and collisionality regimes in a tokamak 199
7.11 Appendix: Hierarchy of moment equations 202
7.12 Appendix: Plasma viscosity tensor in the magnetic field: parallel viscosity, gyroviscosity, and perpendicular viscosity 205
7.13 Appendix: Closure relations for the flux surface averaged parallel viscosity in neoclassical (banana and plateau) regimes 210
References 211

8. The General Fishbone Like Dispersion Relation 219

F. Zonca

8.1 Introduction 219

8.2 Motivation and outline 219
8.3 Fundamental equations 220
8.3.1 The collisionless gyrokinetic equation 221
8.3.2 Vorticity equation 222
8.3.3 Quasi-neutrality condition 225
8.3.4 Perpendicular Ampère's law 225
8.4 Studying collective modes in burning plasmas 226
8.4.1 Ideal plasma equilibrium in the low-β limit 226
8.4.2 Approximations for the energetic population . . . 227
8.4.3 Characteristic frequencies of particle motions . . . 227
8.4.4 Alfvén wave frequency and wavelength orderings . 228
8.4.5 Applications of the general theoretical framework 229
8.5 The general fishbone like dispersion relation 232
8.5.1 Properties of the fishbone like dispersion relation 232
8.5.2 Derivation of the fishbone like dispersion relation 235
8.6 Special cases of the fishbone like dispersion relation 239
8.6.1 Toroidal Alfvén Eigenmodes (TAE) 239
8.6.2 Alfvén Cascades 240
8.7 Summary and discussions 241
References 242

9. What is a Reversed Field Pinch 247

D. F. Escande

9.1 Introduction 247
9.2 Short description 248
9.3 Usefulness of the RFP configuration for fusion science and dynamo physics 251
9.4 Attractivity of the RFP configuration for a reactor 252
9.5 Challenges ahead 254
9.6 Lawson criterion 255
9.7 Intuitive model of magnetic self-reversal 256
9.8 Intuitive description of the dynamo 257
9.9 Necessity of a helical deformation 259
9.10 MHD simulations 260
9.10.1 From single to multiple helicity 260
9.10.2 Single helicity 262
9.10.3 Multiple helicity 264
9.10.4 Quasi single helicity 266

9.11 Experimental results 266
9.11.1 Multiple helicity 267
9.11.2 Quasi single helicity 267
9.11.3 Upgrade of the RFX device 268
9.11.4 From double to single magnetic axis 269
9.12 Analytical description of the single helicity RFP 271
9.12.1 Helical Grad-Shafranov equation 271
9.12.2 Parallel Ohm's law 272
9.12.3 Pinch-stellarator equation 274
9.12.4 Single helicity ohmic RFP states 275
9.12.5 Calculation of the dynamo 278
9.13 Conclusion 280
References 282

Index 287

Chapter 1

The Atmospheric Wave–Turbulence Jigsaw

Michael E. McIntyre
Department of Applied Mathematics & Theoretical Physics, University of Cambridge, UK
www.atm.damtp.cam.ac.uk/people/mem

1.1 Introduction

It was a huge honour to be asked to give the Marshall Rosenbluth Memorial Lecture. Having never worked on plasma physics, though, I also feel some diffidence! The closest I've ever come has been involvement in some peculiar MHD problems that promise an improved understanding of the solar tachocline — more about confining a magnetic field within a plasma than a plasma within a magnetic field. Before proceeding I want to thank Drs Laurène Jouve and Chris McDevitt for producing the first draft of this chapter following my lecture. Chris also kindly lent assistance with the source files and graphics. However, the final responsibility for this chapter and any errors it may contain is mine alone.

What I do know about is the kind of fluid dynamics that has helped us to understand the Earth's atmosphere and oceans. We still have an enormous phase space, albeit with fewer degrees of freedom than for plasma physics. Thanks to countless observations and to the peculiarities of flow heavily constrained by Coriolis effects and stable density stratification, great progress has been made in penetrating the nonlinear dynamics. Indeed, and quite surprisingly, researchers into atmosphere–ocean dynamics have gained insight in a way that cuts straight to strong nonlinearity, avoiding the standard paradigms. And an even greater surprise, to me at least, has been what Pat Diamond, Paul Terry and others have been saying in recent years, namely that some insights from atmosphere–ocean dynamics are relevant to some aspects of plasma behaviour in tokamaks and stellarators — henceforth "tokamaks" for brevity — especially the self-organizing zonal or quasi-zonal flows that appear so important for plasma heat confinement. See, e.g., [1], [2], [3] and references therein.

The standard paradigms avoided include those of weak nonlinearity, strong scale separation, "cascades" in the strict sense of being local in wavenumber space, and indeed homogeneous turbulence theory in all its flavours. Ever since the 1980s when infrared remote sensing from space began to give us global-scale views of, especially, stratospheric fluid flow, it has become apparent that we are dealing with a highly inhomogeneous "wave–turbulence jigsaw puzzle" with no scale separation but with weakly and strongly nonlinear regions closely adjacent and intimately interdependent [4], [5]. One of the tools that have helped us to make sense of this has been the finite-amplitude "wave–mean interaction theory" developed over many years and recently summarized in a beautiful new book by my colleague Oliver Bühler [6].

For instance a typical phenomenon, once completely mysterious but now well understood, and understood in a very simple way, is the self-organization and the peculiar persistence and quasi-elasticity of the great atmosphere–ocean jet streams. They can persist over surprisingly large distances. If ordinary, domestic-scale jets behaved similarly, you could blow out your birthday candles from the far end of the room. There are "anti-frictional" effects that prevent the great jets from spreading out dissipatively, tending to re-sharpen their velocity profiles if something smears them out. As recorded in the famous books by Edward N. Lorenz, the father of chaos theory, and Victor P. Starr, the pioneer of postwar global upper-air data analysis, this behaviour used to be called "negative viscosity" and regarded as a profound enigma [7], [8]. Such was the state of things when I became Jule G. Charney's postdoc at MIT in the late 1960s.

The most conspicuous jets include the Gulf Stream, the Kuroshio Current, and the atmospheric jet streams that are typically found at airliner cruise altitudes, fast-flowing rivers of air a a few kilometres deep and a few hundred kilometres wide, roughly speaking. As airline operators know very well, these atmospheric jet streams — which are among the most comprehensively observed of natural phenomena — can persist for thousands of kilometres and can have wind speeds sometimes exceeding even 100 ms^{-1} or 200 knot. The cores of these thin jets may meander, river-like, with large amplitudes, but nevertheless form resilient, flexible barriers tending to inhibit the turbulent transport of material across them — by contrast with the high-speed advective transport along them — making these jets almost the "veins and arteries of the climate system". Equally spectacular, though far less well understood, are the prograde jets and associated transport barriers in the visible weather layer of the planet Jupiter.

I find myself wondering whether tokamak zonal flows are closer to the terrestrial or Jovian cases. Our relatively poor understanding of Jupiter makes this a hard question to answer at present. For Jupiter there are relatively few observational constraints, apart from the wind fields derived from cloud-top motions and the peculiar straightness of the prograde jets that makes them, as it were, so conspicuously unearthly. It might be good news if the zonal jets in big tokamaks were more Jupiter-like than Earth-like, because less meandering might mean better confinement.

It cannot be too strongly emphasized that, for all the prolific literature, our understanding of the Jovian problem is indeed in its infancy. For one thing, progress has been impeded by a tendency to forget that in the real planet, as distinct from many models of it that have been studied, there is no solid surface or phase change sufficiently near the visible surface to support the type of baroclinic instabilities that excite terrestrial atmospheric jets. There is a "convective thermostat" mechanism — see [9] and references therein, also footnote 4 of [10] — that pretty much precludes any major role for such baroclinic instabilities. And our understanding of the most basic aspect of all, namely the coupling between Jupiter's weather layer and the underlying deep-convection layer, is very poor indeed. Here I think there are some interesting paradigm changes in progress. But let me come back to Earth and to things we know much more about.

1.2 On eddy-transport barriers

The turbulent transport inhibition or "eddy-transport barrier" effect at jet cores goes hand-in-hand with the anti-frictional effects. It depends on the strong, self-maintaining horizontal shears adjacent to the jet core as well as on the *potential-vorticity gradients* — see below — that are concentrated at the jet core. The importance of shear was pointed out in [11]. So although we used to speak of "potential-vorticity barriers", the tendency in the atmosphere–ocean community these days is to call them "eddy-transport barriers". Their dynamics involves both wave propagation and turbulence — strongly nonlinear and strongly inhomogeneous spatially, with no spatial scale separation. As already hinted, this is far beyond the reach both of homogeneous turbulence theory and of weakly-nonlinear wave or weak-turbulence theory, also called "wave-turbulence" theory.

The eddy-transport-barrier effect has been demonstrated again and again from observations, from laboratory experiments, and from high-resolution numerical models. An early and very striking observational demonstration came from studies of the radioactive debris from atmospheric nuclear tests published in 1968. Using an instrumented aircraft, two material air masses well characterized by differing radioactive properties were observed flowing side by side in close proximity, without mixing, on either side of a jet core [12].

Another striking demonstration came from a laboratory experiment in Harry Swinney's big rotating tank at the University of Texas at Austin [13], [14]; see Fig. 1.1. Dye injected on one side of a jet stayed there, after more than 500 tank revolutions — almost perfectly confined despite the meandering of the jet. The jet core and velocity maximum were found to be almost coincident with the dye boundary.

And again, there has been a huge amount of observational and numerical modelling work in connection with the concern over the ozone hole in the Antarctic stratosphere, where the polar-night jet, or polar-vortex edge, keeps itself sharp and acts as an eddy-transport barrier within which the ozone chemistry proceeds differently from the chemistry outside. This has been studied using intensive observations and a large hierarchy of models for over two decades now; among the many landmarks we may note a remarkable pair of papers by Norton [15] and Waugh and Plumb [16]. My website has a movie from Norton's work that visualizes the barrier effect rather spectacularly — websearch `"dynamics that is significant for chemistry"`.

One might ask why there should be any comparison between the above-mentioned laboratory experiment and the atmosphere. Admittedly both are rapidly-rotating systems, in the sense that Coriolis effects are strong. However, stable stratification is very important in the atmosphere, and in the oceanic examples too, whereas the laboratory experiment used an unstratified fluid. The answer is, I've always thought, a rather surprising one. The kind of dynamics involved in all these systems has the same generic structure, explaining the many qualitative similarities between the systems. Even more surprisingly, the same generic structure is found in the tokamak models of Hasegawa, Mima and Wakatani. As we'll see, it is well illustrated by the simplest such model defined by the Hasegawa–Mima equation

Fig. 1.1 From the laboratory study of Sommeria et al. [13]. For a colour version see Fig. 6.1. Courtesy Dr Joël Sommeria.

(Sec. 1.8 below). And it is this structure that allows the insight into strong nonlinearity.

1.3 The generic dynamics, also called PV dynamics

The generic dynamics is shared by a whole hierarchy of models of stratified, rotating atmosphere–ocean dynamics and their unstratified laboratory counterparts, including some remarkably accurate stratified models that easily explain, for instance, the characteristic cross-sectional structure of jet streams that has long been familiar to observational meteorologists [17], [18]; see Fig. 1.3 below. In all these models one has a single scalar field $Q(\mathbf{x}, t)$ that is a *material invariant* for ideal fluid flow — that is, Q materially conserved, i.e. constant on each material particle — expressing the advective nonlinearity in a way that is easy to understand and to visualize. And more than that, the evolution of the Q field captures *everything* about the advective nonlinearity because, to the extent that these models are accurate, the Q field contains all the dynamical information at each instant. So if for example $\mathbf{u}(\mathbf{x}, t)$ is the velocity field, which for definiteness we'll take relative to the rotating Earth, then $\mathbf{u}$ can be deduced at each instant from the Q field alone.

Q, to be defined shortly, is called the *potential vorticity* (PV) of the model, and the mathematical process of deducing the $\mathbf{u}$ field from the Q

field is called *PV inversion.* The hierarchy of models arises because there is an array of different PV inversion operators. They are inverse elliptic operators, hence nonlocal. They differ among themselves for two reasons. One is simply to accommodate the different physical systems that might be of interest, for example the atmosphere, or the laboratory system of Fig. 1.1, or the tokamak. The other is that, for a given physical system, some PV inversion operators are more accurate than others. In atmosphere–ocean dynamics, at least, there is always a tradeoff between simplicity and accuracy. Anyone who is curious as to what a highly accurate inversion operator looks like — the technicalities are nontrivial — may consult a little review that I wrote for *Advances in Geosciences* [19]. The property shared by the whole hierarchy, that the Q field contains all the dynamical information, is sometimes called the "PV invertibility principle".

Using the notations $D/Dt := \partial/\partial t + \mathbf{u}\cdot\nabla$ for the material derivative and $\mathcal{I}$ for the inversion operator, we can write the generic dynamics in great generality as the following pair of equations:

$$DQ/Dt = \text{forcing} + \text{dissipation}\,, \tag{1.1a}$$

$$\mathbf{u}(\mathbf{x},t) = \mathcal{I}[Q_\mathrm{a}(\,\cdot\,,t)]\,, \tag{1.1b}$$

where $Q_\mathrm{a}(\mathbf{x},t)$ is the PV anomaly field relative to a background state at relative rest in the rotating system, $\mathbf{u}\equiv 0$, whose PV is Q_b, say:

$$Q_\mathrm{a} := Q - Q_\mathrm{b} \tag{1.1c}$$

The dot in (1.1b) signals that the inversion operator acts nonlocally on the Q_a field. We have $\mathcal{I}[Q_\mathrm{a}]\equiv 0$ if and only if $Q_\mathrm{a}\equiv 0$ everywhere. Dissipation in (1.1a) may include negative dissipation, i.e. self-excitation.

PV inversion is a diagnostic, as distinct from a prognostic, operation. Diagnostic means that (1.1b) contains neither time derivatives nor history integrals. The single time derivative in (1.1a) is the only time derivative in the problem. The forcing and dissipation or self-excitation terms in (1.1a) will of course depend on the particular physical system and model assumptions, but may often be considered small for practical purposes.

The single time derivative has strange and interesting consequences. It tells us at once that the only possible wave propagation in this kind of system will be one-way propagation, in the sense that the dispersion relation can have only a single frequency branch. These are the famous westward-propagating Rossby or vorticity waves of atmosphere–ocean dynamics and the equally famous electrostatic drift waves of tokamak dynamics, arising

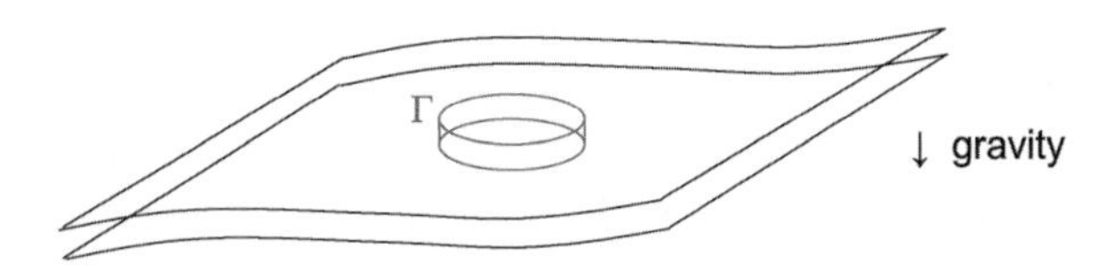

Fig. 1.2 Two stratification surfaces $\theta =$ constant in stratified, rotating flow. The material invariance of Q for ideal-fluid flow, in other words the constancy of absolute Kelvin circulation for an infinitesimal material circuit Γ lying in a stratification surface, exactly captures how the component of vorticity normal to each stratification surface changes under vortex stretching and vortex tilting.

from gradients in the background state $Q_{\rm b}$. The waves seen in Fig. 1.1 are Rossby waves. They are very different from the classical types of waves having pairs of opposite-signed dispersion-relation branches, coming from pairs of time derivatives and reflecting time reversibility. Rossby waves and drift waves have their own peculiar arrow of time.

How do these waves know which way to go? In the atmosphere–ocean case it's in part because they notice which way the Earth, or the laboratory tank, is rotating, along with certain background gradients in the thermodynamic variables. In the tokamak case, they notice which way the azimuthal magnetic field component is pointing, along with the facts that the ions are positively charged and much heavier than the electrons, and that there are background density and pressure gradients. This one-wayness has consequences reaching far beyond small-amplitude wave theory. The single time derivative in (1.1a) is still there, no matter how strongly nonlinear things become.

Of the various definitions of Q the most accurate and general, in atmosphere–ocean dynamics, can be stated as follows. Up to a constant normalizing factor, Q is the absolute Kelvin circulation around an infinitesimally small circuit Γ lying in a stratification surface. That is, Q is proportional to the loop integral $\oint_\Gamma (\mathbf{u} + \mathbf{\Omega}\times\mathbf{x})\cdot d\mathbf{x}$ where $\mathbf{\Omega}$ is the Earth's angular velocity. Stratification surfaces are surfaces of constant θ, where θ is a thermodynamic material invariant for ideal-fluid flow, more generally

$$D\theta/Dt \;=\; \text{forcing} + \text{dissipation}\,. \tag{1.2}$$

For instance we can take θ to be the specific entropy or the so-called potential temperature. Figure 1.2 shows two of the stratification surfaces, which are usually close to horizontal in practice. Also shown are two of the small circuits Γ. The material invariance of Q for ideal-fluid flow, i.e. (1.1a) with right-hand side exactly zero, is an immediate corollary of Kelvin's circulation theorem. In the ideal-fluid limit, (1.2) also has right-hand side zero and the stratification surfaces become material surfaces; so the Γ's can be taken as material circuits and Kelvin's circulation theorem applies.

An equivalent definition of Q can be written in terms of the mass density field ρ, the absolute vorticity $2\boldsymbol{\Omega} + \nabla \times \mathbf{u}$ and the stratification field θ as follows:

$$Q := \rho^{-1}(2\boldsymbol{\Omega} + \nabla \times \mathbf{u}) \cdot \nabla\theta \,. \tag{1.3}$$

Let $\Delta\theta$ be the θ-increment between the two stratification surfaces, taken infinitesimally small, and Δm the mass of the infinitesimally small pillbox-shaped fluid element defined by the pair of Γ's in Fig. 1.2. Using Stokes' theorem we see at once that (1.3) is the Kelvin circulation multiplied by $\Delta\theta/\Delta m$. For ideal-fluid flow, $\Delta\theta$, Δm, the Kelvin circulation, and hence (1.3) are all exact material invariants. Yet another demonstration starts with the vorticity equation, i.e. the curl of the momentum equation, taking its scalar product with $\nabla\theta$ to annihilate the baroclinic vector product $\nabla\rho \times \nabla p$ involving the pressure p, noting that θ is a thermodynamic function of ρ and p alone. That is the route taken in most textbooks.

In the unstratified laboratory case, almost the same picture applies except that the two stratification surfaces in Fig. 1.2 are replaced by the top and bottom boundaries of the tank. The layer now has finite thickness, a function of radius in the case of Fig. 1.1. But because there is no stratification the rapid rotation of the tank keeps the flow approximately two-dimensional — the so-called Taylor–Proudman effect — causing the material circuits Γ to move in parallel.

Historically, the central importance of Q to atmosphere–ocean dynamics was first recognized by Carl-Gustaf Rossby in his seminal papers of 1936, 1938 and 1940 [20], [21], [22]. Rossby's 1938 paper recognized the exact equivalence to the Kelvin circulation and thus showed, in principle, how to define Q exactly both for continuous stratification and for layered systems, including the tank system of Fig. 1.1. Rossby also presented hydrostatic approximations to (1.3) for use with weather data. The formula (1.3) was itself first published by Hans Ertel in 1942 [23], after visiting Rossby at MIT in 1937. The PV invertibility principle, implicit in Rossby's work, was articulated with increasing explicitness from the late 1940s onward through the work of Jule G. Charney [24], Aleksandr Mikhailovich Obukhov [25], and Ernst Kleinschmidt [26].

1.4 Cyclone and jet structure

To complete the generic picture we need a qualitative feel for what PV inversion is like. Illustrative formulae are given in Sec. 1.8; but one way to gain generic insight is to say that on each stratification surface the operator $\mathcal{I}$ delivers a horizontal velocity field $\mathbf{u}$ qualitatively like the electric field $\mathbf{E}$ in a horizontally two-dimensional electrostatics problem, but rotated through a right angle, with $Q_{\rm a}$ in the role of minus the excess charge density — not quite the tokamak case, but qualitatively similar especially as regards rotating the $\mathbf{E}$ vector. ("Horizontal" corresponds to a cross-sectional plane of the tokamak, with the background magnetic field $\mathbf{B}$ nearly "vertical".)

In realistic, stably-stratified atmospheric models, the electrostatic analogy is "layerwise-two-dimensional" insofar as one takes the horizontal component of $\mathbf{E}$ on each stratification surface and ignores the vertical component. The rotation of $\mathbf{E}$ about the vertical to get $\mathbf{u}$ is in a clockwise sense when viewed from above, i.e. compass-wise, in the northern hemisphere, corresponding to upward $\mathbf{B}$. However, thanks to Coriolis effects the field corresponding to the electrostatic potential needs to be calculated in three dimensions, inverting an elliptic operator resembling a modified three-dimensional rather than two-dimensional Laplacian, e.g. Eq. (1.6c) below. Thus the flow on one stratification surface is influenced by the notional charge distributions $Q_{\rm a}$ on other such surfaces over a significant depth of the atmosphere.

Figure 1.3 refines this rough picture by showing the result of an *accurate* three-dimensional inversion, taken from the 1985 review by Hoskins et al. [27]; q.v. for technical detail. The stratification surfaces θ = constant are the light curves that tend toward the horizontal at the periphery. They are more crowded above the tropopause, i.e. in the stratosphere which, as its name suggests, is more stably stratified, increasing the magnitude of $|\nabla\theta|$ in (1.3). The heavy curve represents the tropopause, dipping down from altitudes around 10 km down to 5 km, corresponding to pressure-altitudes (right-hand scale) ranging from around 250 to 500 millibar. The vertical scale is stretched in the conventional way, to make the structure visible. In this case, which uses realistic atmospheric parameter values, the horizontal extent of Fig. 1.3 is 5000 km. So the stratification surfaces are actually very close to being horizontal. The exact (Rossby–Ertel) Q field (1.3) has

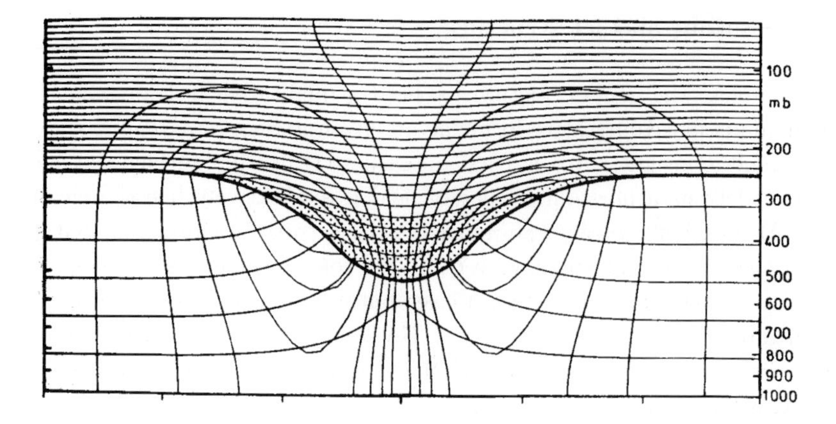

Fig. 1.3 Vertical section through an axisymmetric model of an upper-air cyclone; see text. Calculation by Dr A. J. Thorpe, from the review [27]. The entire structure comes from inverting a single positive PV anomaly located in the central stippled region. The heavy curve is the model tropopause, and the closed contours show the wind-speed profile, typical of atmospheric jets. Stronger jets go with steeper, or even reversed or "folded", tropopause slopes, but their structure is otherwise similar.

an axisymmetric positive anomaly[1] located in the central stippled region. In the electrostatic analogy we can think of it as a concentration of negative charge, making **E** point inward.

The remaining light curves in Fig. 1.3 are the contours of constant wind speed, into the paper on the right and out of the paper on the left as expected from the inward-pointing **E** vector in the layerwise-two-dimensional electrostatic analogy. The maximum wind speed occurs at the tropopause around 8 km altitude, showing a jet structure that is very typical. In this case the maximum wind speed has a rather modest value, just over 20 ms^{-1}. Stronger PV anomalies produce stronger jets with the same structure except that the tropopause, defined as the PV anomaly boundary, tends to slope more steeply and indeed can become vertical or overturned, producing what is famously called a "tropopause fold", e.g. Fig. 9(b) of [27].

[1] As explained in the review [27], accurate inversion operators $\mathcal{I}$ require the anomaly $Q_{\rm a}$ — there referred to as an "IPV anomaly" — to be defined relative to values on the same stratification surface $\theta =$ constant, or "isentropic surface" in atmospheric-science language. This term arises because θ can be taken to be the specific entropy. So "IPV anomaly" means "isentropic anomaly of PV". In the example of Fig. 1.3 the positive anomaly is due mainly to the large magnitude of the factor $\nabla\theta$ in (1.3), i.e. to the presence of stratospheric rather than tropospheric air, in the central stippled region.

I should explain that in order to do the inversion in this kind of problem one has to prescribe the mass under each stratification surface. That is how one tells the model to have a stratosphere with larger values of $|\nabla\theta|$ than the troposphere beneath [27]. One also has to impose what is called a *balance condition*. In this case it is enough to say that the flow is in hydrostatic and cyclostrophic balance. Cyclostrophic means that horizontal pressure gradients are in balance with the Coriolis force plus the centrifugal force $|\mathbf{u}|^2/r$ of the relative motion where r is horizontal distance from the symmetry axis. Thus pressures p are low at the centre of the structure. Although it is left implicit in (1.1b), inversion delivers the p, ρ and θ fields as well as the $\mathbf{u}$ field, as the invertibility principle says it must. In more complicated, non-axisymmetric cases, accurate balance conditions can still be imposed over a surprisingly large range of parameter values, but at the cost of becoming technically much more complicated; see [19] and references therein.

Notice the power of the invertibility principle. The entire structure in Fig. 1.3, long familiar and easily recognizable from the zoological annals of observational meteorology — under such names as "upper-air cutoff cyclone" or "cutoff low", e.g. Fig. 10.8 of [18] or Fig. 8 of [27] — follows from having just a single positive PV anomaly on stratification surfaces intersecting the tropopause. "Cutoff" refers to the way in which such PV anomalies are formed in reality, by a mass of high-Q stratospheric air being advected into lower-Q surroundings and then wrapping itself up into a cyclonic vortex (e.g. the near-circular patch of air over the Balkans in Fig. 1.4 below, also examples in [27]). The idea that a PV anomaly can wrap *itself* up makes perfect sense if the invertibility principle holds.

As already suggested, the jet structure illustrated in Fig. 1.3 is typical and very robust, over a large range of jet speeds and tropopause steepnesses. Although the example in Fig. 1.3 is idealized as being axisymmetric, one gets the same jet structure in more complicated cases whenever there is a concentrated gradient of Q on stratification surfaces, with high values adjacent to low values. The way in which such "isentropic" gradients arise — and observations repeatedly show that they are commonplace — is precisely through the strong nonlinearity I have been hinting at. As I'll explain, the way in which the concentrated gradients arise can be viewed as a rather simple kind of strongly-nonlinear *saturation*. The oft-observed jet structure is telling us that, in reality, saturated states are often approached.

It is worth pointing out for later reference that even without relative motion $\mathbf{u}$ there may well be pre-existing *large-scale* gradients in Q_{b}, the background PV distribution. For the tokamak they come from the radial density and pressure gradients. For realistic atmosphere–ocean dynamics they come from the fact that, in the formula (1.3), horizontal stratification surfaces pick out the vertical component, f say, of the planetary vorticity $2\boldsymbol{\Omega}$. At latitude λ we have to good accuracy

$$f = 2|\boldsymbol{\Omega}| \sin \lambda\,; \tag{1.4}$$

f is called the Coriolis parameter. Its northward gradient is conventionally denoted by β and also by the phrase "beta effect", not to be confused with plasma beta.

Before leaving this topic I should point out that if one wants to cover a wider range of significant cases then one has to count as part of the Q field the distribution of θ at the Earth's surface. This last behaves somewhat like a Dirac delta function in the vertical distribution of Q, and is important in some dynamical processes. A prime example is the baroclinic instability and the associated wrapping-up and "frontogenesis" in the surface-θ field that's so important in the terrestrial atmosphere [28], even though absent in the Jovian. The Earth's large-scale surface-θ gradient across latitudes provides an effective *southward* PV gradient, opposite to that of the beta effect. The opposing gradients mediate a powerful shear instability, the baroclinic instability in question. See §1.6 below. The resulting frontogenesis in the surface θ distribution is another variation on the theme of saturation. This baroclinic instability is the atmosphere–ocean counterpart to plasma drift wave self-excitation by resistive instability, negative dissipation in (1.1a).

1.5 Strong nonlinearity is ubiquitous

Why should concentrated isentropic gradients of PV be so commonplace along with their inversion signatures, the great jet streams, and indeed surface fronts as well, and what justifies associating them with a saturation process? What produces the accompanying anti-frictional or "negative viscosity" effects that used to be thought so mysterious? The answer lies in the idea of *inhomogeneous PV mixing* — more precisely, in the idea of spatially inhomogeneous, layerwise-two-dimensional, turbulent PV mixing, the mixing of PV along stratification surfaces in some regions but not in others. Mixing is a strongly nonlinear process because it involves not weak distortions or resonant-triad interactions but, rather, drastic advective rearrangements of large-scale into smaller-scale PV fields. It has the potential to weaken PV gradients in some places on stratification surfaces, and to strengthen them in others to form jets and eddy-transport barriers; see Secs. 1.6–1.7 below.

Neither mixing nor eddy-transport barriers need be perfect. Indeed in many cases the vortex dynamics produces new coherent structures on different spatial scales, as illustrated in Fig. 1.4. On the other hand, whenever perfect or near-perfect mixing is achieved within some region, we have a rather simple kind of strongly nonlinear saturation because, once we have a well-mixed region, further mixing has little further effect.

Equation (1.1a) with its advective nonlinearity tells us that we can, indeed, reasonably regard Q as a mixable quantity, to that extent like a chemical tracer consisting of notional charged particles. (Moreover, there is an exact "impermeability theorem" stating that the notional particles behave as if trapped on each stratification surface [30] even if air is crossing the surface diabatically.)

Today the reality of PV mixing, and the tendency toward piecewise-saturated, piecewise-well-mixed states in many cases, in the real atmosphere, has extremely strong observational support. We can routinely map real atmospheric Q fields through the highly sophisticated (spacetime Bayesian) observational data-assimilation technology that underpins operational weather forecasting. The results support Eq. (1.1a) with small right-hand side as a key to understanding the ubiquity of real jets, upper-air cyclones, surface fronts and many other meteorological phenomena.

Examples like that of Fig. 1.4 are scrutinized in [29] and [31]. Other examples come from higher altitudes, in the winter stratosphere. They have been intensively studied in connection with the ozone-hole problem already mentioned. Again and again, we see large regions on stratification surfaces within which Q is roughly constant as a result of mixing, bordered by concentrated gradients marking the eddy-transport barriers at jet cores. We can see the mixing taking place in more and more detail, over an increasingly large range of spatial scales as computer power increases. Reference [5] includes a movie from state-of-the-art data-assimilation, for a much-studied stratospheric case. These situations are at an opposite extreme from those assumed in classical homogeneous turbulence theory. In a nutshell, reality is highly inhomogeneous.

Numerical experiments tell the same story in a variety of cases, e.g. [32] and [33], as does the laboratory experiment of Fig. 1.1. In the latter case the laboratory data were good enough to enable the experimenters to map the PV field. The resulting PV maps [13], [14] closely resemble the dye pattern

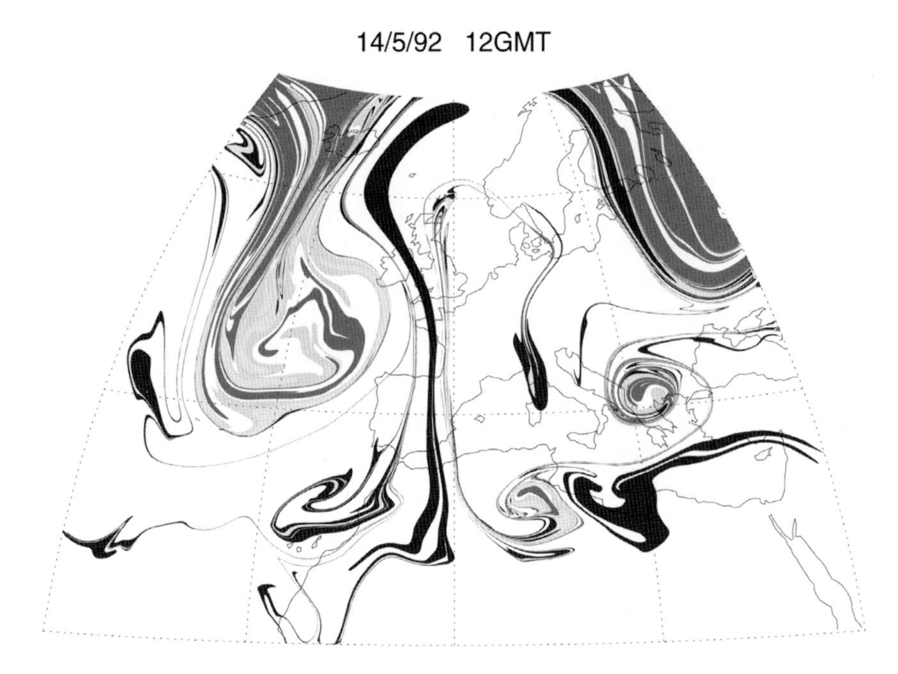

Fig. 1.4 Estimated map of Q, the exact PV defined by Eq. (1.3), on a stratification surface near 10 km altitude. For a colour version see Fig. 6.2. From [29]. The computation assumes that material invariance of Q is a good approximation over a 4-day time interval, and uses a state-of-the-art advection algorithm and weather-forecasting data to trace the flow of high-Q stratospheric air (shaded) and low-Q tropospheric air (clear). The main boundary between stratospheric and tropospheric air marks a jet core showing large-amplitude meandering, from Greenland toward Spain and then back to northern Norway. The leakage of stratospheric air into the troposphere signals intermittent attrition of the eddy-transport barrier at the jet core. The different chemical signatures of the stratospheric and tropospheric air are easily detectable and have been demonstrated in measurement campaigns, even for fine filamentary structures like those shown [16]. The high-Q anomaly over the Balkans illustrates the "cutoff" or self-wrapping-up process that occurs when sufficiently large masses of stratospheric air overcome the barrier. The wrapping-up produces structures like that in Fig. 1.3.

in Fig. 1.1, with well-mixed regions on either side of the jet core. Even though the visible wavy pattern might suggest the validity of linearized or weakly-nonlinear Rossby-wave theory, the suggestion is misleading because the fluid motion has already nonlinearly rearranged its PV distribution to be close to piecewise uniform, uniform to either side of the jet core, with a concentrated PV gradient at the core. Strong nonlinearity did most of its work before the dye was injected. At the instant shown in Fig. 1.1, piecewise regional mixing continues on either side of the jet but has become invisible, since there were no further dye injections. As far as the dynamics is concerned, the mixing process has saturated.

Figure 1.1 well illustrates what I mean by the inhomogeneous "wave–turbulence jigsaw". We have a wavy, weakly nonlinear jet-core region adjacent to strongly nonlinear mixing regions on either side of the jet core. The different regions are coupled dynamically to each other, in a manner to be further analysed in Sec. 1.11 below. The jigsaw has to fit together dynamically as well as geometrically.

A recent numerical model study [34] finesses these inhomogeneous PV-mixing ideas through numerical experiments showing the co-development of jets, well-mixed areas and coherent vortices, with the vortices actively contributing to the mixing, an idealized version of Fig. 1.4. The discussion of strongly-nonlinear coherent structures in Diamond et al. (this volume) hints that we might end up with a similar picture for real drift-wave turbulence in big tokamaks.

1.6 Rossby waves and drift waves

Figure 1.5 shows the generic propagation mechanism for Rossby and drift waves. The background rotation $\boldsymbol{\Omega}$ and azimuthal magnetic field $\mathbf{B}$ point toward the viewer. Propagation occurs whenever the Q field has a background gradient. We consider a background Q field $Q_{\mathrm{b}} = Q_{\mathrm{b}}(y)$, where in atmosphere–ocean cases we take y as pointing northward, in the direction of increasing Q_{b}, and in tokamak cases as pointing radially toward decreasing background pressure and mass density. The y direction is toward the top of the figure and the x direction toward the right; please note that this is not the usual coordinate convention for tokamaks. For the stratified atmosphere $dQ_{\mathrm{b}}/dy > 0$ is the isentropic gradient, i.e., the gradient along a stratification surface as already mentioned.

With no disturbance the constant-Q contours would be straight and parallel to the x axis. We imagine that a small disturbance makes them wavy, as shown. We assume ideal-fluid motion, i.e., zero on the right of Eq. (1.1a). Then the wavy Q contours are also material contours. Linearized wave theory requires the undulations to be gentle: their sideways slopes must be much smaller than unity. We have a row of PV anomalies $Q_{\mathrm{a}} = Q - Q_{\mathrm{b}}$ of alternating sign, as suggested by the plus and minus signs enclosed by circular arrows. In the electrostatic analogy — be careful — plus means a *negative* charge and vice versa. (If only the world were made of antimatter, it would be easier to be lucid here.)

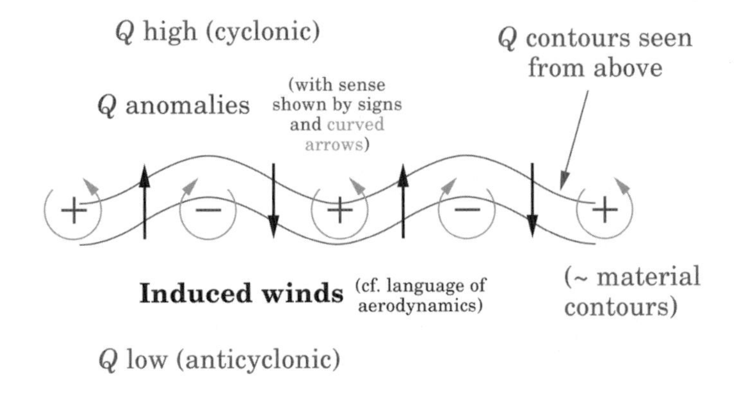

Fig. 1.5 Plane view of an x-periodic Rossby wave, where x points toward the right in the figure. The plus and minus signs respectively indicate the centres of the cyclonic and anticyclonic PV anomalies due to southward and northward air-parcel displacements across a basic northward PV gradient. For a tokamak drift wave, "cyclonic" means a vorticity vector pointing in the $+\mathbf{B}$ direction, toward the viewer in this case. The plus signs then correspond to *negative* net charge, and minimal density, and in the background gradients high Q corresponds to low background pressure and density.

Inversion gives a velocity field $\mathbf{u}$ ($\propto \mathbf{E}\times\mathbf{B}$ for the tokamak) whose north-south component is a quarter wavelength out of phase with the north-south displacement field, as suggested by the big dark arrows in Fig. 1.5. (In the meteorological literature the velocity field resulting from a PV inversion is sometimes called the wind field "induced" by the PV anomalies.[2])

So to understand Rossby-wave and drift-wave propagation — generically, and not just in textbook cases — one need only use one's visual imagination to turn Fig. 1.5 into a movie. With velocity a quarter wavelength out of phase with displacement, the pattern will start to propagate to the left. But invertibility, Eq. (1.1b), says that the velocity pattern must remain phase-locked to the displacement pattern! So the propagation continues indefinitely. With the signs shown, the propagation is in the negative x direction — the famous westward, or quasi-westward, or high-Q-on-the-right phase propagation, relative of course to any mean flow. We may think of the Q contours as possessing a peculiar Rossby-wave "quasi-elasticity". This shows up spectacularly in Norton's movie mentioned in Sec. 1.2.

[2] Here meteorologists are following the language of aerodynamics dating from the pioneering days of Frederick Lanchester and Ludwig Prandtl, where three-dimensional *vorticity inversion* using a Biot–Savart integral has long been a basic conceptual tool. As an aerodynamicist would put it, aeroplanes stay up thanks to the downward velocity "induced" by the trailing wingtip vortices.

If we do a thought-experiment in which all the signs are changed in Fig. 1.5, replacing westward phase propagation by eastward, then we are describing the effect of the large-scale gradient in surface θ noted at the end of Sec. 1.4. The simplest and most powerful baroclinic-instability mode can be thought of as a pair of counterpropagating Rossby waves on the opposing interior and surface PV gradients, phase-locked with the help of the mean vertical shear [27].

The propagation mechanism works similarly for uneven Q contour spacing, including the case of a jet with most of the Q contours bunched up at the jet core. This means that jets can act as Rossby waveguides, with quasi-elastic cores. Figure 1.1 is an example. Some atmospheric examples are discussed in [31]. Explicit toy-model solutions will illustrate the same point in Sec. 1.11.

If the Q contours in Fig. 1.5 were deforming irreversibly rather than gently undulating, then we would say that the Rossby waves are *breaking*. Indeed, for strong reasons grounded in wave–mean interaction theory we may regard such irreversible deformation as the defining property of wave breaking [6], [35]. In the atmosphere, at least, there is no doubt that the breaking of Rossby waves is Nature's principal way of causing PV mixing and its typical consequences, anti-frictional jet sharpening and eddy-transport-barrier reinforcement. In many cases the associated radiation stress or wave-induced momentum transport is an essential part of how the wave–turbulence jigsaw fits together [19], [10], [36]. One way of seeing more precisely how it fits together is through what is called the *Taylor identity*; see (1.9) and (1.10)ff. below.

In Sec. 1.11 we'll see that jet-guided Rossby waves have a strong tendency to break on one or both sides of the jet, leaving the jet core intact. That is, there is a systematic tendency for PV mixing to occur preferentially on the flanks of a jet — sharpening the jet anti-frictionally, reinforcing the eddy-transport-barrier effect, and keeping the wave–turbulence jigsaw highly inhomogeneous. We may think of jets almost as *self-sustaining* elastic structures; the only help they need is for their Rossby waves to be excited now and again, for instance by baroclinic instabilities.

1.7 The PV Phillips effect

There is an even simpler, and independent, argument suggesting that the spatial inhomogeneity or regionality I keep talking about is generically likely. Imagine a large-scale, initially uniform PV gradient subject to random disturbances. Suppose that these disturbances produce PV mixing along stratification surfaces, the mixing being slightly stronger in some regions than others. The regions where the mixing is stronger will have their overall PV gradients weakened. But those regions will then have weaker Rossby-wave quasi-elasticity, and will be even easier to mix. Other things being equal, there is a positive feedback that tends to push PV contours apart in some regions and bunch them together in others.

I like to call this the "PV Phillips effect", after Owen M. Phillips' original suggestion in 1972 that the same thing happens with vertical gradients of θ, i.e. with stable stratification [37]; see also [10], [38] and references therein. In that case the suggestion was beautifully verified in a non-rotating, stratified laboratory tank experiment [39]. One starts with the uniform stratification created by a smooth vertical gradient of salinity. Salinity is a convenient laboratory counterpart to θ, or rather to $-\theta$ because more salinity means more density and less buoyancy. One stirs the tank with smooth vertical rods. This imposes no vertical scale of motion smaller than the depth of the tank. Nevertheless the stratification rearranges itself into horizontal layers. Layers with weak stratification, relatively small $|\nabla\theta|$, are sandwiched between horizontal interfaces with strong stratification, relatively large $|\nabla\theta|$. The vertical scale depends on how vigorously one stirs. With sufficiently vigorous stirring one can of course homogenize the whole tank. Otherwise, we have — guess what — yet another example of the tendency for real turbulence to be highly inhomogeneous and for eddy-transport barriers to form.

The original Phillips effect and its PV counterpart are summarized in Fig. 1.6, courtesy of Drs Jouve and McDevitt. In the original Phillips effect the relevant wave elasticity is the elasticity associated with internal gravity waves — the waves that owe their existence to the stable stratification $\nabla\theta$.

Notice by the way that the positive-feedback argument does not depend on whether the mixing can be described as Fickian eddy diffusion, i.e. as a random walk with *short* steps. The argument transcends any such restrictive modelling assumptions.

In the case of PV the positive feedback tends to be reinforced by the jet shear effects [10], [11], contributing to eddy-transport-barrier formation as suggested in Sec. 1.2. Wherever PV contours bunch together, inversion gives jetlike velocity profiles hence shear. It turns out that the shearing of small-scale disturbances is just as important as the Rossby-wave quasi-elasticity felt by larger-scale disturbances. There is a smoothing effect or *scale effect*,

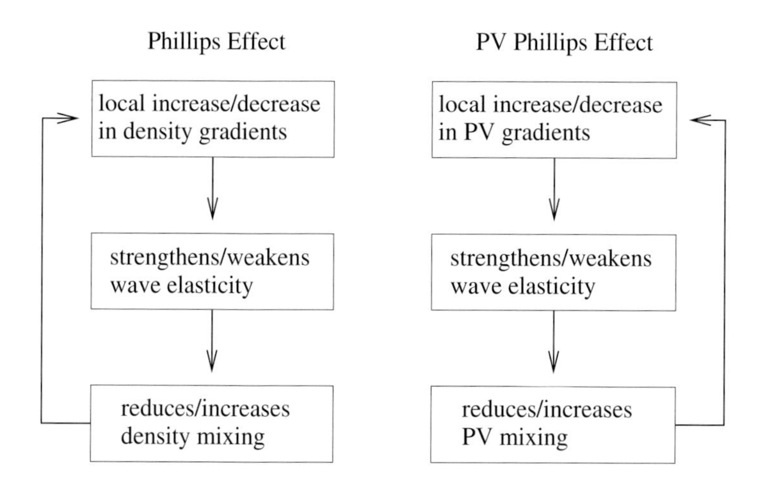

Fig. 1.6 Schematic of the positive feedback loops for the original Phillips effect and its PV counterpart. Courtesy of Drs Jouve and McDevitt. In the PV case the positive feedback is reinforced by jet shear, leading to the formation of eddy-transport barriers such as that illustrated in Fig. 1.1. Further reinforcement can come from the preferred phase speeds of disturbances, as discussed in Sec. 1.11.

coming from the inverse-Laplacian-like character of the inversion operator, that weakens the **u** field of the smallest-scale PV anomalies. So the small-scale behaviour tends to be passive-tracer-like, as indeed was suggested by the filamentary structures in Fig. 1.4.

1.8 Some simple inversion operators

To check our insights we often use models with simplified but qualitatively reasonable inversion operators $\mathcal{I}$. The most important of these models are the so-called *quasi-geostrophic* models. They are not quantitatively accurate but are conceptually important because their dynamics still has the generic form (1.1a)–(1.1c) and, at a good qualitative level — better than that of the crude electrostatic analogy of Sec. 1.4 — they describe phenomena such as the scale effect, Rossby-wave propagation, and Rossby-wave breaking and other strongly nonlinear phenomena such as vortex interactions and so-called "cascades".

PV inversion, which generically is a mildly nonlinear operation, albeit a smoothing operation because of the scale effect, becomes strictly linear in these models. This allows free use of the superposition principle and helps to expand the repertoire of mathematically precise illustrative solutions.

The advective nonlinearity is the only nonlinearity. The quasi-geostrophic models come in a number of versions, including single-layer, multi-layer, and continuously stratified. The standard single-layer or "shallow-water" version is isomorphic to the standard Hasagawa–Mima tokamak model.

The term quasi-geostrophic comes from the balance condition used to contruct the inversion operator $\mathcal{I}$, geostrophic balance, in which we entirely neglect the relative centrifugal and other small terms in the momentum equation. Geostrophic balance means simply a balance between Coriolis forces and horizontal pressure gradients. It is valid as an asymptotic approximation in the limit of small Rossby number $\text{Ro} := f^{-1} \| \hat{\mathbf{z}} \cdot \nabla \times \mathbf{u} \|$ where f is the Coriolis parameter as before, $\hat{\mathbf{z}}$ is a unit vertical vector, and $\| \ \|$ denotes a typical magnitude.

In these models it is convenient to use modified definitions of the PV, to be denoted by q, with background q_b. Within the asymptotic approximation schemes that lead to the models, which originated in the independent pioneering work of Charney [24] and Obukhov [25] and are described in many textbooks, we may regard the velocity field as purely horizontal and nondivergent to leading order. An $O(\text{Ro})$ correction is implicit, allowing weak vertical motion and horizontal divergence. So to leading order we can introduce a streamfunction $\phi(\mathbf{x}, t)$, which is a suitably Coriolis-scaled pressure anomaly such that, to leading order, $\mathbf{u} = \mathbf{u}_\text{g} := \hat{\mathbf{z}} \times \nabla\phi = (-\phi_y, \phi_x, 0)$, expressing geostrophic balance. Suffixes x and y denote partial differentiation. In terms of the corresponding "geostrophic material derivative" $D_\text{g}/Dt := \partial/\partial t + \mathbf{u}_\text{g} \cdot \nabla$ the dynamics takes the form

$$D_\text{g} q/Dt = \text{forcing} + \text{dissipation} , \tag{1.5a}$$

$$\mathbf{u}_\text{g}(\mathbf{x}, t) = \mathcal{I}[q_\text{a}(\,\cdot\,, t)] := \hat{\mathbf{z}} \times \nabla \mathcal{L}^{-1} q_\text{a} , \tag{1.5b}$$

$$q_\text{a} := q - q_\text{b} , \tag{1.5c}$$

where again "dissipation" includes negative dissipation, i.e. self-excitation, and where $\mathcal{L}$ is a linear elliptic operator given by the horizontal Laplacian $\nabla_H^2 = \partial_{xx} + \partial_{yy}$ plus extra terms that vary from model to model. These operators have well-behaved inverses $\mathcal{L}^{-1}$ if reasonable boundary conditions are given. Of course (1.5b)–(1.5c) amount to saying that in each case the definition of q is $q_\text{b} + \mathcal{L}\phi$. However, saying it via (1.5b)–(1.5c) emphasizes that these models are indeed examples of the generic dynamics.

Examples include the following three. The first two are single-layer, two-dimensional models with $\phi = \phi(x, y, t)$, involving a fixed lengthscale L_D to be specified shortly. The third is three-dimensional with $\phi = \phi(x, y, z, t)$, taking account of continuous background stratification:

$$\text{Model 1:} \qquad \mathcal{L}\phi := \nabla_H^2\phi - L_\mathrm{D}^{-2}\phi, \tag{1.6a}$$

$$\text{Model 2:} \qquad \mathcal{L}\phi := \nabla_H^2\phi - L_\mathrm{D}^{-2}\tilde{\phi}, \tag{1.6b}$$

$$\text{Model 3:} \qquad \mathcal{L}\phi := \nabla_H^2\phi + \frac{1}{\rho_0}\frac{\partial}{\partial z}\left(\frac{\rho_0 f_0^2}{N^2}\frac{\partial\phi}{\partial z}\right). \tag{1.6c}$$

Model 2 is exclusive to the tokamak, having no atmosphere–ocean counterpart beyond its conformity to the generic dynamics, whose most important implication, for our purposes, is PV mixability. The tilde in model 2 denotes the departure from a zonal or x-average: $\tilde{\phi} := \phi - \langle\phi\rangle$.

In models 1 and 3, the extra terms added to the Laplacian represent hydrostatic balance together with vortex stretching by the implicit, $O(\mathrm{Ro})$ vertical motion, essentially the ballerina effect from the accompanying horizontal convergence. The continuous background stratification in model 3 is represented approximately in terms of background profiles $\rho = \rho_0(z)$ and $\theta = \theta_0(z)$, with the buoyancy frequency $N(z)$ defined in terms of the gravity acceleration g by $N^2 := g d\ln\theta_0/dz$. Coriolis effects are represented by a domain-average Coriolis parameter $f = f_0 =$ constant. In model 1 the lengthscale L_D, called the Rossby deformation length, is f_0^{-1} times the gravity-wave speed for the layer, which has a free top surface. More precisely, one uses the notional gravity-wave speed that would apply if f_0 were zero, measuring the hydrostatic free-surface gravitational elasticity.

Since model 1 is often called the Hasegawa–Mima model or sometimes, for the sake of historical justice, the Charney–Obukhov–Hasegawa–Mima model, it is reasonable to call model 2 with the tilde a "generalized", or "modified", or "extended" Hasegawa–Mima model, e.g. [40], [41]. Model 1 in its atmosphere–ocean applications is given the self-explanatory name "shallow-water quasi-geostrophic model" and sometimes, less transparently, "equivalent barotropic model". Notice that we can turn model 3 into model 1 by assuming a fixed vertical structure with a single vertical scale H. Then a scale analysis applied to the extra term with the vertical derivatives gives $L_\mathrm{D} = NH/f_0$, related to the notional internal-gravity-wave speed NH for $f_0 = 0$. Typical extratropical L_D values range between $\sim 10^3$ km–10^2 km for the atmosphere and $\sim 10^2$ km–10^1 km for the oceans, with H in the ballpark of say 5–10 km.

Model 3 requires a boundary condition $\phi_z = 0$ at a flat lower boundary, say $z = 0$, idealizing the Earth's surface. (In otherwise-unbounded domains we take evanescent boundary conditions.) For model 3 one can show from hydrostatic balance that $\phi_z = g/f_0$ times the anomaly in $\ln\theta$ (ϕ_z having dimensions of velocity, like ϕ_x and ϕ_y). So the nonvanishing θ anomalies at $z = 0$, which as already mentioned are critical to baroclinic instability and frontogenesis, correspond to nonvanishing ϕ_z at $z = 0$. As hinted at the end of Sec. 1.4, we can regard this situation as equivalent to $\phi_z = 0$ at $z = 0$ together with a compensating delta function in the last term of (1.6c), coming from a jump discontinuity in ϕ_z [42].

In model 2 the ballerina effect is replaced by spinup via the magnetic Lorentz force $\mathbf{u} \times \mathbf{B}$ when ions converge, $\mathbf{u}$ being the ion flow component normal to a background magnetic field $\mathbf{B} = |\mathbf{B}|\hat{\mathbf{z}}$. Geostrophic balance is replaced by the leading-order force balance $\mathbf{u} \times \mathbf{B} \approx -\mathbf{E}$, implying that the disturbance streamfunction $\tilde{\phi}$ is now $|\mathbf{B}|^{-1}$ times the disturbance to the electrostatic potential. In other words the pressure-gradient force is replaced by the electrostatic force and the Coriolis parameter by $|\mathbf{B}|$, suitably normalized — most conveniently as the ion gyrofrequency ω_{ci} given by $|\mathbf{B}|$ times the ions' charge-to-mass ratio. Model 2 assumes that ion temperatures are much lower than electron temperatures so that the ion pressure gradient is relatively unimportant in the ion flow dynamics.

Also implicit in the dynamics of model 2 are assumptions that a typical disturbance $\tilde{\phi}$, while treated as two-dimensional in the cross-sectional plane of the tokamak, actually has a finite wavenumber component $k_{\parallel}$ parallel to $\mathbf{B}$, allowing the electrons to adjust quasi-statically along the $\mathbf{B}$ lines encircling the tokamak. More precisely, with thermalized, hence Boltzmann-distributed, electrons at fixed temperature T_e there is a mutual adjustment between $\tilde{\phi}$ and the electron density anomaly $\tilde{n}_e$ such that $\tilde{n}_e \propto \tilde{\phi}$. For an insightful discussion see [43]. Together with quasineutrality this mutual adjustment gives rise to a notional isothermal sound speed c_{s} based on electron temperature but ion mass, corresponding to the notional shallow-water gravity-wave speed in model 1; c_{s}^2 is of the order of the electron-to-ion mass ratio times the square of the electrons' thermal speed v_e. The corresponding L_{D} value $c_{\mathrm{s}}/\omega_{ci}$ is conventionally denoted by ρ_{s} and is small in comparison with tokamak dimensions. It would be equal to the ions' gyroradius if the ions were heated up to temperature T_e, as indeed happens in models more realistic than model 2. Also implicit in model 2 is an assumption that the zonal-mean flow $\langle\phi\rangle$ does, by contrast, have $k_{\parallel} = 0$, suppressing the mutual adjustment between $\tilde{\phi}$ and $\tilde{n}_e$ and giving rise to the tilde in Eq. (1.6b).

In more realistic tokamak models $\langle \ \rangle$ becomes an average over an entire torus-shaped magnetic flux surface encircling the tokamak.

The Rossby number is replaced by the small parameter $\omega_{ci}^{-1} \| \hat{\mathbf{z}} \cdot \nabla \times \mathbf{u} \|$. For the background PV gradient the atmosphere–ocean β, with dimensions vorticity/length, is replaced by ω_{ci}/L where L is the lengthscale of the background electron-pressure gradient, with signs as in Fig. 1.5. Also, "dissipation" in Eq. (1.5a) includes the negative dissipation from resistive self-excitation of drift waves [44], predominantly at scales of the order of $L_{\mathrm{D}} = \rho_{\mathrm{s}} = c_{\mathrm{s}}/\omega_{ci}$ and coming from a slight time delay in the mutual adjustment between $\tilde{\phi}$ and $\tilde{n}_e$.

Omitted from the list above are various two-layer and higher multi-layer atmosphere–ocean models that also conform to the generic dynamics, and for which the PV-mixing paradigm is also robust. They are essentially stacks of shallow water quasi-geostrophic models and are popular because they capture some aspects of continuous stratification, model 3, but with a reduced computational burden. There is a rigidly-bounded two-layer model that might, however, be worth exploration as a more consistent version of model 2 for the tokamak. In its atmosphere–ocean interpretation it represents a physically consistent thought-experiment with two L_{D} values, one finite and the other infinite, that might be put into correspondence with the tokamak's finite-$k_{\|}$ and zero-$k_{\|}$ modes. It has been intensively studied, e.g. [45] and references.

Such a two-layer model might well, on the other hand, fail to improve on model 2 especially if PV mixing turns out to be important in real tokamaks. This is because mixing due to advection by a chaotic disturbance velocity field $\tilde{\mathbf{u}}$ is insensitive to sign changes $\tilde{\mathbf{u}} \to -\tilde{\mathbf{u}}$. So mixing by finite-$k_{\|}$ disturbances in a real tokamak should be well able to robustly generate and maintain jets in the flux-surface-averaged, zero-$k_{\|}$ velocity field $\langle \mathbf{u} \rangle$, a scenario that is implicit in model 2 despite its being heavily idealized.

Taking $q_{\mathrm{b}} = \beta y +$ constant as the background PV, with constant gradient β, we can easily check that models 1–3 possess waves that propagate in the manner sketched in Fig. 1.5. For instance both model 1 and model 2 linearized about relative rest have the same elementary wave solutions $\tilde{\phi} \propto \exp(ikx + ily - i\omega t)$ with the same, single-branched dispersion relation

$$\omega = \frac{-\beta k}{k^2 + l^2 + L_{\mathrm{D}}^{-2}}, \tag{1.7}$$

where the denominator comes from $q_{\mathrm{a}} = \tilde{q} = \mathcal{L}\tilde{\phi} = -(k^2 + l^2 + L_{\mathrm{D}}^{-2})\tilde{\phi}$ in the linearized (1.5a) with right-hand side zero, $\partial\tilde{q}/\partial t + D_{\mathrm{g}} q_{\mathrm{b}}/Dt =$

$\partial(\mathcal{L}\tilde{\phi})/\partial t + \beta\partial\tilde{\phi}/\partial x = 0$. Notice the scale effect: wave propagation is weakened as scales become smaller and $k^2 + l^2$ larger. In the long-wave limit $(k^2 + l^2) \to 0$ the phase velocity asymptotes to $-\beta L_{\rm D}^2$, which in the tokamak case coincides with the background diamagnetic drift velocity, not of ions but of electrons.

The laboratory flow in Fig. 1.1 has a rigid lid and is well described by model 1 with $L_{\rm D} = \infty$. This is ordinary (Euler, inertial) two-dimensional vortex dynamics except that there is a nontrivial background $q_{\rm b} = \beta y +$ constant. This comes from a gently sloping cone-shaped tank bottom, where now y is radial distance toward the tank centre.

For model 1 in an unbounded domain with finite $L_{\rm D}$ one has the explicit inversion formula

$$\phi(\mathbf{x}, t) = \mathcal{L}^{-1} q_{\rm a} := -\frac{1}{2\pi}\iint K_0\left(\frac{|\mathbf{x}-\mathbf{x}'|}{L_{\rm D}}\right) q_{\rm a}(\mathbf{x}', t)\, dx'dy' \tag{1.8}$$

where $|\mathbf{x}-\mathbf{x}'|^2 = (x-x')^2 + (y-y')^2$ and where K_0 is the modified Bessel function [25]. Its exponential decay shows that both linear and nonlinear interactions become very weak at distances significantly greater than $L_{\rm D}$. Aphoristically speaking, we have action at a distance but not too great a distance, which helps to explain the surprising robustness of the generic dynamics [19] which, in effect, through the balance condition, assumes that gravity-wave propagation is instantaneous. In model 2 this corresponds to instantaneous acoustic propagation, $c_{\rm s} = \infty$.

1.9 Pseudomomentum and the Taylor identity

Models 1–3 and their multi-layer counterparts all possess what are called pseudomomentum theorems and Taylor identities. All these theorems and identities stem from the seminal 1915 work of Sir Geoffrey Ingram Taylor [46] and its further development in the 1960s by Jule G. Charney and Melvin E. Stern [47] and by Francis P. Bretherton [48]. (I'm almost a direct intellectual descendant because Taylor was Bretherton's PhD advisor and Bretherton was mine.) Taylor's results apply to models 1 and 2. Charney, Stern and Bretherton extended them to model 3, leading in turn to a milestone 1969 paper by Robert E. Dickinson [49]. Dickinson's paper took what I regard as the decisive step toward solving the "negative viscosity" enigma for the real atmosphere, though further work was needed to see how simply it could be understood using the concept of Rossby-wave breaking. For more history see [10].

In their usual forms the pseudomomentum theorems depend on linearization and so apply to small-amplitude waves only. The Taylor identity, by contrast, is valid at any amplitude, and so applies to the whole wave–turbulence jigsaw.

Consider an arbitrary zonal-mean state $\langle q \rangle = q_{\rm b} + \langle q_{\rm a} \rangle$. The non-background part $\langle q_{\rm a} \rangle$ might for instance represent a jet flow. Take the disturbance part $\tilde{q} = \mathcal{L}\tilde{\phi}$ of (1.6a) or (1.6b). Multiply it by $\tilde{\phi}_x$ and take the zonal mean. The $L_{\rm D}$ terms disappear, because $\langle \tilde{\phi}_x \tilde{\phi} \rangle = \langle \frac{1}{2}(\tilde{\phi}^2)_x \rangle = 0$. From the horizontal Laplacian we have $\langle \tilde{\phi}_x \tilde{\phi}_{xx} \rangle = \langle \frac{1}{2}(\tilde{\phi}_x^2)_x \rangle = 0$ and so, noting also that $\langle \tilde{\phi}_{xy} \tilde{\phi}_y \rangle = \langle \frac{1}{2}(\tilde{\phi}_y^2)_x \rangle = 0$ and writing $\tilde{v} = \tilde{v}_{\rm g} = \tilde{\phi}_x$, $\tilde{u} = \tilde{u}_{\rm g} = -\tilde{\phi}_y$, dropping the suffix g from now on, we have

$$\langle \tilde{v}\tilde{q} \rangle = -\partial \langle \tilde{u}\tilde{v} \rangle / \partial y \,. \tag{1.9}$$

This is the Taylor identity for models 1 and 2. It is valid at any amplitude and is indifferent to whether the dynamics is ideal-fluid or not: Eq. (1.5a) was never used. It was derived by Taylor for the nondivergent barotropic case $L_{\rm D} = \infty$. As just shown, however, it extends trivially to any value of $L_{\rm D}$. It tells us that in these models the eddy flux of PV, including any contributions due to strongly nonlinear processes like wave breaking and PV mixing, is directly tied to the eddy flux of momentum and hence to the self-sharpening, anti-frictional jet dynamics. The negative-viscosity enigma has vanished in a puff of insight! It is the mixable quantity PV that tends to go down its mean gradient — not momentum, which there is no reason to suppose is mixable.

For model 3 one replaces the momentum-flux convergence on the right of (1.9) by its counterpart in the yz plane, the convergence of an effective momentum flux whose vertical component is minus what oceanographers call the *form stress* across an undulating stratification surface due to correlations between pressure fluctuations and stratification-surface slopes. This gives the Taylor identity for model 3:

$$\langle \tilde{v}\tilde{q} \rangle = -\frac{\partial \langle \tilde{u}\tilde{v} \rangle}{\partial y} + \frac{1}{\rho_0}\frac{\partial}{\partial z}\left(\frac{\rho_0 f_0^2}{N^2} \left\langle \frac{\partial \tilde{\phi}}{\partial x}\frac{\partial \tilde{\phi}}{\partial z} \right\rangle \right). \tag{1.10}$$

To see that the last term contains the form stress, within the round brackets, note that $\langle \tilde{\phi}_x \tilde{\phi}_z \rangle = -\langle \tilde{\phi}\tilde{\phi}_{zx} \rangle$ and that $\rho_0 f_0 \tilde{\phi}$ is the pressure fluctuation while, thanks to hydrostatic balance, $-f_0 N^{-2} \tilde{\phi}_z$ is the stratification-surface displacement and $-f_0 N^{-2} \tilde{\phi}_{zx}$ its slope. For historical reasons this effective momentum flux has often been defined with a perverse sign convention and labelled the *Eliassen–Palm flux*. By the usual conventions, $-\langle \tilde{u}\tilde{v} \rangle$ ought to be minus the momentum flux or plus the stress, and similarly for the vertical, form-stress term. Like (1.9), the identity (1.10) holds whether the dynamics is ideal-fluid or not and for disturbances of any amplitude whatever, wavelike, or turbulent, or both.

The small-amplitude pseudomomentum theorems were also derived by Taylor and Bretherton albeit in slightly disguised form. As things panned out historically, full clarity (in the atmosphere–ocean community) had to await introduction of what are usually called the "transformed Eulerian-mean equations" [50], shortly after which the conceptual connection with theoretical-physics principles — translational invariance, quasi-particle gases and so on — was finally made, helped by a correspondence I had with Sir Rudolf Peierls. An in-depth discussion is given in [6].

Here we define the pseudomomentum P per unit mass for small-amplitude fluctuations $\tilde{q}$ about the translationally-invariant mean state $\langle q\rangle$, for all three models, as

$$P := -\tfrac{1}{2}\langle\tilde{q}^2\rangle/\langle q\rangle_y\,. \tag{1.11}$$

The sign convention is chosen to make P agree with the usual ray-theoretic pseudomomentum or quasimomentum, i.e. wave action times wave vector. We expect a minus sign precisely because of the one-wayness of Rossby and drift waves.

Linearizing (1.5a) with right-hand side zero, about the mean state $\langle q\rangle$, we easily find, on multiplication by $\tilde{q}$ and taking the zonal mean, the disturbance potential-enstrophy equation $\partial_t\frac{1}{2}\langle\tilde{q}^2\rangle = -\langle q\rangle_y\langle\tilde{v}\tilde{q}\rangle$ for all three models. The Taylor identity says that we can turn the potential-enstrophy equation into a conservation theorem if we divide it by $-\langle q\rangle_y$ and use the fact that, at small amplitude, we may consistently neglect the rate of mean-state evolution $\langle q\rangle_{yt}$. Thus for models 1 and 2, for instance,

$$\frac{\partial P}{\partial t} + \frac{\partial\langle\tilde{u}\tilde{v}\rangle}{\partial y} = 0 \tag{1.12}$$

which is indeed in conservation form, as is the corresponding result for model 3 for which we need only replace the second term of (1.12) by minus the right-hand side of (1.10):

$$\frac{\partial P}{\partial t} + \frac{\partial\langle\tilde{u}\tilde{v}\rangle}{\partial y} - \frac{1}{\rho_0}\frac{\partial}{\partial z}\left(\frac{\rho_0 f_0^2}{N^2}\left\langle\frac{\partial\tilde{\phi}}{\partial x}\frac{\partial\tilde{\phi}}{\partial z}\right\rangle\right) = 0\,. \tag{1.13}$$

Of course one can usefully repeat these derivations with the forcing or dissipation terms explicitly included on the right of (1.5a) whenever one has a particular model for those terms, such as a viscous term, or infrared radiative damping, or drift-wave self-excitation. Then one has sources and sinks of pseudomomentum P. Such sources and sinks do not necessarily require external forces to be exerted. They require only that waves be generated or dissipated somehow. This underlines the fact that pseudomomentum is not the same thing as momentum, *despite* sharing the same flux terms.

The small-amplitude pseudomomentum conservation theorems are sometimes called Charney–Drazin theorems, even though they originated in the 1915 work of Taylor [46]. The 1961 work of Charney and Drazin [51] restricted attention to steady, nondissipating waves on a special mean flow $\langle u\rangle(z)$ with vertical shear only, and vanishing horizontal shear $\langle u\rangle_y$ and vanishing Reynolds stress $\langle\tilde{u}\tilde{v}\rangle$, in model 3. Their theorem, Eqs. (8.13)ff. of [51], though influential in its time, said only that, in this special steady-waves case, the $\partial/\partial z$ term of (1.13) vanishes — the counterpart to saying that $\partial\langle\tilde{u}\tilde{v}\rangle/\partial y$ vanishes for the steady-waves case of (1.12). Unlike Taylor, they did not consider time-dependent waves and so did not find any results like (1.12) and (1.13) involving $\partial P/\partial t$.

The reader interested in deeper aspects of the theory is recommended to read the penetrating account in Bühler's book [6]. For instance the assumption $\langle q\rangle_{yt} = 0$ used above, in order to derive (1.12) and (1.13), is consistent with the mean state being not only translationally but also temporally invariant. Then pseudoenergy as well as pseudomomentum is conserved for ideal-fluid flow. Generically, the pseudoenergy E in a particular frame of reference can be defined as $\langle u\rangle P$ plus the positive-definite wave-energy of the linearized disturbance equations. This holds for arbitrarily-sheared mean flows $\langle u\rangle$.

Domain-integrated P and E conservation immediately give us the well known Rayleigh–Kuo–Charney–Stern and Fjørtoft shear stability theorems. The first implies stability whenever P is inherently one-signed, i.e. whenever $\langle q\rangle_y$ is one-signed, and the second whenever a translating frame of reference can be found that makes $\langle u\rangle P$ positive definite, i.e. $\langle u\rangle\langle q\rangle_y$ negative definite, even if $\langle q\rangle_y$ is two-signed. This last points to the role of counterpropagating Rossby waves in shear instabilities mentioned in Sec. 1.6. States in which $\langle q\rangle_y$ is only just one-signed (e.g. Fig. 1.1, also "PV staircases", next section) are sometimes called "states of Rayleigh–Kuo–Charney–Stern marginal stability", or for brevity "states of Rayleigh–Kuo marginal stability". Beautiful generalizations and extensions of all the stability theorems were discovered by V. I. Arnol'd; see e.g. [36], [52], and references therein.

Bühler's book [6] also makes clear to what extent one can generalize small-amplitude results like (1.12) and (1.13) to finite amplitude, for atmosphere–ocean models at least. The finite-amplitude counterparts, though conceptually important, are computationally impractical in the cases that interest us here because they require retention of Lagrangian flow information — more precisely, the shapes of originally-zonal material contours entering Kelvin's circulation theorem. That is no great problem for waves that are not breaking, like the waves in Fig. 1.5, but becomes hopelessly complicated in turbulent zones where the waves are breaking, meaning that the material contours are deforming irreversibly. It is perhaps worth noting that this difficulty has, nevertheless, been circumvented in a recent proof of a finite-amplitude version of the Rayleigh–Kuo–Charney–Stern stability theorem that is even more general than Arnol'd's version, allowing fully-turbulent wave breaking and PV mixing [36], in models 1–3. (But the Fjørtoft theorem then fails: it is easy to find counterexamples.)

It is Kelvin's circulation theorem applied to originally-zonal material contours that accounts for the otherwise mysterious fact that, even though momentum and pseudomomentum are different physical quantities related to different translational symmetry operations, they share the same off-diagonal flux terms. Kelvin's circulation theorem is exactly the ideal-flow constraint preventing what Diamond et al. (this volume) call "the slippage of a quasi-particle gas" of drift waves relative to the zonal flow.

We may think of the momentum-flux convergences that appear in (1.12), (1.13) and their generalizations, including the generalizations to finite disturbance amplitude, as effective wave-induced forces felt by the mean state. Such effective forces may or may not be equal to the actual mean-flow acceleration $\partial\langle u\rangle/\partial t$. If they are, then we have simply

$$\partial_t\left(\langle u\rangle - P\right) = 0 \tag{1.14}$$

for ideal-fluid flow. It can be shown that this always holds in model 2, and in model 1 for $L_{\mathrm{D}} = \infty$. It is not true more generally because the response to the effective mean force normally includes implicit $O(\mathrm{Ro})$ mean motions and mass fluxes in the y direction, whose Coriolis forces contribute to $\partial\langle u\rangle/\partial t$. In the case $L_{\mathrm{D}} = \infty$, and in model 2 for all L_{D}, such mean mass fluxes are shut down by the kinematic constraints. $L_{\mathrm{D}} = \infty$ models are two-dimensionally nondivergent to sufficient accuracy since with gravity $g = \infty$ we have replaced the free upper surface with a rigid lid. Otherwise it is simplest to think directly in terms of eddy fluxes of PV, focusing on the left-hand sides of the Taylor identity (1.9) or (1.10), and using PV inversion to calculate $\partial\langle u\rangle/\partial t$. Inversion implicitly takes full account of the implicit $O(\mathrm{Ro})$ mean motions and mass fluxes.

For models 1–3 there are actually four momentum-like quantities that are liable to be conflated but need to be distinguished, namely (1) momentum, (2) Eulerian pseudomomentum (of which P is a simple example), (3) Lagrangian pseudomomentum (definable at finite amplitude), and (4) Kelvin impulse. Adding to the potential for confusion, the word impulse often *means* momentum in some European languages.

Bühler's book [6] keeps these distinctions clear at all stages. It also lays out in full detail exactly how (1.14) generalizes to finite amplitude, and to atmosphere–ocean models more accurate than quasi-geostrophic, and spells out the price to be paid for such generalizations in terms of retaining the accurate Lagrangian flow information required to keep track of the shapes of material contours.

It is a nontrivial question whether or not model 2 admits any corresponding finite-amplitude results. The reason is that, in the atmosphere–ocean cases, the application of Kelvin's circulation theorem to originally-zonal material contours requires the implicit $O(\mathrm{Ro})$ mean motions and mass fluxes to be taken into account — that is, it requires consideration of the $O(\mathrm{Ro})$ corrections to the leading-order velocity field $\hat{\mathbf{z}}\times\nabla\phi$. My current conjecture is that, since the zonal-mean kinematics of model 2 would appear to constrain such corrections to be zero, at least in the zonal mean, there may well be a finite-amplitude counterpart to (1.14) though, if so, there is still the question of how far the Lagrangian information required might limit its usefulness in practice. However, all this would need to be checked in detail and preferably by someone who understands the plasma physics better than I do.

1.10 PV staircases

Focusing again on our main theme of strong nonlinearity, we note that the PV Phillips effect, Fig. 1.6, suggests the possible relevance of idealized saturated states consisting of perfectly mixed zones cut out of a background y-profile $q_{\mathrm{b}}(y)$ having a constant gradient β (again, not to be confused with plasma beta). We take $q_{\mathrm{b}}=\beta y$ as before. The graph of q against y then looks like a staircase cut out of a sloping hillside. So these idealized saturated, marginally Rayleigh–Kuo–Charney–Stern stable states are often called "PV staircases".

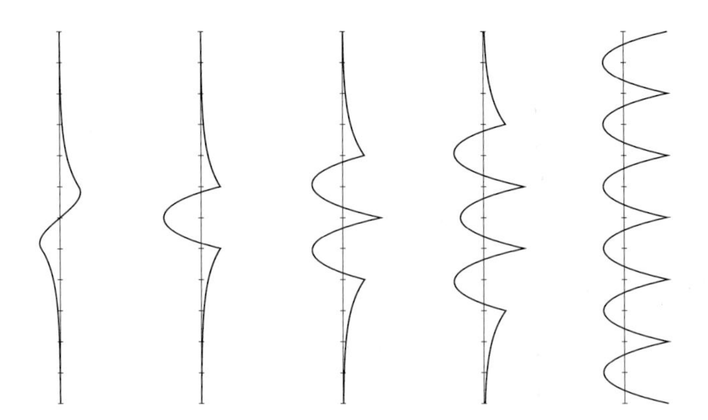

Fig. 1.7 Profiles of ϕ and u for perfect, zonally-symmetric PV staircases with step size L, in model 1. Tick marks are at intervals of $y = \frac{1}{2}L = L_{\rm D}$. From left to right, the first profile is that of ϕ for a single step and the second is the corresponding u profile. The remaining profiles are those of $u = -\phi_y$ for two, three and the limiting case of an infinite number of perfect steps. From [10]; q.v. for mathematical details of the inversions.

The corresponding $q_{\rm a}$ profile is a zigzag. Inverting this we get an array of zonally symmetric jets whose profiles depend on the PV jump $q_{\rm j}$ at each step, as well as on the step size L, i.e. the jet spacing, in units of $L_{\rm D}$. Some examples are shown in Fig. 1.7 above, for model 1 with $L/L_{\rm D} = 2$ and for staircases of one, two, and three steps along with the limiting case of an infinite number of steps. The two profiles on the left are the ϕ and u profiles for the case of one step, an idealized version of the terrestrial winter stratosphere in its usual wintertime state. The others are all u profiles. Recall that $u = -\phi_y$. The solutions are taken from [10], q.v. for mathematical details as well as for caveats regarding the conflicting uses of the term "Rhines scale", whose relation, if any, to the jet spacing L is not as straightforward as is sometimes assumed.[3]

Some researchers think that the PV-staircase idealization provides a good model for Jupiter's weather layer in extratropical latitudes — a persuasive case is made in the review article by Marcus [54] — though, hardly surprisingly, in view of our generally poor understanding of Jovian fluid

[3]One problem, sometimes forgotten, is that Rhines' original concept came from a variant of homogeneous turbulence theory in which an upscale energy "cascade" is all-important, yet the Rossby waves feel only the background PV gradient β, and therefore obey the dispersion relation (1.7). Also, $L_{\rm D} = \infty$ is often assumed, as in Rhines' original work [53]. We shall see shortly that Rossby waves on PV staircases can have dispersion properties that differ substantially from (1.7), and in a way that can drastically reshape the wave–turbulence interactions.

dynamics, this is controversial. Alternatives have been put forward, notably by Dowling [55]. What is not in doubt is that staircase-like structures of one, two and sometimes three steps are well observed and well documented in the Earth's atmosphere [31]. The observed states are not zonally symmetric; what they share with the simple staircase idealization are large, zonally-extensive regions of fairly well mixed PV on stratification surfaces. Those regions are bordered by narrow bands of concentrated PV gradients, the meandering jets already noted. If PV contours on a stratification surface are splayed out by mixing in some places, then they must be bunched up in others. (The only simple alternative consistent with (1.3) and its conservation properties [30] is an overwhelmingly improbable alternative, namely, to wipe out the entire pole-to-pole planetary PV gradient and thus kill the atmosphere's rotation $\boldsymbol{\Omega}$ altogether.) The simplest and clearest example of the splaying-out and bunching-up that I'm talking about is the winter stratosphere [5], [56], often resembling a single-step staircase idealizable as the left-hand case in Fig. 1.7.

As L/L_{D} varies, the jets retain their peaked profiles, corresponding to the infinitely tight bunching of PV contours in this idealization. They change their shapes in other ways. For large L/L_{D} we get isolated jets, still Rayleigh–Kuo–Charney–Stern marginally stable. Figure 1.8 zooms in on one such jet, near which the q_{a} zigzag looks like a simple jump discontinuity, at $y = 0$, say, as shown in the right-hand graph. The zonally-symmetric

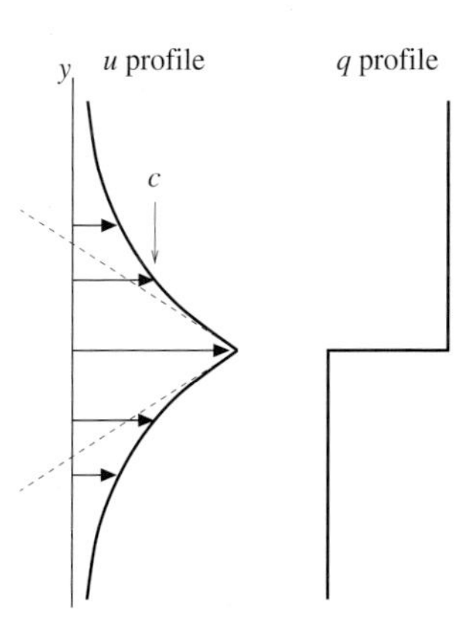

Fig. 1.8 Sketch of PV discontinuity and the corresponding velocity profile obtained by PV inversion. The step size or jet spacing L is much larger than the jet scale L_{D}, and we can ignore β and equate q_{a} and q. The solid velocity profile is from Eq. (1.15). The angle between the dashed lines is governed by the strength of the delta function $q_{\mathrm{j}}\delta(y)$.

inversion problem for u then simplifies to $\mathcal{L}u = u_{yy} - L_{\mathrm{D}}^{-2}u = -\langle q_{\mathrm{a}}\rangle_y = -\langle q\rangle_y = -q_{\mathrm{j}}\delta(y)$ and delivers the velocity profile shown in the left-hand graph,

$$u = \langle u\rangle = u_{\mathrm{j}} \exp\left(-|y|/L_{\mathrm{D}}\right) \tag{1.15}$$

where $u_{\mathrm{j}} := \frac{1}{2}L_{\mathrm{D}}q_{\mathrm{j}}$. The dashed lines in Fig. 1.8 mark the jump in the slope u_y at the jet peak, determined by the strength of the delta function $q_{\mathrm{j}}\delta(y)$ in the inversion problem. Integration of $u_{yy} - L_{\mathrm{D}}^{-2}u = -q_{\mathrm{j}}\delta(y)$ across $y = 0$ gives $\left[u_y\right]_{y=0-}^{y=0+} = -q_{\mathrm{j}}$.

Staircase models are no more than an idealization, for one thing being much more deterministic than current models of the fluctuating zonal flows in tokamaks. However, it may be interesting to speculate whether, for strong nonlinearity, the staircase idealization has some relevance to the saturation problem in real tokamaks.

There is an ill-understood dependence on how the turbulence is excited. In the vast literature on numerical forced-dissipative versions of model 1 (the "beta-turbulence" literature, mostly on model 1 with $q_{\mathrm{b}} = \beta y$ and $L_{\mathrm{D}} = \infty$), it seems that some of the numerical experiments produce sharp staircases, e.g. [57], while others produce only a faint "washboard" (e.g. B. Galperin, personal communication). This could be partly due to the fact that differing choices of artificial forcing will disrupt PV mixing to differing extents.

Most of these studies use a prescribed stochastic external force field, rather than self-excitation. We have the same difficulty in the Jupiter problem — a lack of clarity as to what kind of forcing or self-excitation best mimicks reality. There is also great uncertainty as to when or whether up-scale energy "cascades" are involved and when, by contrast, the passive shearing of repeated PV-anomaly injections is more important [58], [59], [60]. In "cascade" language this last produces an enstrophy cascade but not an energy cascade. It also disrupts simple PV mixing.

The laboratory flows exemplified by Fig. 1.1 are forced in several ways that all contrive to preserve the material invariance of PV to good approximation, giving PV mixing a chance to predominate. In the experiments that produced Fig. 1.1 and similar flows, the forcings are non-stochastic, though spatially variable. In a variety of runs the outcome, in these experiments, always seems to be robustly a flow like that in Fig. 1.1, a prograde jet that is "staircase-like" in the same sense as the flow in Fig. 1.8.

In these experiments, forcing and dissipation are kept far weaker than in most laboratory and numerical experiments. The tank in Fig. 1.1 is unusually large — 86.4 cm in diameter — and can be rotated unusually fast without disintegrating, up to 4 revolutions per second. Figure 1.1 used 3 per second, $|\mathbf{\Omega}| = 18.8\,\mathrm{s}^{-1}$. The wave–turbulence interactions that shape the flow are almost free of forcing and dissipation, and strongly nonlinear. See [13] and [14] for more details of these remarkable experiments.

The only other laboratory experiments with comparable weakness of forcing and dissipation, of which I'm aware, are those of Read et al. [61]. They use an ingenious quasi-stochastic forcing method in an extremely large tank, 13m diameter, though at much smaller $|\mathbf{\Omega}|$. Meandering jets are obtained, and temporary local reversals of PV gradients "indicative of Rossby wave breaking", though in the absence of PV maps like those in [13] and [14] it is hard to make a closer assessment. It appears, however, that in these experiments the turbulence is too weak to bring the system close to a well developed staircase-like regime. In this connection some recent numerical experiments by R. K. Scott and D. G. Dritschel are noteworthy, in that with increased computing power they can reach further into the weakly-excited, weakly-dissipative, strongly-nonlinear parameter ranges conducive to staircasing [33].

T. E. Dowling and co-workers have argued that Jupiter may be different again, possibly close to a state that is marginal not by the Rayleigh–Kuo–Charney–Stern stability criterion but, rather, by that of Arnol'd's Second Theorem. The q profiles would then be "hyperstaircase-like" to the extent that they have reversed PV gradients. There is some indirect, but I think persuasive, observational evidence that Jupiter's weather layer may be close to such a state [55]. But the question of what forcing or self-excitation mechanisms might lead to it does not yet seem to have been clearly answered — though an extremely promising candidate, direct external forcing by moist (thunderstorm) convection, has been put forward from time to time and is now under intensive investigation, e.g. [62], [63], and references therein. At first sight this would seem to suggest a leading role for the passive shearing of repeated PV-anomaly injections [58]–[60], the simplest mechanism potentially able to overcome PV-mixing tendencies and produce a hyperstaircase.

1.11 Jet self-sharpening and meandering: a toy model

The single isolated jet in Fig. 1.8 is the simplest of all Rossby waveguides. Its dispersion relation differs substantially from (1.7), and in a way that will prove interesting. It signals a kind of wave–turbulence, linear–nonlinear coupling that differs greatly from the standard homogeneous-turbulent Rhines picture.

Linearizing (1.5a) with right-hand side zero about the profiles in Fig. 1.8, with $\mathcal{L}$ defined as in (1.6a), and continuing to drop the suffix g, we have

$$\tilde{q}_t + \langle u \rangle \tilde{q}_x + \langle q \rangle_y \tilde{\phi}_x = 0 \tag{1.16}$$

where $\langle u \rangle$ is given by (1.15), and $\langle q \rangle_y = q_{\mathrm{j}} \delta(y)$ as before. For a disturbance of the standard form $\tilde{\phi} \propto \hat{\phi}(y) \exp\{ik(x - ct)\}$, with phase velocity c and wavenumber k along the jet, we have from (1.6a)

$$\tilde{q} = \mathcal{L}\tilde{\phi} = \tilde{\phi}_{yy} - (k^2 + L_{\mathrm{D}}^{-2})\tilde{\phi} \, . \tag{1.17}$$

So, with $\hat{q} := \hat{\phi}_{yy} - (k^2 + L_{\mathrm{D}}^{-2})\hat{\phi}$, (1.16) implies when $k \neq 0$ that

$$(\langle u \rangle - c)\, \hat{q} + q_{\mathrm{j}} \delta(y) \hat{\phi} = 0 \, , \tag{1.18}$$

integration of which across $y = 0$ gives

$$(u_{\mathrm{j}} - c) \left[\hat{\phi}_y \right]_{y=0-}^{y=0+} + q_{\mathrm{j}} \hat{\phi} \big|_{y=0} = 0 \tag{1.19}$$

for the jump in $\hat{\phi}_y$ across $y = 0$. Away from $y = 0$, we have $\langle q \rangle_y = 0$ hence $\hat{q} = 0$ and therefore

$$\hat{\phi}(y) \propto \exp\{-(k^2 + L_{\mathrm{D}}^{-2})^{1/2} |y|\} \, . \tag{1.20}$$

For (1.19) and (1.20) to be compatible we must have, with $u_{\mathrm{j}} = \frac{1}{2} L_{\mathrm{D}} q_{\mathrm{j}}$,

$$c = u_{\mathrm{j}} \{1 - \left(1 + L_{\mathrm{D}}^2 k^2\right)^{-1/2}\} \, , \tag{1.21}$$

which may be compared with (1.7), which has no square root in its denominator $L_{\mathrm{D}}^{-2} + k^2 + l^2$. The square root in (1.21) arises from the infinite bunching of PV contours at the jet core, giving dispersion properties distinctly different from those where the background PV contours are evenly spread out, as assumed in (1.7) and in its limiting case $L_{\mathrm{D}}^2 \to \infty$ used in the standard Rhines-scale argument.

In (1.21) notice especially that, as k varies monotonically between 0 and ∞, c varies monotonically between 0 and u_{j}. That is, for this jet there are always "critical lines" — that is, y values such that $\langle u \rangle = c$ — on each flank of the jet as indicated in Fig. 1.8. Thus, whenever the jet is disturbed, the undulations of its core will inevitably give rise to PV mixing on either side. In a frame of reference moving with the wave at speed c, one sees Kelvin-cat's-eye flow patterns in the jet flanks, which must twist up any stray PV contours like spaghetti on a fork. Whenever the undulations increase in amplitude, the widths of the mixing regions expand, tending on average to keep the PV contours bunched up in the core and maintaining the eddy-transport barrier there, in the typical way.

The presence of more than one zonal wavenumber k complicates the kinematical details, going over to Kelvin sheared-disturbance kinematics [58] for broad spectra. But none of this can suppress the PV mixing in the jet flanks. This is the typical anti-frictional jet self-sharpening process and, broadly speaking, is the kind of kinematics already encountered in Fig. 1.1 and in the other examples mentioned. It is typical of the real, highly inhomogeneous wave–turbulence interactions observed in the atmosphere and oceans — see also [64] for a good discussion of some oceanic examples — with linear, wavelike regions closely adjacent to fully nonlinear, wave-breaking regions. It is well verified in fully nonlinear numerical experiments, e.g. [65], [32]. As already emphasized, it is at an opposite extreme from a homogeneous-turbulence scenario using (1.7) as the Rossby-wave dispersion relation. Further discussion and further illustrations can be found for instance in [19], [34], and [36].

The long-wave behaviour is relevant to the issue of meandering. If we take $k^2 \ll L_\mathrm{D}^{-2}$ in (1.21) we find phase and group velocities

$$c \approx \tfrac{1}{2} u_\mathrm{j} L_\mathrm{D}^2 k^2 \, , \qquad c_\mathrm{group} \approx \tfrac{3}{2} u_\mathrm{j} L_\mathrm{D}^2 k^2 \tag{1.22}$$

that match the jet's surrounding flow velocity $\langle u \rangle \approx 0$ for sufficiently small k^2. This says that a meandering jet flows through its surroundings almost like a river. Large-scale meanders have very little propensity to propagate upstream or downstream. The same thing holds even for large-amplitude meanders, and for generalizations of model 1 beyond quasi-geostrophic theory, as shown in [66]. The meandering behaviour of a nonlinear version of our isolated-jet model is strikingly reminiscent of the large-scale, large-amplitude meandering of real atmosphere–ocean jets. The nonlinear model even succeeds in mimicking the cutoff behaviour that sheds Gulf-stream rings. Again, further discussion is in [19].

It is worth remarking that the long-wave dispersion properties just described do not depend on having an idealized, perfectly sharp jet, with infinitely close bunching of its PV contours — at least not within linear theory. As long as $\langle q \rangle_y = \langle u \rangle_{yy} - L_\mathrm{D}^{-2} \langle u \rangle$ to sufficient accuracy — which means continuing to neglect the planetary-scale background contribution β — and as long as the jet velocity profile $\langle u \rangle$ approaches zero on either side of the jet core, we always find phase speeds c that also approach zero as $k^2 \to 0$. This can be seen by inspection of (1.16)–(1.17). As long as k^2 can be neglected in (1.17), the disturbance problem (1.16) evidently has a solution $\hat{\phi} \propto \langle u \rangle$ with $c = 0$. This fact was pointed out in [67] and extensively exploited for planetary-scale Rossby waves on a strongly-sheared stratospheric polar-night jet in [68]. The physical reason is that the long-wavelength meanders tend to take place with all the jet's PV contours moving as one, as long as they are sufficiently bunched to avoid further mixing by the wave breaking on either side.

How much of this carries over to model 2? The general long-wave solution just described fails; but in the idealized sharp-jet case we get precisely the same dispersion properties, (1.21)–(1.22), as in model 1, despite having a very different jet velocity profile. In model 2 the sharp-jet velocity profile is given by the pair of dashed lines on the left of Fig. 1.8. (Model 2, with the same PV jump q_{j}, requires us to set $L_{\mathrm{D}} = \infty$ in the local inversion problem for the jet profile, i.e. to set $L_{\mathrm{D}} = \infty$ in $\langle u \rangle_{yy} - L_{\mathrm{D}}^{-2} \langle u \rangle = \langle u \rangle_{yy} = -q_{\mathrm{j}} \delta(y)$, which is satisfied by the pair of dashed lines.)

The derivation (1.16)–(1.22) of the linear wave theory applies word for word and symbol for symbol, provided that u_{j} still denotes the maximum jet velocity. With $\hat{q} = 0$ everywhere except for the delta function at $y = 0$, advection $\langle u \rangle \partial_x$ by the the mean flow $\langle u \rangle$ has no role away from $y = 0$.

The dashed lines in Fig. 1.8 are of course only a local approximation within a full staircase with large step size L; it is easy to show by restoring the β contribution to q_y that the dashed lines are really part of a set of larger-scale parabolic profiles, qualitatively like the profiles in Fig. 1.7 though shifted to the left. Again, details are given in [10]. All that matters for the present discussion is that the range of $\langle u \rangle$ values expands when we go from model 1 to model 2, and in particular that model 2 still has the property noted above that critical lines $\langle u \rangle = c$ will always be present.

1.12 Concluding remarks

The most important idea to have emerged from our excursion into atmosphere–ocean dynamics is, I'd argue, the apparent robustness of the inhomogeneous PV-mixing paradigm in strongly nonlinear two-dimensional and layerwise-two-dimensional turbulence, whose weakly-excited, weakly-dissipative parameter regimes exhibiting PV mixing are now beginning to be reached thanks to today's computing power (e.g. [33]). It is perhaps worth saying that model 2, the extended Hasegawa–Mima model, seems to point toward a similar conclusion for real tokamaks. Model 2 is heavily idealized, to be sure — for instance in assuming unrealistically low ion temperatures — but a process as robust as inhomogeneous PV mixing might well have counterparts in more realistic plasma models.

At first sight model 2 presents us with a strangely inconsistent-looking mélange of the real tokamak's zero-$k_\parallel$ and nonzero-$k_\parallel$ modes of motion. Zonal averages $\langle\phi\rangle$ in model 2 are associated with zero $k_\parallel$ because $\langle\cdot\rangle$ is a surrogate for averaging over one of the magnetic flux surfaces within the real tokamak — these toroidal surfaces being approximately fixed in space, providing a peculiar spiral scaffolding that constrains, among other things, the mutual adjustment between $\tilde{n}_e$ and $\tilde{\phi}$ in the nonzero-$k_\parallel$ modes. The zonal average $\langle\phi\rangle$ within a single two-dimensional cross-section of the torus is close to the zonal average within another such cross-section, both being close to the full flux-surface average especially in cases where each spiral field line intersects most neighbourhoods on the flux surface. By contrast, the fluctuating velocity fields $\tilde{\mathbf{u}}(\mathbf{x}, t)$ about such a zonal average tend to have nonzero values of $k_\parallel$, subjecting them to the mutual adjustment between $\tilde{n}_e$ and $\tilde{\phi}$ and making them different in different cross-sections.

However, as already suggested in Sec. 1.8, a key point about advective *mixing* processes is that they are insensitive to details in the fluctuating velocity fields responsible for the mixing. Thus, for instance, if in model 2 a two-dimensional fluctuating velocity field $\tilde{\mathbf{u}}(\mathbf{x}, t)$ were to be replaced by $-\tilde{\mathbf{u}}(\mathbf{x}, t)$, then mixing could still be expected to occur. A random walk remains a random walk when all the displacements are multiplied by -1. Thus if model 2 exhibits inhomogeneous PV mixing, then its entire behaviour could be a surprisingly good surrogate for the real, three-dimensional behaviour.

Of course the PV field is not a passively-advected scalar, even though the scale effect (Sec. 1.7) tends to make the PV almost passive and the velocity field effective at mixing, to a remarkable extent in many observed atmosphere–ocean cases. These often have the character of breaking Rossby waves, involving quasi-passive mixing by a $\mathbf{u}$ field whose scale is generally larger than the rapidly-shrinking scale of the advected PV features. It seems possible, then, that similar ideas might help toward understanding the dynamics of strongly nonlinear drift-wave turbulence.

In summary, we might reasonably expect that the zonally-averaged — in reality flux-surface-averaged — zonal flows in tokamaks will be strongly affected by inhomogeneous PV mixing even though the details of the mixing process must change as we go around the torus. It would be of great interest to see in detail what model 2's PV fields actually look like in the strongly nonlinear, weakly-excited, weakly-dissipative cases accessible to today's computers. To my knowledge there are no such fields published, though zonal-flow generation in model 2 has been clearly demonstrated in reference [41]. If we were to zoom in to look at the fine details, would the PV look anything remotely like Fig. 1.4 above?

Note added May 2013: Jupiter now looks like a different case again. From current work with my graduate student Stephen Thomson an impression is growing that Jupiter has an altogether stronger "scaffolding" in the form of deep, relatively steady, zonally symmetric jets generated in the massive cloud-free convection layer that underlies the weather layer. It now seems likely that such steady deep jets exist in reality and guide the upper, weather-layer jets, suppressing what would otherwise be their tendency to meander. This is despite Jupiter's weather layer being in a regime similar to that discussed in Sec. 1.11, in which long-wavelength meandering would be very easy to excite, in the absence of the deep jets.

Acknowledgments

I am very grateful to Laurène Jouve and Chris McDevitt for producing the first draft and to Steve Cowley, Pat Diamond, Philippe Ghendrih, Alexey Mishchenko, Gabriel Plunk, and Timothy Stoltzfus-Dueck for patiently explaining some aspects of tokamak and stellarator dynamics, and for help with the plasma literature and its notational conventions. Tim Dowling, Peter Read, Richard Wood, and Stephen Thomson have helped me with recent developents in Jovian dynamics, real and idealized. Huw Davies helped with recent terrestrial developments, and Steven Balbus and Toby Wood with solar.

References

[1] P. H. Diamond, S.-I. Itoh, K. Itoh, and T. S. Hahm. Zonal flows in plasmas — a review. *Plasma Phys. Control. Fusion*, 47:R35–R161, 2005.

[2] P. W. Terry. Suppression of turbulence and transport by sheared flow. *Rev. Mod. Phys.*, 72:109–165, 2000.

[3] J. A. Krommes. The gyrokinetic description of microturbulence in magnetized plasmas. *Ann. Rev. Fluid Mech.*, 44:175–201, 2012.

[4] M. E. McIntyre and T. N. Palmer. Breaking planetary waves in the stratosphere. *Nature*, 305:593–600, 1983.

[5] M. E. McIntyre. A tale of two paradigms, with remarks on unconscious assumptions. Haurwitz Memorial Lecture presented to the American Meteorological Society, 19th Conference on Atmospheric and Oceanic Fluid Dynamics and 17th Conference on the Middle Atmosphere, available from www.atm.damtp.cam.ac.uk/people/mem/#haurwitz, 2013. Slide 29 shows a state-of-the-art movie of PV in the real stratosphere, courtesy Dr A. J. Simmons of the European Centre for Medium-Range Weather Forecasts.

[6] O. Bühler. *Waves and Mean Flows*. University Press, Cambridge, 2009. Also 2013: paperback with corrections, in press.

[7] E. N. Lorenz. *The Nature and Theory of the General Circulation of the Atmosphere*. World Meteor. Org., Geneva, 1967.

[8] V. P. Starr. *Physics of Negative Viscosity Phenomena*. McGraw-Hill, 1968.

[9] J. H. Rogers. *The Giant Planet Jupiter*. Cambridge University Press, 1995.

[10] D. G. Dritschel and M. E. McIntyre. Multiple jets as PV staircases: the Phillips effect and the resilience of eddy-transport barriers. *J. Atmos. Sci.*, 65:855–874., 2008.

[11] M. N. Juckes and M. E. McIntyre. A high resolution, one-layer model of breaking planetary waves in the stratosphere. *Nature*, 328:590–596, 1987.

[12] E. F. Danielsen. Stratospheric-tropospheric exchange based on radioactivity, ozone and potential vorticity. *J. Atmos. Sci.*, 25:502–518, 1968.

[13] J. Sommeria, S. D. Meyers, and H. L. Swinney. Laboratory model of a planetary eastward jet. *Nature*, 337:58–61, 1989.

[14] J. Sommeria, S. D. Meyers, and H. L. Swinney. Experiments on vortices and Rossby waves in eastward and westward jets. In A. R. Osborne, editor, *Nonlinear Topics in Ocean Physics*, pages 227–269, Amsterdam, 1991. North-Holland.

[15] W. A. Norton. Breaking Rossby waves in a model stratosphere diagnosed by a vortex-following coordinate system and a technique for advecting material contours. *J. Atmos. Sci.*, 51:654–673, 1994.

[16] D. W. Waugh and R. A. Plumb. Contour advection with surgery: a technique for investigating finescale structure in tracer transport. *J. Atmos. Sci.*, 51:530–540, 1994.

[17] H. Riehl. *Jet Streams of the Atmosphere.* Colorado State University, Fort Collins, CO., USA, 1962. Tech Paper No. 32.

[18] E. Palmén and C. W. Newton. *Atmospheric Circulation Systems.* Academic Press, 1969.

[19] M. E. McIntyre. Potential-vorticity inversion and the wave-turbulence jigsaw: some recent clarifications. *ADGEO (Advances in Geosciences)*, 15:47–56, 2008.

[20] C.-G. Rossby. Dynamics of steady ocean currents in the light of experimental fluid mechanics. *Mass. Inst. of Technology and Woods Hole Oceanogr. Instn., Papers in Physical Oceanography and Meteorology*, 5(1):1–43, 1936.

[21] C.-G. Rossby. On the mutual adjustment of pressure and velocity distributions in certain simple current systems. II. *J. Mar. Res.*, 2:239–263, 1938.

[22] C.-G. Rossby. Planetary flow patterns in the atmosphere. *Q. J. Roy. Meteorol. Soc.*, 66 (Suppl.):68–87, 1940.

[23] H. Ertel. Ein neuer hydrodynamischer Wirbelsatz. *Met. Z.*, 59:271–281, 1942.

[24] J. G. Charney. On the scale of atmospheric motions. *Geofysiske Publ.*, 17(2):3–17, 1948.

[25] A. M. Obukhov. On the problem of the geostrophic wind. *Izv. Akad. Nauk SSSR, Ser. Geograf. Geofiz.*, 13(4):281–306, 1949. in Russian.

[26] E. Kleinschmidt. Über Aufbau und Entstehung von Zyklonen (1 Teil). *Meteorol. Runds.*, 3:1–6, 1950.

[27] B. J. Hoskins, M. E. McIntyre, and A. W. Robertson. On the use and significance of isentropic potential-vorticity maps. *Q. J. Roy. Meteorol. Soc.*, 111:877–946, 1985. Corrigendum 113, 402–404.

[28] C. D. Thorncroft, B. J. Hoskins, and M. E. McIntyre. Two paradigms of baroclinic-wave life-cycle behaviour. *Q. J. Roy. Meteorol. Soc.*, 119:17–55, 1993.

[29] C. Appenzeller, H. C. Davies, and W. A. Norton. Fragmentation of stratospheric intrusions. *J. Geophys. Res.*, 101:1435–1456, 1996.

[30] P. H. Haynes and M. E. McIntyre. On the conservation and impermeability theorems for potential vorticity. *J. Atmos. Sci.*, 47:2021–2031, 1990.

[31] O. Martius, C. Schwierz, and H. C. Davies. Tropopause-level waveguides. *J. Atmos. Sci.*, 67:866–879, 2010.

[32] J. G. Esler. The turbulent equilibration of an unstable baroclinic jet. *J. Fluid Mech.*, 599:241–268, 2008.

[33] R. K. Scott and D. G. Dritschel. The structure of zonal jets in geostrophic turbulence. *J. Fluid Mech*, 711:576–598, 2012.
[34] D. G. Dritschel and R. K. Scott. Jet sharpening by turbulent mixing. *Phil. Trans. R. Soc. A*, 369:754–770, 2011.
[35] M. E. McIntyre and T. N. Palmer. A note on the general concept of wave breaking for Rossby and gravity waves. *Pure Appl. Geophys.*, 123:964–975, 1985.
[36] R. B. Wood and M. E. McIntyre. A general theorem on angular-momentum changes due to potential vorticity mixing and on potential-energy changes due to buoyancy mixing. *J. Atmos. Sci.*, 67:1261–1274, 2010.
[37] O. M. Phillips. Turbulence in a strongly stratified fluid — is it unstable? *Deep Sea Res.*, 19:79–81, 1972.
[38] P. H. Haynes, D. A. Poet, and E. F. Shuckburgh. Transport and mixing in kinematic and dynamically-consistent flows. *J. Atmos. Sci.*, 64:3640–3651, 2007.
[39] B. R. Ruddick, T. J. McDougall, and J. S. Turner. The formation of layers in a uniformly stirred density gradient. *Deep Sea Res.*, 36:597–609, 1989.
[40] J. A. Krommes and C. B. Kim. Interactions of disparate scales in drift-wave turbulence. *Phys. Rev. E.*, 62:8508–8539, 1990.
[41] S. Gallagher, B. Hnat, C. Connaughton, S. Nazarenko, and G. Rowlands. The modulational instability in the extended Hasegawa–Mima equation with a finite Larmor radius. *Phys. Plasmas*, 19:122115, 2012.
[42] F. P. Bretherton. Baroclinic instability and the short wavelength cut-off in terms of potential vorticity. *Q. J. Roy. Meteorol. Soc.*, 92:335–345, 1966.
[43] T. Stoltzfus-Dueck, B.D. Scott, and J.A. Krommes. Nonadiabatic electron response in the Hasegawa–Wakatani equations. *Phys. Plasmas*, 20, paper no. 082314, 2013.
[44] M. Wakatani and A. Hasegawa. A collisional drift wave description of plasma edge turbulence. *Phys. Fluids*, 27:611–618, 1984.
[45] A. F. Thompson and W. R. Young. Two-layer baroclinic eddy heat fluxes: zonal flows and energy balance. *J. Atmos. Sci.*, 64:3214–3231, 2007.
[46] G. I. Taylor. Eddy motion in the atmosphere. *Phil. Trans. Roy. Soc. Lond.*, A215:1–23, 1915.
[47] J. G. Charney and M. E. Stern. On the stability of internal baroclinic jets in a rotating atmosphere. *J. Atmos. Sci.*, 19:159–172, 1962.

[48] F. P. Bretherton. Critical layer instability in baroclinic flows. *Q. J. Roy. Meteorol. Soc.*, 92:325–334, 1966.

[49] R. E. Dickinson. Theory of planetary wave–zonal flow interaction. *J. Atmos. Sci.*, 26:73–81, 1969.

[50] D. G. Andrews and M. E. McIntyre. Planetary waves in horizontal and vertical shear: the generalized Eliassen–Palm relation and the mean zonal acceleration. *J. Atmos. Sci.*, 33:2031–2048, 1976.

[51] J. G. Charney and P. G. Drazin. Propagation of planetary-scale disturbances from the lower into the upper atmosphere. *J. Geophys. Res.*, 66:83–109, 1961.

[52] M. E. McIntyre and T. G. Shepherd. An exact local conservation theorem for finite-amplitude disturbances to nonparallel shear flows, with remarks on Hamiltonian structure and on Arnol'd's stability theorems. *J. Fluid Mech.*, 181:527–565, 1987.

[53] P. B. Rhines. Waves and turbulence on a beta-plane. *J. Fluid Mech.*, 69:417–443, 1975.

[54] P. S. Marcus. Jupiter's Great Red Spot and other vortices. *Ann. Rev. Astron. Astrophys.*, 31:523–573, 1993.

[55] T. E. Dowling. A relationship between potential vorticity and zonal wind on Jupiter. *J. Atmos. Sci.*, 50:14–22, 1993.

[56] M. Riese, G. L. Manney, J. Oberheide, X. Tie, R. Spang, and V. Küll. Stratospheric transport by planetary wave mixing as observed during CRISTA-2. *J. Geophys. Res.*, 107(D23), paper no. 8179, 2002.

[57] S. Danilov and D. Gurarie. Scaling, spectra and zonal jets in β-plane turbulence. *Phys. Fluids*, 16:2592–2603, 2004.

[58] W. Thomson (Lord Kelvin). Stability of fluid motion — rectilineal motion of viscous fluid between two parallel planes. *Phil. Mag.*, 24:188–196, 1887.

[59] B. F. Farrell and P. J. Ioannou. Structure and spacing of jets in barotropic turbulence. *J. Atmos. Sci.*, 64:3652–3665, 2007.

[60] K. Srinivasan and W. R. Young. Zonostrophic instability. *J. Atmos. Sci.*, 69:1633–1656, 2012.

[61] P. L. Read, Y. H. Yamazaki, S. R. Lewis, P. D. Williams, R. Wordsworth, K. Miki-Yamazaki, J. Sommeria, and H. Didelle. Dynamics of convectively driven banded jets in the laboratory. *J. Atmos. Sci.*, 64:4031–4052, 2007.

[62] A. P. Ingersoll, P. J. Gierasch, D. Banfield, W. R. Vasavada, and the Galileo Imaging Team. Moist convection as an energy source for the large-scale motions in Jupiter's atmosphere. *Nature*, 403:630–632, 2000.

[63] Y. Lian and A. P. Showman. Generation of equatorial jets by large-scale latent heating on the giant planets. *Icarus*, 207:373–393, 2010.
[64] C. W. Hughes. The Antarctic Circumpolar Current as a waveguide for Rossby waves. *J. Phys. Oceanogr.*, 26:1375–1387, 1996.
[65] M. D. Greenslade and P. H. Haynes. Vertical transition in transport and mixing in baroclinic flows. *J. Atmos. Sci.*, 65:1137–1157, 2008.
[66] J. Nycander, D. G. Dritschel, and G. G. Sutyrin. The dynamics of long frontal waves in the shallow water equations. *Phys. Fluids*, 5(5):1089–1091, 1993.
[67] P. G. Drazin and L. N. Howard. The instability to long waves of unbounded parallel inviscid flow. *J. Fluid Mech.*, 14:257–283, 1962.
[68] A. J. Simmons. Planetary-scale disturbances in the polar winter stratosphere. *Q. J. Roy. Meteorol. Soc.*, 100:76–108, 1974.

Chapter 2

A Review of the Possible Role of Constraints in MHD Turbulence

Annick Pouquet[1,2,3]

[1] *Visiting Research Scientist,*
Laboratory for Atmospheric and Space Science,
Colorado University, Boulder CO 80304, USA
[2] *Adjunct Professor,*
Department of Applied Mathematics,
Colorado University, Boulder CO 80304, USA
[3] *Senior Scientist, Emeritus,*
National Center for Atmospheric Research,
P.O. Box 3000, Boulder CO 80307, USA

2.1 Introduction

2.1.1 *The context*

Astrophysical and geophysical flows are highly turbulent in general, be it only because they extend on a large ratio of interacting scales. Hence, one expects dissipative processes to play a secondary role as far as energy exchanges among scales are concerned, even though at the smallest scales where extreme events such as solar flares and coronal mass ejections can occur, heating prevails. Thus, it is thought that the overall dynamics results from nonlinear coupling between Fourier modes, waves, eddies and coherent structures in interaction. There have been many observations of turbulence in such flows. Our close environment such as the magnetosphere, the solar wind [22, 146, 157] and the Sun itself, as well as the interstellar medium (ISM) [40] are particularly well documented for their turbulence properties; in the ISM, the high Mach number implies that interactions with the density and self-gravity play determining roles as well. Observations also indicate that magnetic fields are main agents of the acceleration of charged particles, contributing to shape auroral emissions and magnetospheric-ionospheric coupling, as in the case of Jupiter [128].

When such fields are present, they are often (but not always) in quasi-equipartition with either the gas pressure and/or the velocity. If plasma (kinetic) effects play a determining role at small scale, as observations show clearly [130], the magnetohydrodynamics (MHD) regime – whereby the displacement current is neglected at low velocity compared to the speed of light – is important because it covers the dynamics of the large scales which hold the energy. Indeed, with an energy spectrum $E(k) \sim k^{-m}$, with $1 \leq m \leq 3$ for energy and dissipation to be defined and interactions to be local, then most of the energy is contained at large scale described by the MHD limit. For that reason, however unrealistic this description becomes at small scales, the large-scale dynamics of complex magnetized flows can be described successfully by MHD. Beyond scaling laws, turbulent flows are known to develop characteristic structures: plasma sheet turbulence [153] and turbulence in a geomagnetic storm [129] were observed using remote sensing, such as with the CLUSTER configurations of four satellites, and such observations have yielded a wealth of detailed data on these complex flows, with for example coherent vortices [145] and vortex filaments in the magneto-sheath [3]. Note also that the interplay between small-scale waves, such as Langmuir waves, and turbulent eddies contribute to forging the shapes of the wave-packets, as observed in the solar wind [62].

Even at high Reynolds number, the energy eventually gets dissipated if not in shocks in the supersonic case, then in intermittent and quasi-singular structures (vortex filaments, current sheets) that are as thin as the dynamics permits; roughly speaking, of the order of $\sqrt{R_V}$, as for example for a one-dimensional Burgers shocks or a Harris current sheet described locally by a hyperbolic tangent profile, where R_V is the Reynolds number to be defined below. The precise shape of those dissipative structures, the fact that they roll-up or not (see [61, 108] for observations of such rolls in the solar wind), and hence presumably the rate at which energy is dissipated, in a finite time or not, does depend on interactions between large-scale and small-scale processes. There are special cases for which there is a proof for the lack of singularity in a finite time, such as neutral fluids in two space dimensions (2D), or fluids described by the Lagrangian-averaged approach in three space dimensions (3D)[29, 63, 64]; but in general, the available amount of energy in the system is transferred to the small scales where it is dissipated at the rate at which it is transferred (otherwise a run-off situation would occur). This is what has been observed rather unambiguously using direct numerical simulations, for neutral fluids [66, 68], and in MHD both in 2D [14, 113] and in 3D [91].

Turbulent flows are known for their complex behavior, both in time (think of strange attractors) and in space. This is due to the nonlinearity of the advection term, and in magnetohydrodynamics (MHD), of the Lorentz force and Ohm's law. These quadratically nonlinear terms, under the assumption of incompressibility, lead, in Fourier space to a convolution which couples $\mathbf{k}$, $\mathbf{p}$ and $\mathbf{q}$ modes together such that $\mathbf{k} = \mathbf{p} + \mathbf{q}$ is fulfilled; these are the so-called triadic interactions which conserve energy and any other quadratic invariants I_Q (see below) in each of their individual exchanges; hence one can truncate the system and still conserve the energy and the I_Qs.

In the absence of dissipation, nothing can arrest this coupling of modes, with the invariants being transferred both to large scales (as modeled by eddy noise or negative turbulent diffusivities for example, and as observed in inverse cascades in the forced dissipative case) and to small scales (as modeled by positive eddy viscosity and eddy resistivity, and leading to what is called a direct cascade). As far as jargon is concerned, let us reserve the vocable "cascade" for a transfer with a *constant* flux across scales in the so-called inertial range, as opposed to transfer which is simply referring to nonlinear coupling with a possible scale-dependency of the flux. In this latter case, this may arise from Ekman friction for rotating fluids or from anisotropic Joule damping in the quasi-static limit of MHD at low magnetic Reynolds number [150, 151]. Also note that, in the ideal (non-dissipative) case, the cascades are flux-less.

2.1.2 *The Kolmogorov law*

The question we want to address now is: what is the distribution of energy among modes in a turbulent fluid? Anything? Or are there constraining rules? Phenomenology can come to the rescue: in the absence of any other time-scale but the advection time or nonlinear time $\tau_{NL} = \ell/u_\ell$ with u_ℓ the velocity at scale ℓ, in the fluid case, there is a solution when one assumes, following Kolmogorov, that the energy spectrum depends only on the energy injection rate $\epsilon \equiv DE_T/Dt$ and the wavenumber k, henceforth the so-called K41 law [70]:

$$E(k) \sim \epsilon^{2/3} k^{-5/3} .$$

This law is well verified in the laboratory and in observations of the atmosphere and the ocean, in certain parameter regimes, although there are corrections due to intermittency, i.e. the presence of strong small-scale localized structures in the form of vortex filaments. One can extend the K41 law to higher-order moments of the structure function of a given field $\mathbf{u}$ (say, the velocity field) computed on a distance $\mathbf{r}$ between two points, and assuming homogeneity:

$$\delta\mathbf{u}(\mathbf{r}) = \mathbf{u}(\mathbf{x}+\mathbf{r}) - \mathbf{u}(\mathbf{x}), \tag{2.1}$$

with $u_L = \mathbf{u} \cdot \hat{r}$ the longitudinal structure function of $\mathbf{u}$, the unit vector along distance $\mathbf{r}$ being defined as $\hat{r} = \mathbf{r}/|r|$. In the case of a self-similar field for classical (K41) turbulence, one has:

$$\langle \delta u_L(r)^p \rangle \sim r^{\zeta_p}, \quad \zeta_p = p/3\,;$$

this law is not verified by numerical data, nor by atmospheric data. Corrections to the self-similar scaling stem from intermittency, or the predominance at high-order of quasi-singular (dissipative) structures.

One could also consider large-scale shear, in which case the same type of dimensional analysis gives $E(k) \sim k^{-1}$ or, in MHD, one can think of the Alfvén time associated with the propagation of waves along an imposed uniform magnetic field $\mathbf{B}_0$, namely $\tau_A = \ell/B_0$. The simplest solution could be to ignore these time scales and say that interactions will take place on the nonlinear time-scale as in the standard fluid case, i.e. the eddy turn-over time τ_{NL}.

Is this the case?

This paper is devoted to a rapid review of where we stand now on this thorny problem. Thorny because it is difficult to measure spectra in astrophysics whereas it is easy to do so, at least in principle, with Direct Numerical Simulations (DNS) under the assumption of isotropy. However, the Reynolds numbers achieved in DNS in three space dimensions are moderate, hence the extent of the range of scales where such laws can apply omitting the influence of the forcing and of the dissipation, is not very large. And this issue is thorny because, when there is more than one time-scale involved, dimensional analysis alone cannot give us the answer.

Moreover, in the case of forced flows achieving a statistically steady state, the time-scale associated with the forcing itself (white noise, red-noise with some coherence time, or constant) may play a role as well. So could the existence or not of a uniform imposed magnetic field, as well as the aspect ratio of the computational box and the presence of boundary layers. Also, it is found in numerous studies of two-dimensional fluid turbulence [34, 106, 149] that when adding friction at large scale, this can modify both the aspect of the coherent structures that form in the inverse cascade and the scaling law for the distribution of energy (such structures can be detected using wavelet algorithms, see [78] and references therein). Similarly, there are observations, both in the solar wind [112] and for solar flares [1], that indicate variations with time/space of spectral exponents for MHD turbulence.

Finally, one needs to stress that the value achieved by such a spectral index in MHD turbulence is not simply a matter of a theoretical nature or numerology. It can also have consequences, such as on the resulting heating rate of coronal events as a function of the axial field [127].

2.1.3 *The equations*

We give for reference the dynamical equations for MHD. They read, for an incompressible fluid with $\mathbf{v}$ and $\mathbf{b}$ respectively the velocity and magnetic fields in Alfvénic units, assuming a uniform density equal to unity:

$$\frac{\partial \mathbf{v}}{\partial t} + \mathbf{v} \cdot \nabla \mathbf{v} = -\nabla \mathcal{P} + \mathbf{j} \times \mathbf{b} + \nu \nabla^2 \mathbf{v}\,, \tag{2.2}$$

$$\frac{\partial \mathbf{b}}{\partial t} = \nabla \times (\mathbf{v} \times \mathbf{b}) + \eta \nabla^2 \mathbf{b}\,. \tag{2.3}$$

Note that $\mathbf{b}$ is, dimensionally, a velocity as well, the Alfvén velocity. $\mathcal{P}$ is the total pressure, $\nabla \cdot \mathbf{v} = \nabla \cdot \mathbf{b} = 0$, and ν and η are respectively the kinematic viscosity and magnetic diffusivity as stated before. With $\nu = 0,\ \eta = 0$, the energy E_T, the cross helicity H_C and the magnetic helicity H_M, defined as

$$E_T = E_V + E_M = \left\langle v^2 + b^2 \right\rangle /2, \quad H_C = \langle \mathbf{v} \cdot \mathbf{b} \rangle /2\ , \quad H_M = \langle \mathbf{A} \cdot \mathbf{b} \rangle /2\,,$$

where $\mathbf{b} = \nabla \times \mathbf{A}$ with $\mathbf{A}$ the magnetic potential, are conserved (factors of $1/2$ in the helicities are here for convenience, to agree with definitions given in [42]). Note that in the strict two dimensional case (2D, $v_z \equiv 0,\ b_z \equiv 0$), magnetic helicity is identically zero and this invariant is replaced by $\left\langle A_z^2 \right\rangle$ (and higher order moments associated with so-called Casimirs, which are not preserved by the Fourier truncation, though, beyond second order).

The kinetic energy spectrum is defined as the Fourier transform of the velocity two-point correlation function. Once homogeneity, isotropy and incompressibility have been taken into account, only two defining functions remain: $E_V(k)$ is proportional to the kinetic energy, with $\int E_V(k)dk =$

$E_V = \frac{1}{2}\langle v^2 \rangle$, and the kinetic helicity, $H_V(k)$, stems from the anti-symmetric part of the velocity gradient tensor; helicity is a pseudo-scalar, with $\int H_V(k)dk = \frac{1}{2}\langle \mathbf{v} \cdot \omega \rangle$; H_V is an invariant in the absence of viscosity in the pure fluid case but not in MHD.

One can also introduce the Elsässer variables $\mathbf{z}^{\pm} = \mathbf{v} \pm \mathbf{b}$; defining $D_{\pm}/Dt = \partial_t + \mathbf{z}^{\pm} \cdot \nabla$, one can then obtain a more symmetric form of Eqs. (2.2) and (2.3) and write:

$$\frac{D_{\mp}\mathbf{z}^{\pm}}{Dt} = -\nabla \mathcal{P} + \frac{\nu + \eta}{2} \nabla^2 \mathbf{z}^{\pm} + \frac{\nu - \eta}{2} \nabla^2 \mathbf{z}^{\mp} . \tag{2.4}$$

There are no self-advected non-linearities z^+z^+ or z^-z^-, hence the common conjecture that in some sense MHD turbulence is weaker than its fluid counterpart, the effect being associated with Alfvén waves (corresponding to $\mathbf{z}^{\pm} = 0$). Finally, the kinetic and magnetic Reynolds numbers are defined as

$$R_V = U_0 L_0/\nu, \quad R_M = U_0 L_0/\eta ,$$

where U_0, L_0 are the characteristic velocity and length scale; one could define similarly

$$R_{\pm} = Z_0^{\mp} L_0/\nu^{\pm} ,$$

with $Z_0^{\pm} = U_0 \pm B_0$ and $2\nu_{\pm} = \nu \pm \eta$, with B_0 being the magnitude of $\mathbf{b}$, a characteristic large-scale magnetic field. We shall assume $\nu = \eta$ in the following, hence $R_V = R_M$ and the magnetic Prandtl number $P_M = \nu/\eta = 1$ unless otherwise stated; however $Z_0^{\pm}$ can be quite different, depending on the amount of cross-correlation in the flow. Of course, when $\nu = \eta$, hence $\nu_- = 0$, the $\pm$ Reynolds number defined above reduce to one, R_+.

In the next section, we analyze the case of ideal (non-dissipative) dynamics in three space dimensions, and examine energy spectra in §2.3. The role of the exact flux laws that can be written in MHD is reviewed in §2.4 and an example of lack of universality in MHD is dealt with in §2.5. The next two sections examine the role played by the degree of alignment between the velocity and the magnetic field (§2.6), and by magnetic helicity (§2.7). In the penultimate section, one deals briefly with how models of turbulence can help, and finally §2.9 is the conclusion. Note that some of the concepts to be delineated below are well known, see e.g. [13] (see also [28, 123–126], and references therein), and the implications for astrophysical turbulence have been reviewed for example in [49, 146, 157].

2.2 What can we learn from statistical mechanics

A long-standing concept is that the dynamics of turbulent flows is fully encompassed in the nonlinearities of the primitive equations, provided we are not in a weak turbulence regime where waves may be the dominant effect, at least at early times, and provided as well that the Reynolds number is very high, i.e. that, for the most part, dissipative effects are negligible. Note however that, through mode coupling, there is no way to arrest the cascade to small scales and therefore there exists a scale, called the dissipative scale ℓ_D, at which dissipation sets-in, no matter how small the viscosity is; it is found simply by equating the time associated with dissipation, $\tau_{diss} = \ell_D^2/\nu$, and the relevant (say, nonlinear) time at that scale. This leads to what is called reconnection of field lines that are broken by dissipation (see e.g., [74, 86, 88, 113, 115, 134, 148]). It gives rise, in astrophysics and plasma physics, to energetic phenomena such as coronal mass ejections and solar flares (see [104] for a large-scale reconnection event observed in the magnetosphere).

In such a framework, one can consider the dynamics of a truncated ensemble of modes under their nonlinear dynamics. It was shown by T. D. Lee [75], for three-dimensional fluids and MHD, that one can write in fact statistical equilibria for such systems, taking into consideration the only constraint, in the absence of dissipation, of the energy conservation since it is preserved by the truncation (for the one-dimensional problem, see [23, 67] and references therein). In the simplest case, this leads to equipartition among Fourier modes, and thus to an energy spectrum in dimension d, $E(k) \sim k^{d-1}$, i.e. proportional to the number of modes in each (isotropic) Fourier shell.

It was soon realized that, when there is more than one invariant (in fact, the generic case for fluids and in MHD as well), strange things can happen: namely, that when the relative importance of these constraints, due to invariants, differ, the resulting spectra can vary profoundly. On this basis, Kraichnan postulated an inverse cascade of energy toward large scales in 2D Navier-Stokes , assuming that adding viscosity to the system would allow for an energetic balance but would not alter the nonlinear dynamics (see [72]). It was shown recently for the Kolmogorov spectrum in 3D ideal fluids obtained at intermediate times and intermediate scales [33], that the small-scale k^2 equipartition spectrum acts as an effective eddy viscosity to these intermediates scales, even though the equipartition spectrum is progressing, slowly, toward the large scales. Even in the presence of waves, one can observe these equilibria although it takes a substantially longer time as the linear term grows in strength, for example in rotating turbulence as the Rossby number decreases [93] (note that the so-called "1/f" noise can also develop in turbulent flows for long times, showing the importance of memory effects, see e.g. for a recent study [38] and references therein). A similar conclusion of intermediate time-intermediate scale turbulent-like behavior in ideal fluids is reached in MHD in two dimensions [43, 73, 152], although the power-law index is not necessarily a K41 law (see §2.3). Presumably, the same happens in three dimensions as well (see also [58]).

So, what are these statistical equilibria in 3D MHD? They were derived in [42]; with α, β and γ the Lagrange multipliers associated with the E_T, H_M and H_C invariants, namely

$$\alpha E_T + \beta H_M + \gamma H_C\,;$$

note that β does not have the same physical dimension as α and γ. These equilibria read:

$$H_J(k) = -\frac{8\pi\beta}{\alpha^2\Gamma^4}\frac{k^2}{\mathcal{D}(k)}\,; \quad H_V(k) = \frac{\gamma^2}{4\alpha^2}\,H_J(k)$$

$$H_M(k) = \frac{1}{k^2}\,H_J(k)\,; \qquad H_C(k) = \frac{\gamma\Gamma^2}{2\beta}\,H_J(k) \tag{2.5}$$

$$E_M(k) = -\frac{\alpha\Gamma^2}{\beta}\,H_J(k)\,; \quad E_V(k) = \left(1 - \frac{\beta^2}{4\alpha^2\Gamma^2}\frac{1}{k^2}\right)E_M(k) \tag{2.6}$$

where $H_J = \int k^2 H_M(k)dk$ is the current helicity, and

$$\alpha > 0, \quad \Gamma^2 = 1 - \frac{\gamma^2}{4\alpha^2} > 0$$

$$\mathcal{D}(k) = \left(1 - \frac{\beta^2}{\alpha^2\Gamma^4}\frac{1}{k^2}\right) > 0 \; \forall k \in [k_{\min}, k_{\max}]\,. \tag{2.7}$$

Conditions on these multipliers and relations with the minimum wavenumber in the system are such that realizability is ensured (see [42] for details). By realizability is meant positivity of the energy spectrum, or a Schwarz inequality relating energy and helicity spectra, namely $H(k) \leq kE(k)$ with maximal helicity in the case of an equality, corresponding to alignment (parallel or anti-parallel) of say the velocity and the vorticity.

All Fourier spectra are proportional, i.e. equivalent, to within constants based on the Lagrange multipliers i.e. on the relative importance of the three invariants imposed to the truncated system of modes at $t = 0$. This is true except for the kinetic energy, which has to recover its non-MHD formulation when $\mathbf{b} \equiv 0$. More importantly perhaps, $\forall k$, one has $E_V(k) \leq E_M(k)$, a point already noted in [143]. In fact, defining the relative modal energy and helicity $E_R(k)$ and $H_R(k)$, we have $E_R(k) \leq 0$, and $H_R(k) * H_M(k) \leq 0$, or more specifically:

$$E_R(k) \equiv E_V(k) - E_M(k) = -E_M(k)\frac{\beta^2}{4\alpha^2\Gamma^2} \leq 0\,, \tag{2.8}$$

$$H_R(k) \equiv H_V(k) - k^2 H_M(k) = -\Gamma^2 k^2 H_M(k)\,. \tag{2.9}$$

Examining a bit further these statistical equilibria (see also [141]), one observes that there can be an inverse cascade of magnetic helicity for proper values of β with respect to the other two multipliers α, γ (note that $H_{C,M}$ are pseudo-scalar and that, unlike energy, they can have either $\pm$ sign). Moreover, equipartition ($E_R \equiv 0$, $H_R \equiv 0$) does not occur except for zero magnetic helicity ($\beta \equiv 0$), or for maximal correlation in the initial state, as well as for very large systems ($k \to \infty$). It should be noted here that a lack of equipartition is viewed consistently in the solar wind [83], with most often a slight excess of magnetic energy. It can be interpreted as due to strong v-b correlations, arising from the source. However, since a similar trend is also seen in numerical simulations and in models of MHD turbulence, in each case the slight excess of E_M can likely be attributed to ideal dynamics prevailing at high Reynolds number. Other remarks, perhaps a bit unexpected, are that:

- In the absence of correlations ($\gamma \equiv 0$, $\Gamma^2 \equiv 1$), of course $H_C(k) \equiv 0$, but so is the kinetic helicity, $H_V(k) \equiv 0$, which can thus be viewed in ideal MHD as being due totally to the amount of alignment between the velocity and the magnetic field (or between the magnetic field and the magnetic potential). In that case, the kinetic energy spectrum has its fluid value, $E_V(k) = 4\pi k^2/\alpha$ (the magnetic helicity playing no role), whereas the magnetic energy is $E_M(k) = E_V(k)/\mathcal{D}(k)$; so that, for wavenumbers approaching $|\beta|/\alpha$, magnetic energy becomes very large, and so does magnetic helicity. This is at the origin of the inverse cascade excitation of large scales in MHD turbulence, as postulated in [42]. For $\gamma \neq 0$, such condensation of magnetic excitation can happen as well.
- In the opposite case of maximal correlation, equipartition is recovered ($E_R(k) \equiv 0$, $H_R(k) \equiv 0$), as expected.

- Furthermore, note that, rewriting in part equations (5)–(7), we have:

$$H_C(k) = -\frac{\gamma}{2\alpha} E_M(k) \quad \forall k \ . \tag{2.10}$$

In other words, when $\gamma \neq 0$, the correlation spectrum in ideal MHD is proportional to the magnetic energy spectrum, the latter being constrained by a Schwarz inequality to be larger than the magnetic helicity, namely $E_M(k) \geq kH_M(k)$. Thus, when magnetic helicity undergoes an inverse cascade to large scales, magnetic energy has to grow to large scales as well, and in the force-free (fully helical) case, it will have a spectrum determined by that of magnetic helicity; and thus the correlations will grow as well to the large scales. A negative transfer at large scale was observed in some early low-resolution numerical simulations of the MHD equations [122]; this remark could also explain recent observations in the solar wind [140].

It should also be noted that the correlation spectrum does not change sign at any scale, as evident from Eq. (2.10). Such a change of sign can thus only occur through forcing mechanisms or else dynamically at the dissipative scale, as it has been observed and modeled using phenomenological arguments in two-point statistical closures of turbulence [57] (see [50] for 2D DNS). Similarly, kinetic and magnetic helicity are of the same sign at all scales, as Eq. (2.5) indicates; this sign is determined by the initial conditions and the helicity spectra can only change sign beyond the dissipative scale (leaving aside here a discussion involving the case when the magnetic Prandtl number $P_M = \nu/\eta \neq 1$ for which more than one dissipative scale can occur). It is not clear whether this latter point concerning the behavior of $H_M(k)$ at the dissipative scale has yet been verified by data.

- Other cases of ideal MHD flows have been studied, and the invariants change as one either changes space dimension [82, 98], or adds either a uniform magnetic field B_0, or a uniform rotation Ω_0; in the case when $B_0 \neq 0$ and $\Omega_0 \neq 0$, a new invariant appears [136, 137], called the parallel helicity

$$H_P = H_C - \sigma H_M \ ,$$

when these two externally imposed agents are parallel, namely with $\mathbf{\Omega}_0 = \sigma \mathbf{B}_0$, even though H_C and H_M are not invariant anymore in that case (see [138] for the corresponding two-dimensional geometry).

Finally, ideal dynamics is of course well-suited for studying the possible development of singularities in a finite time in MHD, a problem that is open, as it is as well in the three-dimensional fluid case. Although this question is of a mathematical nature, one can help with direct numerical simulations that are accurate so that the invariants are well preserved. One such example of ideal 3D MHD dynamics is given in the following figures. In the computation described succinctly below, there is no dissipation, no forcing and no imposed uniform magnetic field; the magnetic helicity is identically zero, and so is the cross-correlation. Initial conditions for both the velocity and the magnetic field have the same four-fold symmetries as that of the Taylor-Green vortex [17] but extended to MHD [76, 77]. Such a vortex corresponds to the flow between two counter-rotating cylinders as encountered in many experimental configurations. One can compute the evolution of such flows until the dynamics reaches the mesh size $\Delta x = 2\pi/N$ where N is the number of points per dimension. After that time, the partial differential equations are no longer resolved and one deals with the dynamics of a truncated number of Fourier modes which will evolve progressively toward the ideal spectra given in Eqs. (5)–(7).

When fitting the temporal evolution of the energy spectra as:

$$E(k,t) \sim C(t)k^{-n(t)} \exp^{[-2\delta(t)k]} , \tag{2.11}$$

where $\delta(t)$ is called the logarithmic decrement and $n(t)$ is the inertial index, one is assured of the regularity of the MHD equations as long as $\delta \neq 0$. In Figs. 2.1 and 2.2 are displayed the temporal evolution of δ and of n for an ideal MHD run on an equivalent grid of 3072^3 points; the horizontal dash line represents the mesh size which is reached for $t \approx 2.5$; just before that time, a sharp acceleration of the evolution of δ is observed, as in the computation reported in [76] where it was interpreted as the formation of a rotational discontinuity between two approaching current sheets. After that time, the computation of the PDEs is not reliable any longer. A higher-resolution run on an equivalent grid of 6144^3 points has allowed for the further study of this evolution [18]. The dominant structure that emerges consists in the collision and subsequent coupled evolution of two neighboring current sheets. At the last reliable time of the computation (i.e. when the logarithmic decrement is comparable to the grid size), magnetic field lines embedded in this dual system of current sheets are all seen converging to the point of maximum current [18].

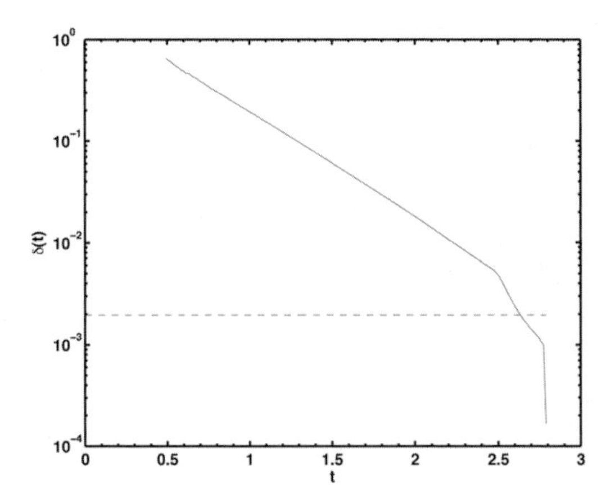

Fig. 2.1 Temporal evolution of the logarithmic decrement $\delta(t)$ (see text, Eq. (2.11)) for the ideal MHD case with, as initial conditions, the so-called I-flow (see [76]). The horizontal dashed line indicates the mesh size Δx, on a run using the equivalent of 3072^3 points, implementing the symmetries of the Taylor-Green flow generalized to MHD.

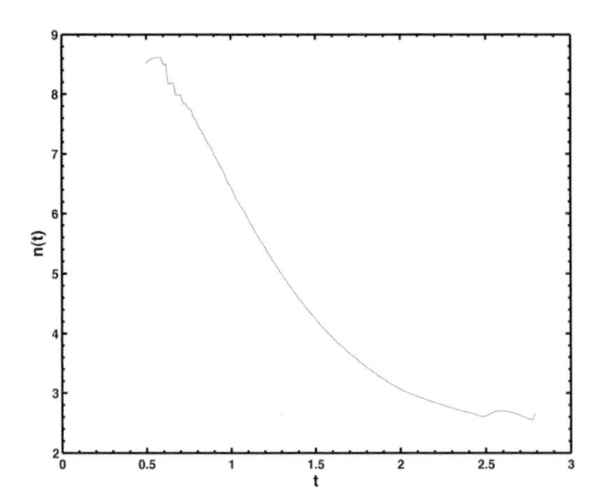

Fig. 2.2 Temporal evolution of the energy spectral index $n(t)$ (see text, Eq. (2.11)) for the same run as in Fig. 2.1. The computation of the MHD equations is reliable until $\delta \sim \Delta x$, indicated by the dash line in Fig. 2.1, or $t \approx 2.5$, time after which one enters the regime of ideal dynamics eventually described by the statistical spectra given in Eqs. (5)–(7).

Note that the spectral index of the total energy spectrum seems to reach values below $n = 3$ (see [73, 152] for recent studies of the corresponding ideal problem in two dimensions). Finally, in Fig. 2.3, are given the Fourier spectra of the total energy, with different successive interval of times displayed in different colors, and with a change in color every $\Delta T \sim 0.4$. At the latest time displayed here, the spectrum begins to show an accumulation at small scale, the premises to the equipartition spectrum at small scales following a k^2 law in the simplest (non-helical) case. The jagged aspect of the spectra at large scale is likely due to resonances between modes because of the symmetries of the initial conditions.

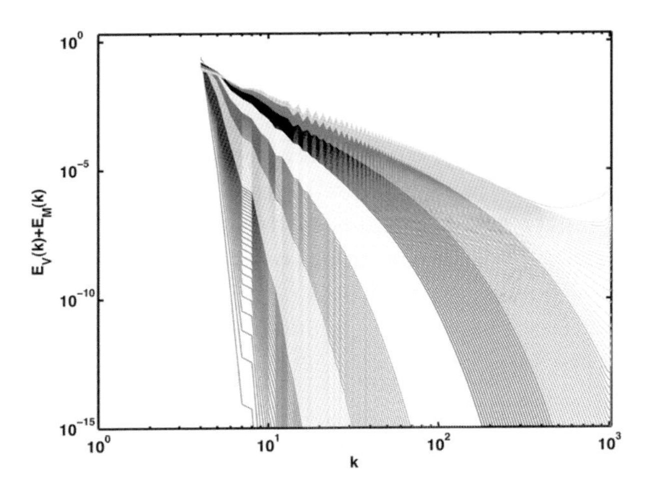

Fig. 2.3 Temporal evolution of the total energy spectrum $E_T(k,t)$ for the same flow as in Figs. 2.1 and 2.2, in logarithmic coordinates. Colors shift from blue to green, red, yellow, black, purple and cyan with increments of roughly $\Delta T = 0.4$, see the color version Fig. 6.3. The first spectrum is at $t = 0$, and the last one is at $t = 2.815$, time at which pile-up of energy in the smallest scale is clearly visible; it would lead at later times to an energy spectrum $\sim k^2$, corresponding to the non-helical ideal MHD.

However, the analysis of ideal dynamics in MHD does not tell us what actually happens in the inertial range at intermediate scales for MHD turbulence in the presence of dissipation, nor does it put any constraint that we can think of in the inertial ranges themselves. So we now turn to analysis and phenomenology, using dimensional arguments based on characteristic time scales, in order to advance in our understanding of MHD dynamics.

2.3 Energy spectra

2.3.1 *Weak MHD turbulence*

Let us assume that $\delta b/B_0 \ll 1$ where δv and δb are velocity and magnetic field fluctuations in the presence of a strong uniform magnetic field $\mathbf{B}_0$. Under the effect of such a strong imposed field, Alfvén waves will propagate, with $\delta v \sim \delta b$. Hence, the eddy turnover time is slow compared to the Alfvén time based on $B_0 \gg \delta b$, and we have a small parameter in the problem, namely:

$$\epsilon_{WT} = \tau_A/\tau_{NL} = \frac{\delta b}{B_0}\frac{\ell_{\|}}{\ell_{\perp}} ,$$

assuming the turbulence becomes quasi bi-dimensional, with $\ell_{\|}$ and $\ell_{\perp}$ referring to directions parallel and perpendicular to $\mathbf{B}_0$. This is a situation familiar in weak turbulence theory: the nonlinear interactions are weak except when the two counter-propagating waves are at resonance, and thus the statistical problem in terms of moments and cumulants of the fluctuating fields can be closed (see e.g. for review [103] and references therein). This was exploited in the case of MHD in [44, 45]. The procedure leads to integro-differential equations for the diverse spectra (kinetic and magnetic energy and helicity) and such equations can be analyzed for their flux-less solutions (corresponding to statistical mechanics of ideal truncated systems), and for their constant flux solutions, corresponding to the weakly turbulent regime. In the latter case, one finds

$$E(k_{\perp}, k_{\|}) \sim k_{\perp}^{-2} f(k_{\|}) \tag{2.12}$$

(henceforth, the WT spectrum), where the perpendicular and parallel directions refer to the imposed magnetic field. Note that, at this order, weak MHD turbulence can be viewed as being degenerate, since there is no coupling between different planes parallel to the magnetic field axis. One advantage of the weak turbulence approach is that not only can one determine the power-law scaling but also the (generalized) Kolmogorov constant appearing in front of the scaling which is known as well; it is found to depend strongly on the degree of alignment between the velocity and the magnetic field [44].

Phenomenology can help us in recovering this spectrum in Eq. (2.12), with "proper" dimensional analysis, by using ϵ_{WT} in a hand-waving argument. In the presence of a strong B_0, the turbulence is weak, hence the nonlinear transfer is weak; or perhaps, better said, it is delayed. By how much? Let us use the only parameter we have at our disposal, namely ϵ_{WT}. Hence let us conjecture, following Iroshnikov (1963) and Kraichnan (1965) [65, 71] that the characteristic time of transfer to small scales τ_{TR} is longer than in the fluid case by $1/\epsilon_{WT}$, hence $\tau_{TR} = \tau_{NL}^2/\tau_A$. Dimensional analysis then gives, with $\tau_{NL} = \ell_{\perp}/u_{\ell_{\perp}}$ using the fact that most of the nonlinear transfer occurs in planes perpendicular to B_0, and $\tau_A = \ell_\parallel / B_0$ following the anisotropic dispersion of Alfvén waves,

$$E(k_\perp, k_\parallel) \sim k_\perp^{-2} k_\parallel^{-1/2} .$$

In the isotropic case ($k_\perp \sim k_\parallel$), and after integration over one direction, one has $E(k) \sim [\epsilon_T B_0]^{1/2} k^{-3/2}$, henceforth the Iroshnikov-Kraichnan (IK) spectrum [65, 71], with $\epsilon_T \equiv DE_T/DT$ the rate of dissipation of the total energy. Hence the IK spectrum is compatible with the theory of weak MHD turbulence.

2.3.2 *When the weak regime breaks down*

Are such spectra observed? This is where trouble begins. Of course, in the weak turbulence integro-differential system of equations, they are obtained analytically, and they are consequently observed when performing a numerical integration of the complex equations emanating from the weak turbulence development [44, 45]. But the weak MHD theory is non-uniform in scale, because the two characteristic time scales vary in different ways with the size of eddies; hence, there exists a scale at which $\epsilon_{WT} \sim 1$ and some sort of strong turbulence takes over, to be determined. This is why it is rather difficult to observe such a spectrum although there are indications that it has been observed in the magnetosphere of Jupiter [131] for which $\delta b / B_0 \sim 0.008$ and one estimates $\epsilon_{WT} \approx 0.06$. Saturn could also be a candidate for such a regime to be observed. Note that this situation of break-down of weak turbulence is well known in geophysical fluid dynamics, where the Ozmidov scale for stratified flows and the so-called Zeman scale for rotating flows have been defined in similar ways (as the scale for which the appropriate $\epsilon_{WT} = 1$). It was shown recently using a computation on a grid of 3072^3 points of rotating turbulence that a Kolmogorov spectrum recovers beyond the Zeman scale; such a high resolution was needed in order to be able to resolve the large scales, the first (weak turbulence) inertial range, the K41 range and the dissipation range as well [94].

In the weak turbulence regime, Alfvén waves are fast and equipartition obtains at all scales, except possibly at the largest scales where magnetic helicity would dominate. However, it is well-known that a defect of equipartition has been observed in solar wind data, with what is called an (inverse) Alfvén ratio $E_M(k)/E_V(k)$ often found to be close to 2, similar to models and DNS of MHD turbulence. We show in Fig. 2.4 such a ratio at the peak of dissipation in several runs done on moderate grid resolution [141]: the

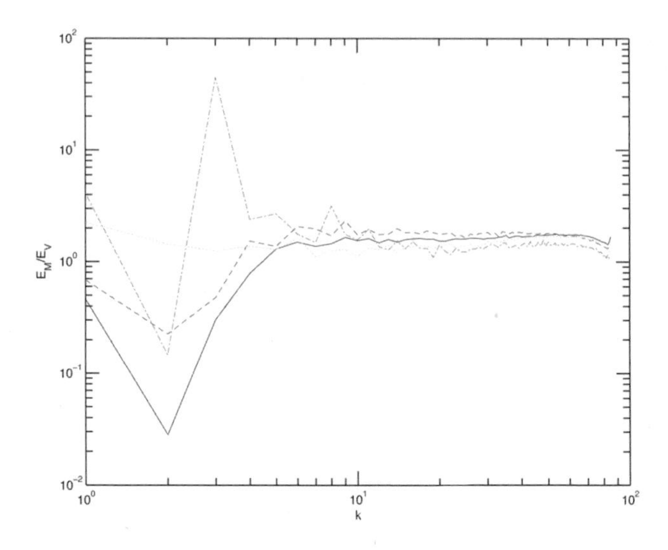

Fig. 2.4 Ratio of spectra of magnetic to kinetic energy at peak of dissipation for several runs in three dimensions (see also [141]). Note the differences at large scale, presumably responsible for different scaling laws, and the constancy in the inertial range, with a slight excess of modal magnetic energy as also observed in the solar wind [35].

spectra are seen to differ at large scale but are close to unity in the inertial range, as already found in [77]. This numerical result is reminiscent of solar wind observations of the Alfvén wave ratio that also appears to be constant in the inertial range [35].

There are impressive observations in astrophysics which exhibit, combining different techniques, a Kolmogorov spectrum for the density (and hence, probably for the energy as well), on many orders of magnitude, from the astronomical unit to the size of the inter-galactic medium [5, 6]. Of course, errors are large, and other spectra could be accommodated as well by the data. Note that one reason why such spectra might be observed is because even though this medium is in general supersonic, there are strong pockets of subsonic flows where K41 spectra can develop [31] (see also §2.8), as indicated by numerical simulations, for example using modeling of the small scales. In the solar wind, it has been known for a long time that the energy spectrum is often quite close to K41 [83] (see also the reviews [22, 146, 157]), but not always by far [112].

One could argue that an anisotropic Kolmogorov law is the solution to MHD turbulence, once one takes into account that the main effect of a strong magnetic field is to render the turbulence quasi bi-dimensional.

This is basically the solution advocated in [53]. The argument used is one of "critical balance" at all scales between Alfvén wave propagation on a time $\tau_A(\ell)$ and nonlinear transfer on a time $\tau_{NL}(\ell)$ at scale ℓ, namely that at all scales, one has $\epsilon_{GS}(\ell) = \tau_A/\tau_{NL} = 1$, leading to $k_\parallel \sim k_\perp^{2/3}/B_0$ (see [49] for a recent review). This scaling is corroborated by some numerical simulations [81] and not by others [77, 101, 141]. Moreover, note that one can extend this phenomenological argument to a constant ratio with scale, but not equal to unity, i.e. $\epsilon_{GS2} = c\,\forall\ell,\ c \neq 1$, in which case other solutions (such as the IK spectrum and the Weak Turbulence (WT) spectrum) can emerge as well [46]. The analysis of data is further complicated by the fact that, for different angles between the imposed field $\mathbf{B}_0$ and the wavevector $\mathbf{k}$, different spectra may occur, as documented in rotating flows [94] and in MHD as well [11], with the slow modes (corresponding to zero-frequency standing waves) and three-dimensional modes evolving differently, the latter having a steeper spectrum. Henceforth, averaging over angles as is done to obtain isotropic spectra may not be justified (see also [12]).

Recent numerical simulations taking care in obtaining convergence as the resolution is increased conclude that the GS model is what MHD will tend to, as the Reynolds number is increased [9]. However, it should be noted that the runs with hyper-diffusivities (dissipation operator with orders higher than a Laplacian) create their strong bottleneck effect, and that from an examination of the data in [9] one may not yet be able to discard IK spectra in these instances.

2.3.3 *Intermittency*

Intermittency deals with rare but strong events in turbulent flows, as well as in other critical phenomena [20]. Because of this phenomenon, the data for the energy spectrum in MHD turbulence is perhaps still a bit ambiguous: after all, the K41 spectral index 5/3 is not very far from the IK solution of 3/2 in the isotropic case, and besides we expect these values to be different because of intermittency corrections due to the presence of strong localized structures, in the form of vortex filaments for three-dimensional fluids, and vortex and current sheets [113, 115, 148], that eventually roll-up in 3D [88], as observed in the magnetosphere [61, 108] (although in that case the roll-up is at times interpreted as being due to Hall currents).

When examining higher-order moments of structure functions, the ambiguity is completely lifted: MHD turbulence does differ rather strongly from fluid turbulence in its scaling laws, independently of isotropy. In two dimensions [118], as well as in three dimensions [91, 99], intermittency exponents are very different from the fluid case, and MHD is more intermittent (in the sense that there is a stronger departure from self-similar linear scaling, although in that latter 3D case, to a lesser extent than for 2D). Such intermittency can be described by somewhat ad-hoc models, following the lead of She and Lévêque [135], as done for example in [60, 114], and a link to self-organized criticality has been made recently for 3D numerical simulations of MHD turbulence [147].

Similar results hold, moreover, in observations of the solar wind [24] and more recently in auroral absorption [142] as well as in X-ray emissions from the solar photosphere in active regions [1], with quite dramatic, order unity, changes in scaling exponents when going from weak flares (for which the intermittency is quite close to that of neutral fluids) to large flares in which case the exponents are closer to the MHD case in 2D, presumably indicative of the presence of a strong quasi-uniform field. Note that one also finds different intermittency exponents for supersonic MHD turbulence when varying for example the Mach number [105].

Is it thus conceivable to claim that only one solution will emerge for energy spectra at high Reynolds number, considering that the Reynolds number of astrophysical flows in which variable intermittency is observed, such as in the Sun, is quite high already? Can these differences only be ascribed to measurements errors (granted, they are large)? In particular in light of the fact that the absolute difference in anomalous scaling indices between strong and weaker solar flares is of order unity? Or is it the difference in physical parameters (compressibility, radiative transfer, stratification, boundaries, geometry) that is the culprit? Perhaps not. So, what can we do?

2.4 Exact laws

There is one thing we know about MHD turbulence: there are exact laws that can be written, under five simplifying hypotheses, and such laws can be of some help. Under the assumptions of incompressibility, homogeneity, stationarity, isotropy and high Reynolds numbers, one can write for the scaling of third-order structure functions defined before in Eq. (2.1), for the Elsässer variables, $\mathbf{z}^{\pm} = \mathbf{v} \pm \mathbf{b}$ [116, 117]:

$$\langle \delta z_L^-(\mathbf{r}) \Sigma_i [\delta z_i^+(\mathbf{r})]^2 \rangle = -\frac{4}{d}\epsilon^+ r \, ;$$

$$\langle \delta z_L^+(\mathbf{r}) \Sigma_i [\delta z_i^-(\mathbf{r})]^2 \rangle = -\frac{4}{d} \epsilon^- r \qquad (2.13)$$

or, in terms of the velocity and magnetic field:

$$\langle \delta v_L \Sigma_i (\delta v_i)^2 \rangle + \langle \delta v_L \Sigma_i (\delta b_i)^2 \rangle - 2 \langle \delta b_L \Sigma_i \delta v_i \delta b_i \rangle = -\frac{4}{d} \epsilon^T r \,, \qquad (2.14)$$

$$-\langle \delta b_L \Sigma_i (\delta b_i)^2 \rangle - \langle \delta b_L \Sigma_i (\delta v_i)^2 \rangle + 2 \langle \delta v_L \Sigma_i \delta v_i \delta b_i \rangle = -\frac{4}{d} \epsilon^C r \,. \qquad (2.15)$$

Such laws stem from the conservation of energy and cross-helicity. A detailed study of the relative importance of these terms has been performed in [156] for a set of numerical simulations. Similar laws exist for Navier-Stokes turbulence , and for the dynamics of passive tracers in fluids. Note that these exact laws have been extended to several cases, for example with anisotropy [48], or including the Hall term [47].

One essential remark is that equations (2.14) and (2.15) are coupled: the invariance of total energy (governed by the imposed flux of energy ϵ_T) and the invariance of cross-correlation, governed by its rate of injection ϵ_C chosen independently of ϵ_T, play dual symmetric and coupled roles; it is particularly visible when considering the Elsässer variables, and their rates $\epsilon_\pm = [\epsilon_T \pm \epsilon_C]/2$. Each rate can impose its own time-scale, say T_E, T_C for the energy and cross-helicity (or $T_\pm$), and thus there is a priori the possibility of breaking the universality that is often assumed in the neutral fluid case. Note that the possible role played by the invariance of kinetic helicity for neutral fluids is discussed in [55], and the case of magnetic helicity in MHD is analyzed separately in §2.7 (see also [119]).

Let us examine a bit further equations (2.14) and (2.15). There are three possibilities a priori: either $\delta v \gg \delta b$ (the kinematic case of the dynamo starting from a weak magnetic field, in which case we certainly expect a Kolmogorov law for the total energy spectrum dominated by its kinetic part), or $\delta v \ll \delta b$ (corresponding to a strongly magnetized force-free plasma), or of course $\delta v \sim \delta b$, which can be called the Alfvénic case. Such partition of phase space has been documented, both in 2D and 3D numerically [141, 143, 144] and is plausible. But what do the exact laws above tell us in each case?

In the kinematic regime, one recovers approximately the scaling for standard fluid turbulence, with $\langle \delta v_L(r) \Sigma_i \delta v_i(r)^2 \rangle = -\frac{4}{3} \epsilon r$ (in dimension three), compatible with the perhaps more familiar expression in terms only of the longitudinal structure function $\langle \delta v_L(r) \rangle^3 = -\frac{4}{5} \epsilon r$, with a Kolmogorov energy spectrum for the velocity. The magnetic field is likely to follow in the early kinematic phase $E_M(k) \sim k^{3/2}$, the Kazantsev solution to the problem (see [21] for review). In the case when $\delta v \ll \delta b$, the opposite is true: the structure function cubic in the magnetic fluctuations scales like r, in which case again a Kolmogorov spectrum might be plausible, this time for the magnetic energy E_M, and thus likely for the total energy dominated by E_M. Finally, when $\delta v \sim \delta b$, we cannot conclude directly from these laws the scaling of energy spectra, but we observe that velocity-magnetic field correlations will play an essential role, as stressed in [116] (see §2.6). In other words one should not factorize the correlators in equations (2.14, 2.15), although that would lead of course to K41.

2.5 Explicit examples of non-universal behavior in MHD

Let us end this section by noting that there are a few known examples of MHD turbulence with different energy spectral indices, either in the context of reduced MHD corresponding to a simplification of the MHD equations when a strong field is imposed [36], or using the full MHD equations with a strong B_0, in the latter case finding either a K41 or an IK spectrum [101].

There is also a recent example where all three spectra (K41, IK, WT) are observed when changing in some controlled and seemingly simple fashion the initial conditions [77]. Three different runs have been performed, each with the same initial velocity field, specifically the Taylor-Green vortex, a configuration corresponding to laboratory experiments between two counter-rotating cylinders, see e.g [97]. Three different initial conditions were taken for the magnetic field, with the same magnetic energy and magnetic helicity (zero in that case) and comparable cross-correlations (between 0 and 4% in relative terms, i.e. when normalized by the energy). Also equal initial kinetic and magnetic energy were taken, and there was no uniform magnetic field. Thus, at $t = 0$, $\tau_A = \tau_{NL}$, where here τ_A here is based on the root mean square field. These computations are performed on equivalent grids of 2048^3 points, imposing the four-fold symmetries of the Taylor-Green flow to both the velocity and the magnetic field, taken initially to have such symmetries as well. Finally, there is no forcing and the magnetic Prandtl number is equal to unity. What distinguishes the runs, which end up with three different spectra, is that, in particular in the case where WT is observed, the magnetic energy grows significantly at the expense of the kinetic energy at the largest modes, leading to $\tau_A(\ell) \ll \tau_{NL}(\ell)$ for $\ell \sim L_0$. One should note also that the critical balance hypothesis, even when generalized to a non-unity value for the ratio of time scales, is not fulfilled in these simulations [77]. The scale-dependence of the time ratio is found to be compatible with the energy spectra scaling obtained for these three runs. In fact, both in the computations in [77] and in the parametric study in [141], the attractive solution seems to be one in which quasi-equipartition between kinetic and magnetic modal energies obtain at all wave numbers except in the largest scales, and it is these large scales that determine *in fine* the power-law solution of the dynamics (see Fig. 2.4, and [35, 141]).

However, the statistical equilibria for these three initial conditions are the same since the invariants are (almost) identical. So, either universality in MHD breaks down, presumably in different sub-classes (possibly K41, IK and WT), or else there are hidden invariants of the problem that we are not taking into account (also see [132] for a discussion of universality). One ought to consider higher-order correlators such as tetrads involving the Lagrangian dynamics of four spatial points [32], or higher-order correlators involving skewness and kurtosis for example. Is this the solution to this conundrum? Or is it that these numerical simulations were done at too low a Reynolds number, although the grids were the largest MHD runs performed at the time, with $k_{\max}/k_{\min} = 700$, where $k_{\max}$, $k_{\min}$ are the maximum and minimum wavenumber of the computation (corresponding respectively to the size of the box and the mesh size)? Since no B_0 is imposed, a moderate resolution does lead to a strong large-scale field, but which cannot be considered as quasi-uniform still, from the stand-point of the smallest resolved scales. Furthermore, it is perhaps important to understand that when one imposes $\mathbf{B}_0 \neq 0$ (or rotation or gravity for that matter), the flow is anisotropic at all scales, even if isotropy is recovered at small scales in some cases, like for rotating turbulence; whereas, when $\mathbf{B_0} \equiv 0$, there is no preferred direction that can be identified, although one can define a local mean $B_{0,loc}$ by averaging the magnetic field $\mathbf{b}$ in a ball of diameter the integral scale of the flow. Isotropy does recover on average at large scale in that second case.

Another possible reason for these computations to differ in their energy scaling is that they are decaying flows and temporal averages cannot be taken, except near the peak of dissipation. Will the forced cases follow the same scaling as the decay case, or not? Will quasi-equipartition at the largest scale be achieved on average, or will the dominance of magnetic energy in one particular case leading to the weak turbulence regime be able to maintain its status for all times? Preliminary results indicate that the non-universality is preserved in the forced case [19], but these are open questions in need of further investigation, the answer to which, though, may depend on the correlation time of the forcing as well.

Finally, note that these exact laws can be used to inverse engineering data and deduce what is the heating rate in the solar wind due to a turbulent cascade, as done for example in [80], a cascade for which there has been ample evidence since the pioneering work done using Voyager data [83] (see also [139]).

2.6 The role of alignment between the velocity and the magnetic field

The considerations of the preceding section were done under the (implicit) hypothesis that the correlations between the velocity and the magnetic field were weak. However, it has been known for a long time that such correlations grow in time (see e.g. [84, 113]). It was shown in [56] using a simple dimensional argument, that attributing spectral indices $m^{\pm}$ to the $E^{\pm}$ spectra, one could deduce that $m^{+} + m^{-} = 3$, with $m^{+} = m^{-} = 3/2$ in the uncorrelated case. The reason why both m^{+} and m^{-} appear in this dimensional analysis is because one Elsässer variable is advected by the other one, hence both fields will appear since there is no self-advection.

This analysis was confirmed in numerical simulations of MHD turbulence both in 2D and in 3D, although in the latter case with moderate resolution. Moreover, not only does the global correlation of the flow, which can be viewed as the ratio of two invariants, viz. H_C/E_T, increases, so does the local correlation which can be defined either as $\rho_I(\mathbf{x}) = \mathbf{v} \cdot \mathbf{b}/[v^2 + b^2]$ or $\rho_A(\mathbf{x}) = \cos(\mathbf{v}, \mathbf{b})$, the latter indicating that this is a geometrical property of the flow, namely the tendency for the velocity and magnetic field to align or not. This local (as opposed to global) alignment was observed in two dimensions [86] as well as more recently in 3D [85], with large plateaux of highly aligned (or anti-aligned) fields and strong gradients in the reconnection regions. It was also shown that other alignments occur, between the velocity and the vorticity, as well as between the magnetic field and its current [133].

The weakening of cross-correlation in the region of strong gradients was already hypothesized phenomenologically and found numerically using second-order closures [56, 57] and confirmed with 2D-DNS soon afterwards [50]. As noted in [155], it may have important consequences for reconnection since in that domain the interactions can be strong (see also [154] for an analysis of solar wind data in terms of the anisotropy of the $\pm$ Elsässer variables).

The weak turbulence analysis gives an unambiguous answer, namely that $m^+ + m^- = 4$, which corresponds in the closure case to $m^+ + m^- = 3$ for the isotropic spectra, as reported in the simulations. It also gives the ratio of the generalized Kolmogorov constants appearing in front of the $E^\pm$ spectra as a function of the imposed degree of alignment in the flow, with $C_+ = C_- \sim 0.585$ at zero correlation, whereas these two constants ~ 0.19 at high correlation [44]. Moreover, the energy fluxes in that latter case differ by more than one order of magnitude.

However, besides these regimes, the question of what are the asymptotic high Reynolds number spectra in MHD turbulence in three dimensions in the presence of strong correlations between the velocity and the magnetic field, is highly debated at the present time. Note that MHD in 2D can be viewed as being weak in some sense since it is the simplest approximation to the case of 3D MHD in the presence of a strong imposed field. The dynamics to lowest order becomes 2D. Or else one can consider the next approximation of the so-called reduced MHD. However, in the plane perpendicular to $\mathbf{B}_0$, interactions can be strong again; and in the absence of a strong imposed field, the large-scale isotropy leads to interactions of packets of turbulence in all directions, feeding isotropy to smaller scales until possibly the magnetic Taylor scale [90].

On the one hand, there is a simple if ad-hoc way to estimate the third-order correlators appearing in Eqs. (2.14), (2.15), as done in [15]: introduce a dimensionless field, such as an angle, and a natural choice is of course the angle θ between $\mathbf{v}$ and $\mathbf{b}$. Thus, the following relation is still dimensionally correct

$$\langle \theta \delta v^2 \delta b \rangle \sim r \ ,$$

and can be viewed as incorporating the effect of the alignment between velocity and magnetic field through θ. Using now the hypothesis of the simplest solution where all variables scale in the same way, one obtains $\delta v \sim \delta b \sim \theta \sim r^{1/4}$, compatible with an anisotropic IK spectrum, $E(k) \sim k_\perp^{-3/2}$ spectrum, as advocated for some time by Boldyrev [15] (see also [25, 52] for another model with similar scaling taking into account the Lagrangian point of view). This spectrum is corroborated by moderate resolution DNS in the presence of a strong uniform magnetic field (see [81] and references therein). Note that, in the forced case, long-time integration can be performed leading to good statistics when averaging over the duration of the statistically steady state. For example, it is shown in [107], using the incompressible reduced MHD equations on a grid of $1024^2 \times 256$ points, a narrow forcing in the perpendicular direction and a wide forcing in the parallel direction, with a resulting global correlation coefficient of ~ 0.8, that the amount of alignment is irrelevant to the final energy spectra which are all found to be proportional to $k_\perp^{-3/2}$, with "pinning" (joining and possibly crossing) at the dissipation scale, and thus compatible with ideal dynamics (see §2.2). Note that another solution, beyond $m^+ + m^- = 4$, $m^+ \neq m^-$, is that in fact, at high Reynolds number, the $E^\pm$ symmetry is recovered and $m^+ = m^- = 2$ for all degrees of alignment between the velocity and the magnetic field, as advocated in [8].

One of the point that may have to be taken into consideration is based on the importance of the actual values of $R^\pm$. Indeed, when taking a very strong global degree of alignment, say positive, one has $Z_0^+ \gg Z_0^-$; hence R_+ could be small unless the viscosity itself is extremely small. It is not clear whether these $\pm$ Reynolds numbers actually play a role in the overall physics, based on the dynamical **v** and **b** fields. A similar argument could be made in fluid helical turbulence when using the $\pm$ variables linked to the eigenmodes of the curl operator. As pointed out in [30], one may be led to the wrong conclusion concerning the dynamics of H_V when considering the Reynolds numbers based on the two $\pm$ helically polarized waves of the problem.

Similarly, one question concerns the rates at which the $\pm$ energies cascade to small scales, let us call them $T_\pm$: do we have $T_+ \neq T_-$? Finally, the presence of a bottleneck at small scale at the onset of the dissipative range could be an issue as well; this accumulation of energy can be rather prominent for fluids, less so for MHD, a point attributed to the greater degree of non-locality of nonlinear interactions in MHD turbulence [2, 27, 39, 89, 92].

At this point, it is difficult to conclude and we can consider this problem of strongly aligned MHD turbulence, as observed in several instances in the solar wind (see e.g. [140] and references therein), as being open from a theoretical and numerical point of view. One way forward may be to use models of turbulence to mimic the effect of higher Reynolds numbers (see §2.8). Models of small-scale MHD turbulence with non-zero cross-helicity have been developed [155] and could lead to improvements in the dynamo problem of generation of magnetic fields and in reconnection regions as well.

2.7 Can the dynamics of magnetic helicity play a role for energy scaling?

Kinetic helicity does not seem to play a significant role in fluid turbulence, although the small-scale vortex filaments have been known for a long time to be fully helical (see [95] for a review). It may be important in the case of rotating flows (see [94] and references therein), or in the dynamics of the atmosphere, where it is measured in so–called supercell storms and hurricanes (see e.g., [96]). The exact law resulting from the flux conservation of kinetic helicity can be found in [55], and that for magnetic helicity was derived in [119]. In terms of the components of the magnetic potential and of the electromotive force $\mathcal{E} = \mathbf{v} \times \mathbf{b}$, it reads:

$$\langle [\mathcal{E}(\mathbf{x}) \times A(\mathbf{x}')]_L \rangle = +\frac{1}{3}\,\tilde{\epsilon}_{H_M} r\,,$$

with $\mathbf{x}' = \mathbf{x} + \mathbf{r}$ and $\tilde{\epsilon}_{H_M} \equiv DH_M/DT$; it can also be written as:

$$\langle v_L(\mathbf{x})\Sigma_i b_i(\mathbf{x})A_i(\mathbf{x}')\rangle - \langle b_L(\mathbf{x})\Sigma_i v_i(\mathbf{x})A_i(\mathbf{x}')\rangle = -\frac{1}{3}\,\tilde{\epsilon}_{H_M} r\,.$$

This gives new constraints on the dynamics of MHD, the consequences of which have not been explored yet. What can be the role, then, in the dynamics of MHD turbulence, of magnetic helicity? We saw in §2.2 that its role is essential: it is the culprit at the origin of the lack of complete equipartition between kinetic and magnetic energy, and its spectrum in the ideal case (when not trivially zero) could be seen as governing all other spatial dynamics except for the kinetic energy spectrum.

Dimensional analysis à la Kolmogorov leads for the magnetic helicity spectrum to a power-law solution $H_M(k) \sim k^{-2}$ [121], a spectrum observed in the inverse cascade of magnetic helicity in turbulence closures as well as in moderate resolution DNS. But in fact other spectra have been observed more recently[79, 91, 102], with $H_M(k) \sim k^{-\kappa}$, with $\kappa \approx 3.3$. The spectrum in [121] was obtained under the assumptions of separation of ranges (direct cascade of total energy, inverse cascade of H_M), and of locality in Fourier space of nonlinear interactions. But it has been found recently that in fact all scales interact for a variety of turbulent fluids [37, 38, 93], even though, in a logarithmic discretization of scales (as opposed to linear), such may not be the case [4]. And if indeed the magnetic helicity spectrum is quite steep, non-local interactions are bound to play a role. Also note that long-time memory effects may lead to recurrence of events, as for example in the case of geo-magnetic reversals [10, 16].

These non-local interactions may render dimensional analysis invalid, and what we need is another way to proceed, and perhaps a paradigm shift. The two safe arguments that can be made rely on ideal dynamics described in §2.2 and on the exact laws analyzed in §2.4, and their extensions to more complex cases. Beyond these two types of results and beyond dimensional analysis à la Kolmogorov, there are a variety of arguments that can be put forward, among which one that may turn to be valuable (see [79, 91, 110]).

The idea is to postulate that the nonlinear dynamics in MHD, beyond preserving the exact laws, is trying throughout the inertial range to achieve a quasi-equipartition between kinetic and magnetic energy and helicity (but not quite because of the ideal constraints discussed in §2.2); in other words, one can expect that there be a proportionality between $E_V(k)/E_M(k)$ and $H_V(k)/H_M(k)$. This argument can be shown as well to be compatible with the dynamo regime [110]. Such a quasi-Alfvénization is observed in the simulations analyzed in [79], using hyperviscosity at small scale. Similarly, in the decaying runs described in [77] using a code that imposed the four-fold symmetries of the Taylor-Green flow, quasi-Alfvénization (or QA) is obtained at all scales but the largest (see Fig. 2 in [77]), whereas the so-called critical balance postulated in [53] is not observed (see Fig. 3, op. cit.). However, one should be careful here: this does not mean that critical balance is ruled out; in fact there is evidence for it in some numerical simulations [81]. It simply means that it may not be the only solution, and that quasi-Alfvénization may well be a relevant concept as well in most of the small-scale inertial range.

2.8 Can models help?

Moore's law gives an increase of resolution by a factor two roughly every six years in three dimensions. Hence, high Reynolds number turbulence will remain un-attainable in the near future. There are several ways around this difficulty, of course. One is to use and impose symmetries of some flows and reach equivalent resolutions that can be as much as 8 times a full DNS (see [17, 69] for fluids, and [77] for MHD). One can also filter the numerical data at small scale, either simply with a higher power of the Laplacian, or else using some approximation to the sub-grid stress tensor (see [87, 100] for reviews of Large Eddy Simulations, or LES).

Another possibility is to use two-point closure models which have been shown to be quite helpful as a source of modeling for high Reynolds number turbulence, as for example the Eddy Damped Quasi Normal Markovian (EDQNM) closure. They yield expressions of transport coefficients that depend on time and on the kinetic and magnetic energy and helicity spectra. These coefficients can be used as a model of the unresolved sub-grid scale interactions; as such, they were recently applied in a variety of conditions, e.g. inhomogeneous flows [41], or MHD flows [7], including in the study of the dynamo effect [120], or when helicity is present [155]. In the EDQNM closure in its simplest formulation, only two time-scales are considered to damp the high-order correlations, namely the eddy turn-over time and the Alfvén time. However, when examining carefully the algorithm that leads to a closure formulation, it was shown in [7] that another time needs to be considered as well, namely one built on the defect of equipartition between velocity and magnetic field, in which case a better model results. Note also that the EDQNM model in the presence of v-b alignment has been written [59] but probably not exploited in the context of modeling of MHD turbulence. This may be an avenue to explore.

Another avenue is to take so-called scalar models, first put forward in MHD in [51] and expanded further since then in several directions (see e.g. [111] for a recent application to the dynamo in a rotating fluid at small magnetic Prandtl number).

Finally, let us mention another filtering methodology which may be quite promising in MHD [109]: in that approach, the invariants are maintained but in a different norm, H_1 instead of L_2 (see e.g. [64]). In other words, within the model, it is $E_{H1} \sim \left\langle |\mathbf{u}|^2 + \alpha^2 |\omega|^2 \right\rangle$ which is conserved, with α an open parameter of the model taken as the scale on which Lagrangian trajectories are averaged. This allows constraining the growth of vorticity and current density, and thus reaching higher Reynolds number at a given resolution than the DNS would allow, by up to a factor 6 or perhaps more [110]. For example, one finds in [110] a clear indication, for an equivalent resolution of $\sim 6000^3$ points, that quasi-Alfvénization is the preferred mode for this given set of initial conditions: the energy ratio $E_M(k)/E_V(k)$ remains remarkably constant (and ≈ 2) throughout the inertial range, and with a ratio of Alfvénic to turn-over time-scales compatible with the scaling of the energy spectrum. This clearly is worth exploring further, at higher Reynolds number as well as in the forced case.

A combination of all these modeling techniques may prove a valuable tool for studying MHD turbulence at Reynolds numbers as encountered in geophysics and astrophysics, in order to explore some of the issues mentioned in the preceding Sections.

2.9 Conclusions

A lot more needs to be done, using every tool we can, to help unravel what is happening in MHD turbulence. Data, both numerical and observational, seems to indicate that, at orders that are higher than the energy spectra, there is a notable difference between fluid and MHD turbulence, due to the dynamics of small scales, vortex filaments in one case and vorticity and current sheets in the other, which do eventually roll-up as observed in the magnetosphere (see e.g. [61]). Whereas it is difficult to measure small discrepancies in scaling exponents, it might be possible to explore the physical differences that lead to critical balance versus quasi-Alfvénization, as discussed in §2.7 in the context of magnetic helicity which may well play a determining role, possibly as important as the degree of alignment between the velocity and the magnetic field. Perhaps, a study of Lagrangian dynamics as performed for fluid turbulence and recently in MHD as well (see [25, 26]) will help understand the detailed properties of small-scale in MHD.

In fact, when discussing the exact laws written in Eqs. (2.14), (2.15), more possibilities could take place beyond the relative ordering of velocity and magnetic fluctuations, in introducing the degree of alignment between **v** and **b** on the one hand, and between **A** and **b** on the other hand. According to which alignment is greater, one regime can dominate over another one. Also worthy of noting is the fact that, in the ideal dynamics of MHD turbulence, the cross-correlation and magnetic helicity spectra are proportional with $H_C \sim k^2 H_M$, with a coefficient of proportionality depending only on the three Lagrange multipliers associated with the three invariants. Since one has postulated and observed an inverse cascade of magnetic helicity to the large scales (see e.g. [42, 121, 123]), then would it be possible to envisage that, for some ratios of the α, β, γ multipliers, the cross-correlation could also grow at large scale (even though, in that case, $k \to 0$), implying a high degree of alignment of the velocity and magnetic field in the large scales of the system? There is some evidence for this in the solar wind [140], but more data needs to be analyzed in detail before we can conclude on this point which may be mature for numerical explorations as well (see also Eq. (2.10)). Also, in the case of the interstellar medium, perhaps, with ALMA for example, more detailed information about the scaling behavior of interstellar turbulence will be available (see e.g. [40]).

On the other hand, structures (such as horse-shoe vortices in channel flows, or convective plumes) arise from the boundaries, and one may wonder what is the role of such boundaries in determining scaling and statistical properties of turbulent flows in general. This question can be extended of course to homogeneous turbulence which naturally develops internal boundary and shear layers, and to the effect of the roughness of such boundaries [54]: do current sheets (and shear layers) provide a similar corrugation of MHD turbulence and lead to specific scaling laws?

Numerically, one of the problem is the lack of adequate resolution in three space dimensions: how high do we need to push the Reynolds number to be convinced that we have an asymptotic solution? Statistical mechanics does tell us that diverse spectra can emerge. Will K41 still emerge at very high Reynolds? Or does the intuition based on the examination of the exact laws stemming from the conservation properties of MHD indicate indeed that three different regimes can occur, as seen in some computations?

Acknowledgments

I am thankful to Marc-Étienne Brachet (ENS, Paris) and Duane Rosenberg (NCAR) for their contributions to the computations described in the Figures 1–3 presented in this paper, and to Julia Stawarz for Figure 4. The computations for Figures 1–3 were performed thanks to a large DOE/INCITE allocation of hours (2011-16013); for Figure 4, the runs were done at NCAR. The National Center for Atmospheric Research is sponsored by the National Science Foundation.

References

[1] V. I. Abramenko et al., *Astrophys. J.* **597**, 1135 (2003).
[2] A. Alexakis, P. D. Mininni and A. Pouquet, *Phys. Rev. E* **72**, 046301 (2005).
[3] O. Alexandrova et al., *J. Geophys. Res.* **111**, A12208 (2006).
[4] H. Aluie and G. L. Eyink, *Phys. Rev. Lett.* **104**, 081101 (2010).
[5] J. W. Armstrong et al., *Nature* **291**, 561 (1981).
[6] J. W. Armstrong, B. J. Rickett and S. R. Spangler, *Astrophys. J.* **443**, 209 (1995).
[7] J. Baerenzung et al., *Phys. Rev. E* **78**, 026310 (2008).
[8] A. Beresnyak and A. Lazarian, *Astrophys. J.* **722**, L110 (2010).
[9] A. Beresnyak, *Mon. Not. R. Astron. Soc.* **422** , 3495 (2012).
[10] R. Benzi and J. F. Pinton, *Phys. Rev. Lett.* **105**, 024501 (2010).

[11] B. Bigot, S. Galtier and H. Politano, *Phys. Rev. E* **78**, 066301 (2008).
[12] B. Bigot and S. Galtier, *Phys. Rev. E* **83**, 026405 (2011).
[13] D. Biskamp, *Nonlinear Magnetohydrodynamics* (Cambridge University Press, 1993).
[14] D. Biskamp and H. Welter, *Phys. Fluids B* **1**, 1964 (1989).
[15] S. Boldyrev, *Phys. Rev. Lett.* **96**, 115002 (2006).
[16] F. Bouchet and E. Simmonet, *Phys. Rev. Lett.* **102**, 094504 (2009).
[17] M. E. Brachet et al., *J. Fluid Mech.* **130**, 411 (1983).
[18] M. E. Brachet, M. Bustamante, P. Mininni, A. Pouquet and D. Rosenberg, *Phys. Rev. E* **87**, 013110 (2013).
[19] M. E. Brachet, G. Krstulovic and A. Pouquet, "Forced dynamics of three-dimensional MHD flows implementing the Taylor-Green symmetries," in preparation (2012).
[20] S. T. Bramwell, P. C. W. Holdsworth and J.-F. Pinton, *Nature* **396**, 552 (1998).
[21] A. Brandenburg and K. Subramanian, *Phys. Rep.* **417**, 1 (2005).
[22] R. Bruno and V. Carbone, Living Review *Sol. Phys.* **2**, 4 (2005).
[23] J. M. Burgers, *Koning. Nederl. Akad. Weten.* **36**, 620 (1933).
[24] L.F. Burlaga, *J. Geophys. Res.* **96**, 5847 (1991).
[25] A. Busse, W.-C. Müller, H. Homann and R. Grauer, *Phys. Plasmas* **14**, 122303 (2007).
[26] A. Busse, W.-C. Müller and G. Gogoberidze, *Phys. Rev. Lett.* **105**, 235005 (2010).
[27] D. Carati et al., *J. Turb.* **7**, 51 (2006).
[28] V. Carbone and A. Pouquet, "An introduction to fluid and MHD turbulence for astrophysical flows: theory, observational and numerical data, and modeling", School on Astrophysical Plasmas, L. Vlahos and P. Cargill Eds., Springer Verlag, pp. 71–128 (2009).
[29] S. Chen, C. Foias, D.D. Holm, E. Olson, E. Titi and S. Wynne, *Phys. Rev. Lett.* **81**, 5338 (1998).
[30] S. Chen, Q. Chen and G. Eyink, *Phys. Fluids* **15**, 361 (2003).
[31] A. A. Chernyshov et al., Astrophys. J. **686**, 1137 (2008).
[32] M. Chertkov, A. Pumir and B. I. Shraiman, *Phys. Fluids* **11**, 2394 (1999).
[33] C. Cichowlas et al., *Phys. Rev. Lett.* **95**, 264502 (2005).
[34] S. Danilov and D. Gurarie, *Phys. Rev. E* **63**, 020203R (2001).
[35] S. Dasso et al., in Solar Wind 10, p. 546 sq., M. Velli, R. Bruno and F. Malara Eds., *Am. Inst. Phys.* (2003).

[36] P. Dmitruk, D. O. Gómez, and W. H. Matthaeus, *Phys. Plasmas* **10**, 3584 (2003).
[37] P. Dmitruk and W.H. Matthaeus, *Phys. Rev. E* **76**, 036305 (2007).
[38] P. Dmitruk et al., *Phys. Rev. E*, **83**, 066318 (2011).
[39] J. A. Domaradzki, B. Teaca, and D. Carati, *Phys. Fluids* **22**, 051702 (2010).
[40] E. Falgarone, J. Pety and P. Hily-Blant, *Astron. Astrophys.* **507**, 355 (2009).
[41] J. S. Frederiksen and S. M. Kepert, *J. Atmos. Sci.* **63**, 3006 (2006).
[42] U. Frisch et al., *J. Fluid Mech.* **68**, 769–778 (1975).
[43] U. Frisch et al., *J. Mécanique Théor. Appl.*, **2**, 191 (1983).
[44] S. Galtier, S. Nazarenko, A. Newell, and A. Pouquet, *J. Plasma Phys.* **63**, 447 (2000).
[45] S. Galtier, S. Nazarenko, A. Newell, and A. Pouquet, *Astrophys. J. Lett.* **564**, L49 (2002).
[46] S. Galtier, A. Pouquet and A. Mangeney, *Phys. Plasmas* **12**, 092310 (2005).
[47] S. Galtier, *Phys. Rev. E* **77**, 015302R (2008).
[48] S. Galtier, *Astrophys. J.* **704**, 1371 (2009).
[49] S. Galtier, "Wave Turbulence," World Scientific, V. Shrira and S. Nazarenko Eds. (2011).
[50] S. Ghosh, W. H. Matthaeus and D. Montgomery, *Phys. Fluids* **31**, 2171 (1988).
[51] C. Gloaguen et al., *Physica D* **17**, 154 (1985).
[52] G. Gogoberidze, *Phys. Plasmas* **14**, 022304 (2007).
[53] P. Goldreich and P. Sridhar, *Astrophys. J.* **438**, 763 (1995).
[54] N. Goldenfeld, *Phys. Rev. Lett.* **96**, 044503 (2006).
[55] T. Gomez, H. Politano and A. Pouquet, ıtPhys. Rev. E **61**, 5321 (2000).
[56] R. Grappin et al., *Astron. Astrophys.* **105**, 6 (1982).
[57] R. Grappin et al., *Astron. Astrophys.* **126**, 51 (1983).
[58] R. Grappin and W. C. Müller, *Phys. Rev. E* **82**, 026406 (2010).
[59] R. Grappin, private communication.
[60] R. Grauer, J. Krug and C. Marliani, *Phys. Lett. A* **195**, 335 (1994).
[61] H. Hasegawa et al., *Nature* **430**, 755 (2004).
[62] S. L. G. Hess, D. M. Malaspina and R. E. Ergun, *JGR* **116**, A07104, 2011
[63] D. D. Holm, J. E. Marsden, and T. S. Ratiu, *Phys. Rev. Lett.* **80**, 4173 (1998).

[64] D. D. Holm, *Chaos* **12**, 518 (2002).
[65] P. S. Iroshnikov, *Soviet Astronomy* **7**, 566 (1964),
[66] T. Ishihara, T. Gotoh and Y. Kaneda, *Ann. Rev. Fluid Mech.* **41**, 165 (2009).
[67] S. Jung, P. Morrison and H. Swinney, *J. FLuid Mech.* **554**, 433 (2006).
[68] Y. Kaneda et al., *Phys. Fluids* **19**, L21 (2003).
[69] S. Kida and Y. Murakami, *Phys. Fluids* **30**, 2030 (1987).
[70] A. N. Kolmogorov, *Dokl. Akad. Nauk SSSR*, **30**, 299 (1941).
[71] R. H. Kraichnan, *Phys. Fluids*, **8**, 1385 (1965).
[72] R. K. Kraichnan and D. Montgomery, *Rev. Prog. Phys.* **43**, 547 (1980).
[73] G. Krstulovic, M.-E. Brachet and A. Pouquet, *Phys. Rev. E* **84**, 016410 (2011).
[74] G. Lapenta, *Phys. Rev. Lett.* **100**, 235001 (2008).
[75] T. D. Lee, *Quart. Appl. Math* **10**, 69 (1952).
[76] E. Lee et al., *Phys. Rev.E* **78**, 066401 (2008).
[77] E. Lee et al., *Phys. Rev.E* **81**, 016318 (2010).
[78] J. W. Lord et al., *Phys. Fluids* **24**, 025102 (2012).
[79] S. K. Malapaka, "A study of magnetic helicity in decaying and forced 3D-MHD turbulence," PhD Thesis, June 2009, Universität Bayreuth.
[80] R. Marino et al., *Astrophys. J.* **677**, L71 (2008).
[81] J. Mason, F. Cattaneo, and S. Boldyrev, *Phys. Rev. E* **77**, 036403 (2008).
[82] W. H. Matthaeus and D. Montgomery, *Ann. N.Y. Acad. Sci.* **357**, 203 (1980).
[83] W. H. Matthaeus and M.L. Goldstein *J. Geophys. Res.* **87**, 6011 (1982).
[84] W. H. Matthaeus, M. L. Goldstein and D. Montgomery, *Phys. Rev. Lett.* **51**, 1484 (1983).
[85] W.H. Matthaeus et al., *Phys. Rev. Lett.* **100**, 085003 (2008).
[86] M. Meneguzzi et al., *J. Comp. Phys.* **123**, 32 (1996).
[87] C. Meneveau and J. Katz, *Ann. Rev. Fluid Mech.* **32**, 1 (2000).
[88] P. D. Mininni, A. G. Pouquet, and D. C. Montgomery, *Phys. Rev. Lett.* **97**, 244503 (2006).
[89] P. D. Mininni, A. Alexakis, and A. Pouquet, *Phys. Rev. E* **74**, 016303 (2006).
[90] P. D. Mininni and A. Pouquet, Phys. Rev. Lett. **99**, 244502 (2007).
[91] P. D. Mininni and A. Pouquet, *Phys. Rev. E* **80**, 025401 (2009).

[92] P. D. Mininni, "Scale Interactions in Magnetohydrodynamic Turbulence," *Ann. Rev. Fluid Mech.* **43**, 377 (2011).
[93] P. D. Mininni et al., *Phys. Rev. E* **83**, 016309 (2011).
[94] P. D. Mininni, D. Rosenberg and A. Pouquet, *J. Fluid Mech.* **699**, 263 (2012).
[95] H. K. Moffatt and A. Tsinober, *Ann. Rev. Fluid Mech.* **24**, 281 (1992).
[96] J. Molinari and D. Vollaro, *J. Atmos. Sci.* **67**, 274 (2010).
[97] R. Monchaux et al., *Phys. Rev. Lett.* **98**, 044502 (2007).
[98] D. Montgomery and L. Turner, *Phys. Fluids* **25**, 345 (1982).
[99] W. C. Müller and D. Biskamp, *Phys. Rev. Lett.* **84**, 475 (2000).
[100] W. C. Müller and D. Carati, *Phys. Plasmas* **9**, 824 (2002).
[101] W. C. Müller and R. Grappin, *Phys. Rev. Lett.* **95**, 114502 (2005).
[102] W. C. Müller, S. K. Malapaka, and A. Busse, *Phys. Rev. E* **85**, 015302(R) (2012).
[103] A. Newell and B. Rumpf, "Wave Turbulence," *Ann. Rev. Fluid Mech.* **43**, 59 (2011).
[104] K. Nykyri et al., *Ann. Geophys.* **24**, 2619 (2006).
[105] P. Padoan et al., *Phys. Rev. Lett.* **92**, 191102 (2004).
[106] J. Paret and P. Tabeling, *Phys. Fluids* **10**, 3126 (1998).
[107] J. C. Perez and S. Boldyrev, *Phys. Plasmas* **17**, 055903 (2010).
[108] T. D. Phan et al., *Nature* **439**, 175 (2006).
[109] J. Pietarila Graham, P. D. Mininni, and A. Pouquet, *Phys. Rev. E* **80**, 016313 (2009).
[110] J. Pietarila Graham, P. D. Mininni and A. Pouquet, "High Reynolds number magnetohydrodynamic turbulence using a Lagrangian model," *Phys. Rev. E* **84**, 016314 (2011).
[111] F. Plunian and R. Stepanov, *Phys. Rev. E* **82**, 046311 (2010).
[112] J. J. Podesta, D. A. Roberts, and M. L. Goldstein, *Astrophys. J.* **664**, 543 (2007).
[113] H. Politano, A. Pouquet, and P. L. Sulem, *Phys. Fluids B* **1**, 2330 (1989).
[114] H. Politano and A. Pouquet, *Phys. Rev. E* **52**, 636 (1995).
[115] H. Politano, A. Pouquet and P. L. Sulem, *Phys. Plasmas* **2**, 2931 (1995).
[116] H. Politano and A. Pouquet, *Geophys. Res. Lett.* **25**, 273 (1998).
[117] H. Politano and A. Pouquet, *Phys. Rev. E* **57**, 21(R) (1998).
[118] H. Politano, A. Pouquet, and V. Carbone, *Europhys. Lett.* **43**, 516 (1998).

[119] H. Politano, T. Gomez, and A. Pouquet, *Phys. Rev. E* **68**, 026315 (2003).

[120] Y. Ponty et al., *New J. Phys.* **9**, 296 (2007).

[121] A. Pouquet, U. Frisch, and J. Léorat, *J. Fluid Mech.*, **77**, 321 (1976).

[122] A. Pouquet and G. S. Patterson, *J. Fluid Mech.* **85**, 305 (1978).

[123] A. Pouquet, *Magnetohydrodynamic Turbulence*, Les Houches Summer School on Astrophysical Fluid Dynamics, Session **XLVII**, 139–227; Eds. J. P. Zahn and J. Zinn–Justin (Elsevier, 1993).

[124] A. Pouquet, *Turbulence, Statistics and Structures: an Introduction*, V^{th} European School in Astrophysics, San Miniato; C. Chiuderi and G. Einaudi (Eds.), Springer–Verlag, Lecture Notes in Physics "Plasma Astrophysics" **468**, 163 (1996).

[125] A. Pouquet, "Hydrodynamic and MHD turbulence: a rapid overview"; *Solar and Astrophysical Magnetohydrodynamics Flows*, K. Tsinganos Ed., Kluwer Acad. Press (Dordretch), NATO Advanced Study Institute Series C **481**, 195 (1996).

[126] A. Pouquet et al., "Lack of universality in MHD turbulence, and the emergence of a new paradigm?" Astrophysical Dynamics: From Stars to Galaxies, Proc. IAU Symposium **271**, Brummell, N. H., Brun, A. S. and Miesch, M., eds. (2010).

[127] A. F. Rappazzo et al., "Coronal heating, weak MHD turbulence and scaling laws," *Astrophys. J.* **657**, L47 (2007).

[128] L. C. Ray et al., *J. Geophys. Res.* **117**, A01205 (2012)

[129] K. A. Riveros et al., *Geofisica Internacional* **47**, 265 (2008).

[130] F. Sahraoui et al., *Phys. Rev. Lett.* **105**, 131101 (2010).

[131] J. Saur et al., *Astron. Astrophys.* **386**, 699 (2002).

[132] A. Schekochihin, S. Cowley and T. Yousef, in IUTAM Book series, **4**, 347, Y. Kaneda (Ed.) (Springer, Dordrecht, 2008).

[133] S. Servidio, W. H. Matthaeus and P. Dmitruk, *Phys. Rev. Lett.* **100**, 095005 (2008).

[134] S. Servidio et al., *Phys. Plasmas* **17**, 032315 (2010).

[135] Z. S. She and E. Lévêque, *Phys. Rev. Lett.* **72**, 336 (1994).

[136] J. Shebalin, *J. Plasma Phys.* **72**, 507 (2006).

[137] J. Shebalin, *Plasma Phys.* **16**, 072301 (2009).

[138] J. Shebalin, *Plasma Phys.* **17**, 092303 (2010).

[139] L. Sorriso-Valvo et al., *Phys. Rev. Lett.* **99**, 115001 (2007).

[140] J. Stawarz et al., *Astrophys. J.* **713**, 920 (2010).

[141] J. Stawarz, A. Pouquet and M.-E. Brachet, *Phys. Rev. E*, **86**, 036307 (2012).

[142] M. V. Stepanova et al., *Adv. Space Res.* **37**, 559 (2006).
[143] T. Stribling and W. H. Matthaeus, *Phys. Fluids B* **2**, 1979 (1990).
[144] T. Stribling and W. H. Matthaeus, *Phys. Fluids B* **3**, 1848 (1991).
[145] D. Sundkvist et al., *Nature* **436**, 825 (2005).
[146] C. Tu and E. Marsch, *J. Geophys. Res.* **95 A4**, 4337 (1990).
[147] V. M. Uritsky et al., *Phys. Rev. Lett.* **99**, 025001 (2007).
[148] D. A. Uzdensky, N. F. Loureiro and A. A. Schekochihin, *Phys. Rev. Lett.* **105**, 235002 (2010).
[149] A. Vallgren and E. Lindborg, *J. Fluid Mech.* **667**, 463 (2011a).
[150] M. Verma, "Variable enstrophy flux and energy spectrum in 2D turbulence with Ekman friction," preprint (2011a).
[151] M. Verma, "Variable energy flux in quasi-static magnetohydrodynamic turbulence," preprint (2011b).
[152] M. Wan et al., *Phys. Plasmas* **16**, 080703 (2009).
[153] J. M. Weygand, *J. Geophys. Res.* **110**, A01205 (2005).
[154] R. T. Wicks et al., *Phys. Rev. Lett.* **106**, 045001 (2011).
[155] N. Yokoi, *J. Turb.* **12**, nb. 27, p. 1–33 (2011).
[156] T. Yousef, F. Rincon, and A. Schekochihin, *J. Fluid Mech.* **575**, 111 (2007).
[157] Y. Zhou, W.H. Matthaeus, and P. Dmitruk, *Rev. Mod. Phys.* **76**, 1015 (2004).

Chapter 3

Dynamics of Structures in Configuration Space and Phase Space: An Introductory Tutorial

P. H. Diamond[1,2], Y. Kosuga[1,2], M. Lesur[1]

[1] *WCI Center for Fusion Theory, N.F.R.I., Korea*

[2] *CMTFO and C.A.S.S., UCSD, USA*

Some basic ideas relevant to the dynamics of phase space and real space structures are presented in a pedagogical fashion. We focus on three paradigmatic examples, namely; G. I. Taylor's structure based re-formulation of Rayleigh's stability criterion and its implications for zonal flow momentum balance relations; Dupree's mechanism for nonlinear current driven ion acoustic instability and its implication for anomalous resistivity; and the dynamics of structures in drift and gyrokinetic turbulence and their relation to zonal flow physics. We briefly survey the extension of mean field theory to calculate evolution in the presence of localized structures for regimes where Kubo number $K \simeq 1$ rather than $K \ll 1$, as is usual for quasilinear theory.

"We all know that the real reason universities have students is in order to educate the professors."

John Archibald Wheeler

"Reward the sneezers who stand up and spread the ideas."

Seth Godin, on viral marketing

3.1 Introduction and basic considerations

The aim of this tutorial is to introduce the ways we think about and describe relaxation, transport and instability in a system consisting of an ensemble of localized structures. Indeed, in simple terms, turbulence is usually better thought of as a 'soup' or 'stew' of 'eddys,' 'vorticies,' 'blobs,' 'clumps,' 'holes,' etc. than as an ensemble of (very) weakly nonlinear waves. However, in MFE, we almost always *calculate* using quasilinear theory, which is based on the idea of turbulence as a random ensemble of waves. We follow this familiar recipe in spite of the fact that at the same time, we often invoke or tacitly assume that the strength of saturated turbulence given by mixing length guesstimates [22], which posit that $\tilde{v} \sim \Delta_c/\tau_c$. Here $\tilde{v}$ is a typical fluctuating velocity, Δ_c is the correlation scale and τ_c is the correlation time. Of course, such a mixing length criteria is simply another way to write Kubo number $K \simeq 1$, where $K = \tilde{v}\tau_c/\Delta_c$. $K \simeq 1$ implies that the particle or fluid element rotation time in a vortex structure is comparable to the lifetime of that structure, while quasilinear theory assumes that the lifetime is short compared to the rotation time (i.e. $\tau_{ac} < \tau_b$). Thus, it is difficult to see how $K \simeq 1$ and quasilinear theory can be mutually compatible, despite perpetual claims to the contrary.

More generally, a structure such as an 'eddy' or a 'blob' can be distinguished from a wave in that an eddy does *not* correspond to the zero of a collective response function (dielectric), i.e. $\epsilon(k, \omega) = 0$, while a wave does. MFE theory is obsessed with the zoology of linear waves and instabilities in spite of the fact that, as noted above, concepts derived from localized structures are frequently more germane to the true, physical dynamics of the real turbulent state. Structures are usually thought to form in the process of nonlinear saturation of the underlying instability. However, as we will discuss, localized structures are sometimes more efficient in *tapping* free energy than linear waves are, and so structures can also act as a mechanism for *triggering* instability and relaxation. We will see that this property follows from the fact that structures scatter energy and momentum differently than waves do, on account of the fact that the structure is *localized* in phase space. In the particular case of drift wave turbulence, consideration of potential vorticity structures provides an illuminating alternative route to understanding zonal flow generation via the conservation of momentum in drift wave-zonal flow interaction.

Given the aphorism that "a picture is worth a thousand words" — and at least as many equations — we deem it useful to present a few relevant

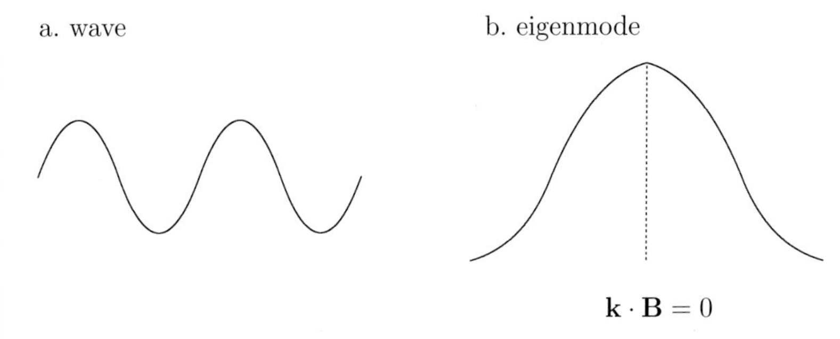

Fig. 3.1 Wave and eigenmode.

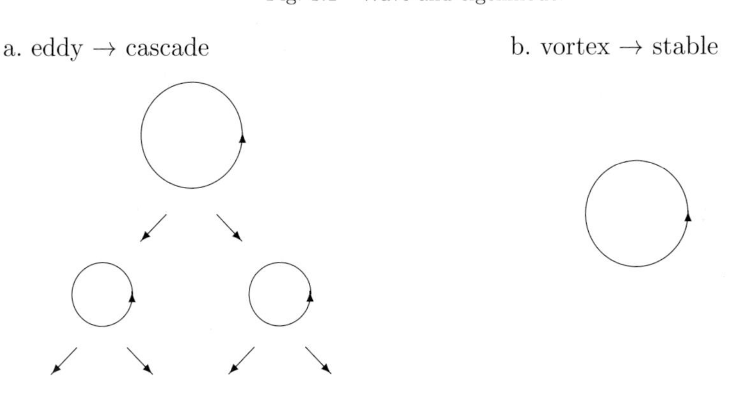

Fig. 3.2 Eddy and vortex.

cartoons here, so as to put flesh on the rather abstract concepts we have been discussing. Figure 3.1a shows a wave and Fig. 3.1b shows an eigenmode localized at a $\mathbf{k} \cdot \mathbf{B} = 0$ surface. These should be compared to cascading eddys, shown in Fig. 3.2a or an isolated coherent vortex, shown in Fig. 3.2b. Also localized structures can interact with mean profiles. One example is shown in Fig. 3.3, where a localized gradient relaxation event (i.e. gradient flattening) generates a blob (i.e. local quantity excess) propagating down the gradient and a void (i.e. local quantity deficit) propagating up the gradient. Structures can form in phase space, as well. Figure 3.4 shows a phase space density hole, in velocity space. Since total phase space density must be conserved, the hole perturbation can grow if the centroid of the hole moves up the gradient. We will refer back to these cartoons from time to time in this paper.

The remainder of this tutorial is organized as follows. In Section 3.2, we discuss the Rayleigh inflection point theorem. After reviewing the classic

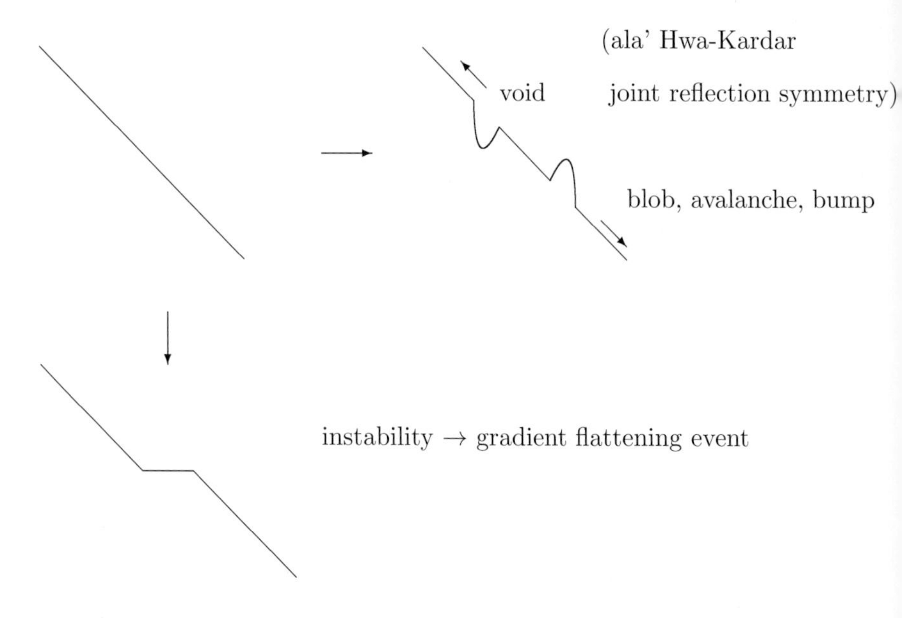

Fig. 3.3 Void and bump.

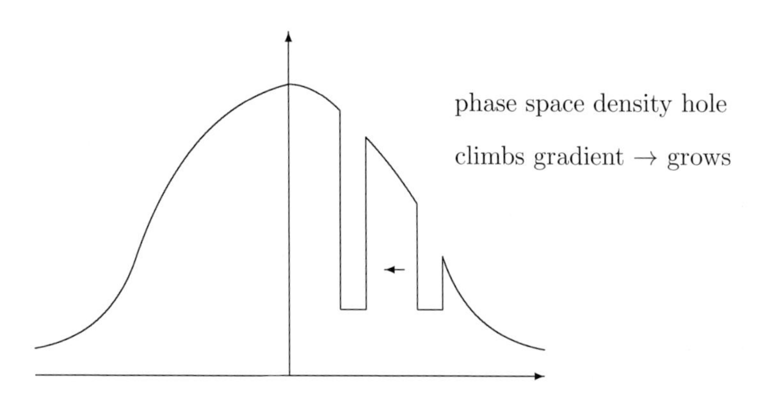

Fig. 3.4 Hole in phase space.

modal approach we present G. I. Taylor's far more physical and intuitive derivation based on a consideration of the displacement of a localized PV blob. As a bonus, from there we immediately derive basic insights into zonal flow generation and the content of the Charney-Drazin non-acceleration theorem for wave-mean flow interaction. In Section 3.3, we re-visit the current driven ion acoustic problem, and discuss a nonlinear, structure-based instability mechanism which is complementary to the text book example based on linear theory. We cast this story in terms of growth of a granulation in phase space. The instability mechanism is nonlinear. We also consider the growth of localized structures in the context of the now classic paradigm of the Berk-Breizman model. In particular, we suggest the possibility of a new nonlinear instability for that model. In Section 3.4, we discuss phase space density structures in the context of drift and gyrokinetic turbulence. That section to some degree unifies the discussions in Section 3.2 and Section 3.3. Special attention is given to the relation of structure dynamics in kinetic drift-zonal flow systems to the Charney-Drazin non-acceleration theorem discussed in Section 3.2. In Section 3.5, we briefly outline the Dupree-Lenard-Balescu theory for calculating transport in the presence of phase space density granulations, where Kubo number $K \sim 1$. That section is short, as the subject is discussed at length in other recent tutorials. Section 3.6 is a brief concluding comment.

3.2 Rayleigh criterion, potential vorticity, and zonal flow momentum

In this section, we first review the traditional 'modal' approach to the Rayleigh criterion, then present a far more physical, 'structure based' approach first given by G. I. Taylor. We then extend this line of thought to extract the essence of several key relations between wave and zonal flow momentum.

3.2.1 *Something old: modal derivation of Rayleigh criterion*

The time honored Rayleigh criterion [16, 28, 42], a necessary condition for inviscid shear flow stability, is relevant to the flow configuration shown in Fig. 3.5.

Here $q = -\nabla\phi^2$ is the potential vorticity, so PV conservation is just $d\nabla^2\phi/dt = 0$. Straightforward linearization then gives the Rayleigh equation for eigenmodes of the shear flow:

$$\left(\partial_y^2 - k_x^2\right)\tilde{\phi}_k + \frac{k_x \partial_y \langle q\rangle}{\omega - k_x V_x(y)}\tilde{\phi}_k = 0\,. \tag{3.1}$$

Multiplying by $\tilde{\phi}_k^*$, integrating from $-a$ to a, noting the no-slip boundary

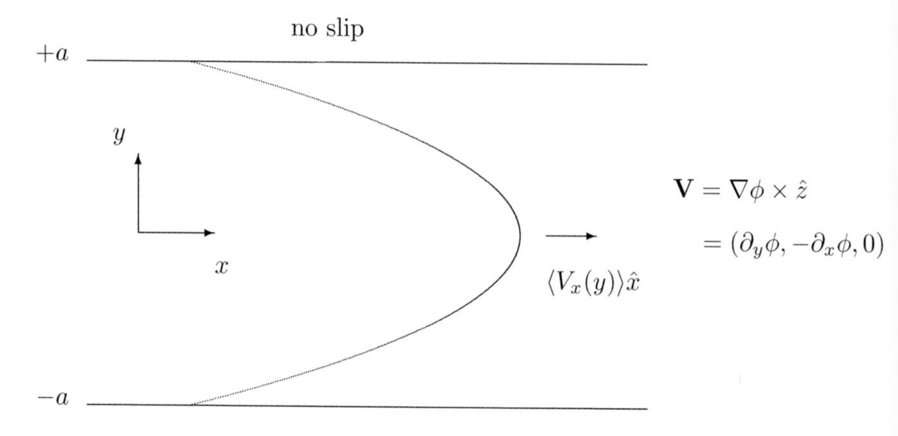

Fig. 3.5 Flow configuration.

condition, and writing $\omega = \omega_r + i\gamma_k$ explicitly, then gives

$$-\int_{-a}^{a} dy\Big(|\partial_y\tilde{\phi}_k|^2 + k_x^2|\tilde{\phi}_k|\Big) + \int_{-a}^{a} dy \frac{k_x\partial_y\langle q\rangle\{(\omega_r - k_x V_x(y)) - i\gamma_k\}}{(\omega_r - k_x V_x(y))^2 + \gamma_k^2}|\tilde{\phi}_k|^2 = 0\,. \tag{3.2}$$

Equation (3.2) is a complex equation, so both the real and imaginary parts of the lefthand side must vanish. For the imaginary part, we have

$$\gamma_k \int_{-a}^{a} dy \frac{k_x\partial_y\langle q\rangle}{(\omega_r - k_x V_x(y))^2 + \gamma_k^2}|\tilde{\phi}_k|^2 = 0\,. \tag{3.3}$$

Thus, if $\langle q\rangle' \neq 0$ everywhere on $[-a, a]$, $\gamma_k = 0$ necessarily, implying stability. For $\gamma_k \neq 0$ (i.e. instability),

$$\int_{-a}^{a} dy \frac{k_x\partial_y\langle q\rangle}{(\omega_r - k_x V_x(y))^2 + \gamma_k^2}|\tilde{\phi}_k|^2 = 0 \tag{3.4}$$

is required. This implies that $\langle q\rangle'$ *must* change sign at some point on the interval $[-a, a]$, i.e. for example, $\langle q\rangle' < 0$ on $[-a, x]$ and $\langle q\rangle' > 0$ on $[x, a]$. Since $\langle q\rangle' = \partial_y\langle\partial_y^2\phi\rangle = \partial_y^2\langle V_x(y)\rangle$, the slope of the vorticity must change somewhere in $[-a, a]$, or equivalently the mean flow profile must have an *inflection point* somewhere on $[-a, a]$. We have arrived at the essence of the famous *Rayleigh inflection point theorem*, namely that a necessary condition for inviscid instability of a shear flow is that the flow must have an inflection point. Fjortoft later showed that, for instability, the inflection point must be a vorticity maximum [16, 42].

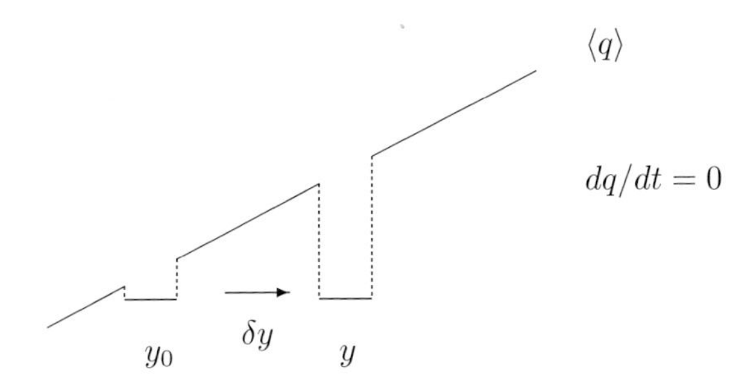

Fig. 3.6 Displacement of PV blob.

3.2.2 *Something newer: G. I. Taylor's stability criterion derived from PV dynamics*

The Rayleigh inflection point theorem (Eq. (4)) is based entirely on manipulations of the equation and so is not very satisfying from a physics perspective. More substantively, Rayleigh's criterion does not address the possibility that excitons or structures other than linear eigenmodes (i.e. waves) may be the optimal ones with which to tap the available free energy in the shear. This brings us to G. I. Taylor's argument from his famous paper of 1915 [38]. Taylor constructed a very physical and intuitive description of the consequence of infinitesimal displacement of a 'blob' or 'slug' of potential vorticity (PV) in a system which locally conserves PV. The class of such systems includes non-dissipative 2D fluids, quasi-geostrophic fluids, and both the Hasegawa-Mima and the Hasegawa-Wakatani system plasmas. The crux of Taylor's argument is on the analysis of consequences for the mean flow when a 'blob' of PV is displaced up or down the mean cross-stream PV gradient, as shown in Fig. 3.6.

To address the effect on the flow, consider the momentum balance equation for inviscid, isentropic displacements

$$\frac{\partial \mathbf{V}}{\partial t} + \mathbf{V} \cdot \nabla \mathbf{V} = -\nabla h \tag{3.5a}$$

where h is enthalpy (i.e. $dh = dp/\rho$). Equivalently,

$$\frac{\partial \mathbf{V}}{\partial t} = -\nabla \left(h + \frac{V^2}{2} \right) + \mathbf{V} \times \boldsymbol{\omega} \,. \tag{3.5b}$$

So, for the mean component,

$$\frac{\partial \langle \mathbf{V} \rangle}{\partial t} = -\partial_y \left(h + \frac{V^2}{2} \right) \hat{y} + \langle \mathbf{V} \times \boldsymbol{\omega} \rangle \tag{3.5c}$$

where we assumed symmetry in $\hat{x}$ direction. For the zonal component:

$$\frac{\partial\langle V_x\rangle}{\partial t} = \langle \tilde{V}_y \tilde{\omega}_z\rangle \tag{3.6a}$$

or equivalently

$$\frac{\partial\langle V_x\rangle}{\partial t} = \langle \tilde{V}_y \tilde{q}\rangle \tag{3.6b}$$

since $\tilde{q} = \tilde{\omega}_z$ for $q = \omega_z + \beta y$, as for quasi-geostrophic (QG) fluids. Note that since $\tilde{\omega}_z = -(\partial_x^2 + \partial_y^2)\tilde{\phi}$, performing the zonal average gives

$$\frac{\partial\langle V_x\rangle}{\partial t} = -\partial_y \langle \tilde{V}_y \tilde{V}_x\rangle\,. \tag{3.6c}$$

Equations (3.6a), (3.6b), and (3.6c) state the famous, and even useful, *Taylor identity.*

$$\langle \tilde{V}_y \tilde{\omega}_z\rangle = \langle \tilde{V}_y \tilde{q}\rangle = -\partial_y \langle \tilde{V}_y \tilde{V}_x\rangle\,. \tag{3.6d}$$

The Taylor Identity states that for a 2D or quasi-geostrophic (QG), etc fluid with conserved vorticity or PV, the cross zonal stream flux of PV equals the along-stream component of the Reynolds force. Taylor's identity establishes that vorticity transport is the process which controls the dynamics of zonal flows and their self-acceleration.

Returning to the issue of stability, we note that to utilize Eq. (3.6b), we must calculate $\tilde{q}$. Now

$$\tilde{q} = (\text{PV of vortex blob at } y) - (\text{Mean PV at } y)\,. \tag{3.7a}$$

Since $\tilde{q}(y)$ is displaced to y from y_0, and since $dq/dt = 0$ with initial fluctuations negligibly small, we have

$$(\text{PV of vortex blob at } y) = \langle q(y_0)\rangle\,. \tag{3.7b}$$

Also, for small $y - y_0 \equiv \delta y$, we can 'Taylor' expand:

$$\langle q(y)\rangle \cong \langle q(y_0)\rangle + (y - y_0)\left.\frac{d\langle q\rangle}{dy}\right|_{y_0}\,. \tag{3.7c}$$

So

$$\tilde{q} \cong -\delta y \left.\frac{d\langle q\rangle}{dy}\right|_{y_0}\,. \tag{3.7d}$$

Since the choice of y_0 is arbitrary, we hereafter drop the subscript. Thus, we finally arrive at

$$\frac{\partial\langle V_x\rangle}{\partial t} = -\langle \tilde{V}_y \delta y\rangle \frac{d\langle q\rangle}{dy} \tag{3.8a}$$

or, taking the fluid parcel displacement $\tilde{\xi} = \delta y$ and noting $\partial_t \tilde{\boldsymbol{\xi}} = \mathbf{V}$,

$$\frac{\partial \langle V_x \rangle}{\partial t} = -\left(\partial_t \frac{\langle \tilde{\xi}^2 \rangle}{2}\right) \frac{d\langle q \rangle}{dy}. \tag{3.8b}$$

Instability of the system to the initial PV slug displacement requires, of course, $\partial_t \langle \tilde{\xi}^2 \rangle > 0$. At the same time, the net momentum of the flow is conserved, so

$$\frac{\partial}{\partial t} \int_{-a}^{a} dy \langle V_x \rangle = -\int_{-a}^{a} dy \left(\partial_t \frac{\langle \tilde{\xi}^2 \rangle}{2}\right) \frac{d\langle q \rangle}{dy} = 0. \tag{3.9}$$

Since $\partial_t \langle \tilde{\xi}^2 \rangle > 0$, total momentum conservation requires that $\partial \langle q \rangle / \partial y$ *must* change the sign at some point on the interval $[-a, a]$. For $q = -\nabla^2 \phi$, this is equivalent to requiring that the flow has an inflection point.

Taylor's derivation of the Rayleigh result is notable in that:

(1) it makes no reference to waves or eigenmodes, but rather is formulated in terms of the displacement of a 'blob' or 'slug' of vorticity. Of course, it is limited to consideration of a small displacement.
(2) it directly links stability to flow evolution, and so is useful for obtaining more general insights into dynamics.

The physical clarity and simplicity of Taylor's derivation make it far more satisfying than Rayleigh's. We should mention here that there were at least two notable follow-ons to Taylor's analysis. First, C. C. Lin [32] showed that a flow profile inflection point is needed to allow the interchange of two vortices without a restoring force, so that the shear layer relaxes. Lin's analysis was based on using the Biot-Savart law to calculate the force on individual vortex elements. Later, V.I. Arnold [2] presented a famous *non-linear* stability analysis which showed that the existence of a flow profile with an inflection point is still a necessary condition for instability.

3.2.3 *Something further: zonal flow evolution and wave-flow interaction*

Recall that Eq. (3.8b) states that

$$\frac{\partial \langle V_x \rangle}{\partial t} = -\left(\partial_t \frac{\langle \tilde{\xi}^2 \rangle}{2}\right) \frac{d\langle q \rangle}{dy} \tag{3.10}$$

and so relates the mean zonal flow evolution to the mean PV gradient and the evolution of the displacement of a fluid element [35]. Note that since the vortex element or PV diffusivity D_q is, by definition,

$$D_q = \partial_t \frac{\langle \tilde{\xi}^2 \rangle}{2} \tag{3.11a}$$

it follows that

$$\frac{\partial \langle V_x \rangle}{\partial t} = -D_q \frac{d\langle q \rangle}{dy} . \tag{3.11b}$$

Equation (3.11b) is an important, general result which states that *a latitudinal diffusive flux of potential vorticity will accelerate a zonal flow.* This is in accord with the general concept that *links zonal flow formation to PV mixing.* Indeed, recently M. McIntyre and R.Wood [45] published a lengthy discussion of this issue which generalizes, but draws heavily upon, Taylor's pioneering insights. Equation (3.11b) prescribes the direction of the zonal acceleration. For $d\langle q \rangle/dy > 0$, the acceleration is westward, while for $d\langle q \rangle/dy < 0$, the acceleration is eastward. In particular, this suggests that the beta effect will drive a westward circulation ($\beta > 0$). In general, any PV mixing process which tends to increase the variance of the latitude of a fluid particle (i.e. $\partial_t \langle \tilde{\xi}^2 \rangle > 0$), will accelerate a zonal flow opposite to $d\langle q \rangle/dy$. Finally, we observe that Eq. (3.11b) can also be obtained using the Taylor Identity (Eq. (3.6d)) and a mean field calculation (i.e. quasilinear theory) for the vorticity flux. We leave this as an exercise for the readers.

Taylor's argument for zonal flow generation is much more fundamental and elegant than modulational instability methods and the other cranks we love to turn [15]. The only two essential elements in his argument are local PV conservation along fluid trajectories and PV mixing, i.e. the net irreversible transport of potential vorticity. This begs the question of *what is the origin of irreversibility in the PV flux?* Equivalently, what is the microscopic mechanism for PV mixing? In the language of MFE methodology, this boils down to calculating the cross-phase in the PV flux $\langle \tilde{V}_y \tilde{q} \rangle$. There are several viable candidates, which include:

(1) Direct dissipation, as by viscosity.
(2) Nonlinear coupling to small scale dissipation by the *forward* cascade of potential vorticity.
(3) Rossby or drift wave absorption at critical layers, where $\omega = k_x \langle V_x(y) \rangle$. This is essentially Landau resonance. Transport or mixing of PV region requires the overlap of neighboring critical layers, leading to stochastization of flow streamlines.

(4) Stochastic nonlinear wave-fluid element scattering, which is analogous to transport induced by nonlinear Landau damping.

Note that the more general concepts are stochasticity of streamlines and forward potential enstrophy cascade to small scale dissipation. Interestingly, when looking at the phenomenon of zonal flow self-organization from the standpoint of PV transport and mixing, it is the *forward enstrophy cascade* which is critical, and not the inverse energy cascade, as is conventionally mentioned!! Finally, we observe that zonal flow acceleration is not necessarily a strongly nonlinear process. PV mixing can occur via wave absorption, and can be manifested in weak turbulence, as an essentially quasilinear process. In this regard, the reader should consult [40].

Equations (3.10) and (3.11b) also lead to a useful and instructive momentum conservation theorem for wave-mean flow interaction. From these, we immediately see that

$$\frac{\partial}{\partial t}\left\{\langle V_x\rangle + \frac{\langle\tilde{\xi}^2\rangle}{2}\frac{\partial\langle q\rangle}{\partial y}\right\} = 0\,. \tag{3.12a}$$

Using the relation

$$\tilde{q} = -\tilde{\xi}\frac{\partial\langle q\rangle}{\partial y} \tag{3.12b}$$

which relates PV perturbation to fluid element displacement, we can then re-write Eq. (3.12a) as

$$\frac{\partial}{\partial t}\left\{\langle V_x\rangle + \frac{\langle\tilde{q}^2\rangle}{2(\partial\langle q\rangle/\partial y)}\right\} = 0\,. \tag{3.12c}$$

It is interesting then to note that for $\partial\langle q\rangle/\partial y = \beta$ and $\tilde{q} = \nabla^2\tilde{\phi}$, we find from Eq. (3.12c)

$$\frac{\partial}{\partial t}\left\{\langle V_x\rangle - \sum_{\mathbf{k}} k_x N_{\mathbf{k}}\right\} = 0 \tag{3.12d}$$

where

$$N_{\mathbf{k}} = \frac{E(k)}{\omega_{\mathbf{k}}} \tag{3.12e}$$

is the wave action density for Rossby waves. $E(k)$ is simply the Rossby wave energy density. We see, then that Eq. (3.12d) is a momentum theorem which ties the zonal flow momentum $\langle V_x\rangle$ to the zonal wave momentum density $k_x N_{\mathbf{k}}$, which is the negative of the pseudomomentum. Equation (3.12d) states that the zonal flow cannot 'slip' relative to the wave momentum density, or equivalently, that the quasi-particle field of the wave packets is frozen into the zonal flow. Equation (3.12d) is a limiting case of the Charney-Drazin non-acceleration theorem, which states that in the absence of sources and sinks, the zonal flow momentum can change *only* if the wave momentum density varies in time. The full Charney-Drazin theorem [11, 13, 42] for the quasi-geostrophic (QG) equation is:

$$\frac{\partial}{\partial t}\left\{\langle V_x\rangle + \frac{\langle\tilde{q}^2\rangle}{2(\partial\langle q\rangle/\partial y)}\right\} = -\nu\langle V_x\rangle + \frac{1}{\langle q\rangle'}\left\{\langle\tilde{f}^2\rangle\tau_c - \mu\langle(\nabla\tilde{q})^2\rangle - \partial_y\langle\tilde{V}_y\tilde{q}^2\rangle\right\}. \tag{3.13}$$

Here $\tilde{f}$ is the stochastic forcing which drives the system, τ_c is its correlation time, μ is the viscosity, and ν is the scale invariant drag. The term $\partial_y\langle\tilde{V}_y\tilde{q}^2\rangle$ accounts for the local convergences and divergences in the flux of potential enstrophy — i.e. turbulence spreading [19]. Physically speaking, Eq. (3.13) states that the sum of the flow momentum and the wave pseudomomentum is conserved up to frictional flow damping and turbulence excitation ($\sim\langle\tilde{f}^2\rangle\tau_c$), dissipation ($\sim\mu\langle(\nabla\tilde{q})^2\rangle$) and spreading. Thus, stationary turbulence can drive a zonal flow *only* by forcing, dissipation or convergence of spreading, all of which break the local freezing-in law for quasi-particle momentum and flow momentum. For stationary flow and turbulence, we find

$$\langle V_x\rangle = \frac{1}{\nu\langle q\rangle'}\left\{\langle\tilde{f}^2\rangle\tau_c - \mu\langle(\nabla\tilde{q})^2\rangle - \partial_r\langle\tilde{V}_r\tilde{q}^2\rangle\right\}. \tag{3.14}$$

Thus, the flow direction is ultimately set by $\langle q\rangle'$ and the spatial distribution of sources and sinks. Note that some finite stand-off distance between source and sink is required for a finite scale shear flow. In simple drift wave turbulence, this is usually provided by the disparity and distance between the electron coupling region of the wave and the ion Landau resonance points, where wave absorption occurs.

Formally, the Charney-Drazin theorem is a consequence of, and is derived from, potential enstrophy density balance. The pseudomomentum density is simply the fluctuation potential enstrophy density divided by $\langle q\rangle'$. Thus, we see that a physical perspective on the pseudomomentum is that the fluctuation potential enstrophy density sets its size, while $\langle q\rangle'$ defines its direction or orientation. Alternatively, note that the pseudomomentum density is a measure of the effective 'roton' intensity field. In this vein, it is easy to see why potential enstrophy flux convergence ($\sim\partial_r\langle\tilde{V}_r\tilde{q}^2\rangle$) affects the zonal momentum balance, since it alters the roton quasi-particle density. Interestingly, drift waves have the character of *both* types of quasi-particles: namely phonons, via the relation $N_{\mathbf{k}} = E(k)/\omega_{\mathbf{k}}$ and rotons, via the proportionality of pseudomomentum to $\langle \tilde{q}^2 \rangle / \langle q \rangle'$ [46]. Ultimately, this dual character is a consequence of the fact that both energy and potential enstrophy are inviscid invariants of the system.

3.3 Dynamics of phase space structures

3.3.1 *Basic concepts*

In this section, we discuss phase space structures in Vlasov turbulence [7, 14, 18]. That there should be a close correspondence between PV dynamics in quasi-geostrophic (QG) and drift wave turbulence on one hand, and Vlasov turbulence on the other hand, is no surprise, since both systems are governed by the incompressible advection of a conserved quantity along Hamiltonian trajectories, with feedback via a Poisson equation. Table 3.1 compares the two systems and is self-explanatory. One point which should be emphasized is that in addition to Hamiltonian advection of a conserved scalar field, both quasi-geostrophic and Vlasov turbulence satisfy a Kelvin's theorem and thus have a conserved circulation. We remark that Donald Lynden-Bell was the first to prove a Kelvin's theorem for the Vlasov equation as an appendix to his pioneering paper [33] on violent relaxation in 1967. Systems governed by PV (i.e. q) or f conservation have several fundamental features in common. In particular, conservation of q, the total PV density and of f, the total phase space density, respectively lead to interesting relations concerning the evolution of a localized blob or structure, as well as its interaction with the mean gradient, $d\langle q \rangle / dy$ or $\partial \langle f \rangle / \partial v$. For PV, we have

$$\frac{d}{dt}(q_0 + \delta q) = 0 \,. \tag{3.15a}$$

Here q_0 is the initial profile including the mean distribution $\langle q \rangle$ as well as the depletion due to structure, Fig. 3.7, and δq is a perturbation. This gives

$$\frac{\partial}{\partial t} \int d^2 x \delta q^2 = -2 \frac{d}{dt} \int d^2 x q_0 \delta q \,. \tag{3.15b}$$

Expanding q_0 around a localized point y_0, we have

$$q_0(y) = q_0(y_0) + (y - y_0) \left. \frac{dq_0}{dy} \right|_{y_0} \equiv q_0(y_0) + \delta y \frac{dq_0}{dy_0} \,. \tag{3.15c}$$

Table 3.1 Comparison of PV and Vlasov system.

	QG system	Vlasov system
Basic Equation	$\partial_t q + \{q, \phi\} - \nu\nabla^2 q = 0$	$\partial_t f + \{f, H\} = C(f)$
Field	PV, $q = \beta y + \nabla^2\phi$ $q = \ln n_0(r) + \phi - \rho_s^2\nabla^2\phi$	distribution function $f(x, v, t)$
Evolution	$\{q, \phi\}$	$\{f, H\}$
Dissipation/Coarse Graining	$-\nu\nabla^2$	$C(f)$
Poisson Eq./Feedback	$q = q_0 + \beta y + \nabla^2\phi$	$\nabla^2\phi = -4\pi n_0 q \int f dv$
Kelvin's Theorem	$\oint (V + 2\Omega a \sin\theta) dl = const$	$\oint \mathbf{v}(s) \cdot d\mathbf{x} = const$
Decomposition	Planetary + Relative $\Rightarrow q = \beta y + \omega$	$f = \langle f \rangle + \tilde{f}$
Physical Element	vortex	granulation $\rightarrow$ hole, clump
Conservation	$\int q da$ (i.e. total charge)	phase volume

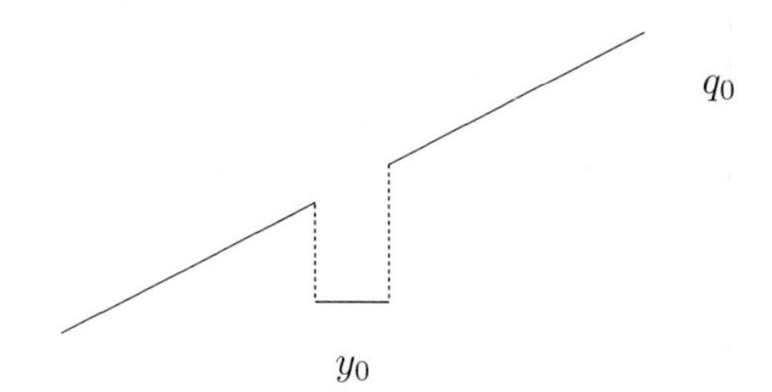

Fig. 3.7 Initial profile with structure.

Then Eq. (3.15b) gives

$$\frac{\partial}{\partial t}\int d^2x \delta q^2 = -2\frac{dq_0}{dy_0}\int d^2x \langle \tilde{V}_y \delta q \rangle . \tag{3.15d}$$

The reader can easily see that after utilizing the Taylor Identity (Eq. (3.6d)) and the mean flow evolution equation with scale independent drag, we recover a limit of the Charney-Drazin theorem. Note that Eq. (3.15d) can be re-written as:

$$\frac{\partial}{\partial t}\int d^2x \delta q^2 = 2\left(\frac{dq_0}{dy}\right)^2 \int d^2x \frac{d}{dt}\frac{\langle \delta y^2 \rangle}{2} . \tag{3.15e}$$

For the corresponding case of a Vlasov plasma structure [18], we have:

$$f = f_0 + \delta f \tag{3.16a}$$

f_0 includes the mean distribution $\langle f \rangle$ as well as the distortion of structure. δf is a perturbation. The growth of the perturbation is:

$$\frac{\partial}{\partial t}\int dv \delta f^2 = -2\frac{d}{dt}\int dv f_0 \delta f \,. \tag{3.16b}$$

Again expanding,

$$f_0 = f_0|_u + (v-u)\left.\frac{\partial f_0}{\partial v}\right|_u = f_0|_u + (v-u)\frac{\partial f_0}{\partial u} \tag{3.16c}$$

we find

$$\frac{\partial}{\partial t}\int dv \delta f^2 = -2\left(\frac{\partial f_0}{\partial u}\right)\frac{d}{dt}\int dv(v-u)\delta f \equiv -2\left(\frac{\partial f_0}{\partial u}\right)\frac{d}{dt}\frac{\langle p \rangle}{m} \,. \tag{3.16d}$$

Here $\langle p \rangle$ is the blob averaged momentum. In deriving Eq. (3.16d), we have assumed that the characteristic oscillation frequency for an individual particle in the localized phase space structure (i.e. "bounce frequency") is high compared to the structure growth rate and the mean relaxation rate, i.e. $\omega_b \gg (\partial_t \delta f)/\delta f, (\partial_t \langle f \rangle)/\langle f \rangle$. Equation (3.16d) is remarkable in that it links local structure growth to the evolution of net, *local* momentum. Depending upon the signs of $\partial f|_0/\partial u$ and the possibilities for $d\langle p \rangle/dt$, structure *growth* or *damping* is possible. The key to this is the nature of possible momentum exchange channels.

3.3.2 *Local structure growth*

In this section, we discuss the dynamics of nonlinear growth of localized Vlasov structures. Before launching into terra nova, we revisit terra firma, so as to place new ideas in the familiar (i.e. boring) context of the same old same old. A staple of Vlasov microstability theory is the current driven ion-acoustic (CDIA) instability [14, 27], which is also important for anomalous resistivity. As shown in Fig. 3.8, CDIA is essentially a battle between inverse electron resonance and ion resonance. The shift in the electron mean drift velocity v_d must satisfy $v_d > c_s$, so $\omega - kv_d < 0$. For

$$\epsilon = \epsilon_r + i\epsilon_{IM} \tag{3.17a}$$

and

$$\gamma_{\mathbf{k}} = -\frac{\epsilon_{IM}}{(\partial\epsilon_r/\partial\omega)|_{\omega_{\mathbf{k}}}} = -\frac{\epsilon_{IM}^e + \epsilon_{IM}^i}{(\partial\epsilon_r/\partial\omega)|_{\omega_{\mathbf{k}}}} \,, \tag{3.17b}$$

the algebra confirms the picture of Fig. 3.8, namely that the shift v_d must be large enough so that inverse electron Landau damping exceeds the positive

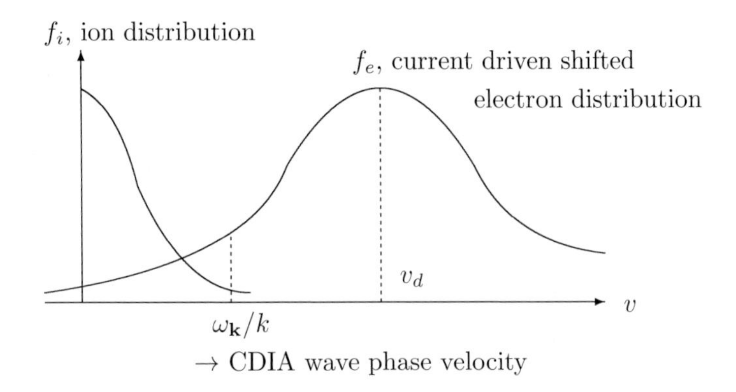

Fig. 3.8 Current Driven Ion-Acoustic (CDIA) instability.

ion Landau damping. As we all learned in kindergarten, this is most easily realized for *minimal* overlap of the ion and electron distribution functions. From the perspective of anomalous resistivity, the requisite momentum exchange is between electrons and *waves*, as opposed to collisional resistivity in which electrons exchange momentum with *ions*, by particle collisions. CDIA saturation requires some sort of *nonlinear* dissipation process, including possibly nonlinear ion Landau damping, in order to dispose of the wave energy.

The situation for localized structures is different, since several possible channels for momentum exchange exist. In general, momentum conservation implies

$$\frac{d}{dt}(\langle p_i\rangle + p_e + p_w) = 0 \tag{3.18}$$

where we assumed that our blob was an *ion* structure (this choice is arbitrary!) and p_e, p_w correspond to electron and wave momentum, respectively. Thus in a one species plasma (dynamically), with u off wave resonance, $d\langle p_i\rangle/dt = 0$ so structure growth is impossible. However, in a two species plasma more interesting things can happen. In particular, for ions:

$$\frac{d}{dt}\langle p_i\rangle = -\frac{1}{2}\frac{m_i}{(\partial f_{0,i}/\partial v)|_u}\partial_t \int dv\delta f_i^2 \tag{3.19a}$$

and for electrons

$$\frac{d}{dt}p_e = -\frac{1}{2}\frac{m_e}{(\partial f_{0,e}/\partial v)|_u}\partial_t \int dv\delta f_e^2 \,. \tag{3.19b}$$

So momentum conservation (with $\partial p_w/\partial t \cong 0$) requires

$$\frac{m_e}{(\partial f_{0,e}/\partial v)|_u}\partial_t \int dv\delta f_e^2 = -\frac{m_i}{(\partial f_{0,i}/\partial v)|_u}\partial_t \int dv\delta f_i^2 \,. \tag{3.19c}$$

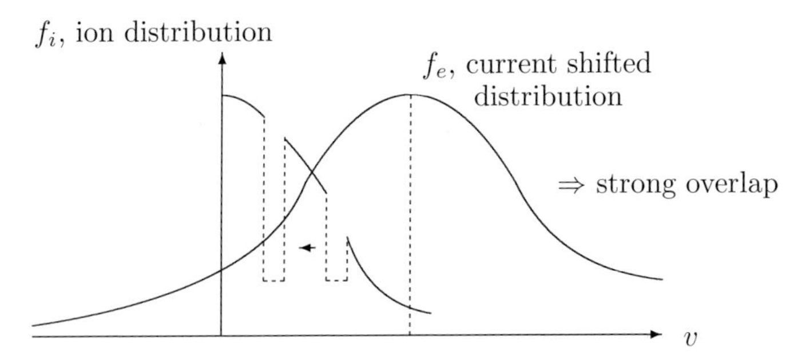

Fig. 3.9 Ion hole growth in current carrying Vlasov plasma.

Thus, we see that if:

$$\left.\frac{\partial f_{i,0}}{\partial v}\right|_u \left.\frac{\partial f_{e,0}}{\partial v}\right|_u < 0 \tag{3.19d}$$

— i.e. the two mean distributions have *opposite* slopes at the structure velocity — growth of δf is possible [18]. This is essentially a statement of availability of free energy — i.e. the presence of a finite v_{de}. Here, growth of localized structures occurs by collisionless inter-species momentum exchange, which move the structure up (for hole) or down (for blob) the mean phase space density gradient. Conservation of phase space density then requires that the fluctuation amplitude grows. See Fig. 3.9 for a demonstration. It is interesting to contrast the conditions for structure growth with the familiar conditions for CDIA wave growth. Structure growth requires *strong* overlap of electron and ion distributions, with opposite slopes. Overlap of $\langle f\rangle$ facilitates inter-species momentum exchange, which can propel a structure up or down the gradient. CDIA, by contrast, requires minimal overlap for instability. Minimal overlap reduces the stabilizing effects of ion Landau damping. Also, we note that Eq. (3.19a) effectively states $\partial p/\partial t \sim \partial_t \int dv \delta f^2/(\partial f_0/\partial v)|_u$, so we see the appearance of $\int dv\delta f^2/(\partial f_0/\partial v)|_u$ as a pseudomomentum. Here, however, we are not necessarily speaking of *linear waves*, but of more general types of fluctuations. Hence this observation attests to the fact that the pseudomomentum is a more general concept. It is comforting, however, to note that plugging the non-resonant linear response in for δf recovers the conventional expression for wave momentum density as derived from small amplitude theory [29].

There is an old injunction to lecturers which advises: 'Don't try to prove everything, but do try to prove at least one thing'. Thus, we now try to calculate the growth rate of an ion phase space structure (hole) driven by an electron current. From Eq. (3.19a), we have

$$\frac{1}{2}\frac{m_i}{(\partial f_{0,i}/\partial v)|_u}\partial_t \int dv\delta f_i^2 = -\frac{d\langle p_i\rangle}{dt} = \frac{dp_e}{dt}\,. \tag{3.20}$$

Now, while ion orbits are trapped, forming a phase space vortex, we can hope to calculate the orbits of the electrons by assuming nearly unperturbed ($u < v_{the}$) and using mean field theory. It is a standard crank to then show

$$\frac{\partial\langle f_e\rangle}{\partial t} = \frac{\partial}{\partial v}D(v)\frac{\partial}{\partial v}\langle f_e\rangle \tag{3.21a}$$

where

$$D(v) = \frac{q^2}{m_e^2}\sum_{k\omega}\langle E^2\rangle_{k\omega}\pi\delta(\omega - kv) \tag{3.21b}$$

is the usual quasilinear diffusion coefficient (but note $\omega \neq \omega(k)$ here!). Then,

$$\begin{aligned}\frac{dp_e}{dt} &= -m_e\int dvD(v)\frac{\partial}{\partial v}\langle f_e\rangle \\ &\simeq -m_e D(u)\left.\frac{\partial f_{0,e}}{\partial v}\right|_u \Delta u \end{aligned}\tag{3.21c}$$

where we have assumed the scattering spectrum to have a phase velocity distribution peaked near u. Δu is the approximate width of the distribution. Then combining Eqs. (3.20) and (3.21c) gives:

$$\partial_t\int dv\delta f_i^2 \cong -2\frac{m_e}{m_i}\left.\frac{\partial f_{0,i}}{\partial v}\right|_u \int dvD(v)\frac{\partial}{\partial v}\langle f_e\rangle \tag{3.21d}$$

$$\cong -2\frac{m_e}{m_i}\Delta u D(u)\left.\frac{\partial f_{0,i}}{\partial v}\right|_u \left.\frac{\partial f_{0,e}}{\partial v}\right|_u . \tag{3.21e}$$

Once again, we see that the need for strong electron and ion distribution function overlap at a location of opposite slope — the condition for available free energy. To extract the key scalings of the growth rate, we note from Eq. (3.21b) that $D(v) \sim \langle\tilde{E}^2\rangle \sim \langle\delta f^2\rangle(\Delta v_T)^2$. Here Δv_T is extent of the phase space structure in velocity. Loosely speaking, it corresponds to a self-trapping width, i.e. the width in velocity of a bunch of resonant particles which define a structure [18]. In resonance broadening theory [17], for a generic 1D plasma, we have:

(1) the spectral auto-correlation time

$$\tau_{ac} = \left[\Delta k\left(\frac{d\omega}{dk} - \frac{\omega}{k}\right)\right]^{-1}$$

which defines the time of self-coherence of a wave packet,

(2) the wave particle correlation time

$$\tau_c = (k^2 D)^{-1/3}$$

which defines the time it takes for a particle to scatter one wavelength, i.e. to decorrelate from the wave by random kicks in velocity,

(3) the $\langle f \rangle$ relaxation time, τ_{relax}.

We always assume the ordering $\tau_{ac} < \tau_c < \tau_{relax}$. In the vein, then, Δv_T is defined by

$$\Delta v_T = \frac{1}{k\tau_c} \,. \tag{3.22}$$

Of course, $\int dv \delta f^2 \sim \Delta v_T \delta f^2$, so the growth rate has the form

$$\gamma \simeq k\Delta v_T \left[-\frac{\partial f_{0,i}}{\partial v}\frac{\partial f_{0,e}}{\partial v} \right]_u F(\text{mess}) \,. \tag{3.23}$$

Here $F(\text{mess})$ is a complicated function related to the details of the phase space structure, and is of no instructive value. Note that the growth rate is nonlinear, i.e. $\gamma \sim k\Delta v_T \sim \omega_b \sim (q\phi/m_i)^{1/2}/\Delta x$. Here ϕ is the self-potential of the phase space structure, as determined by Poisson's equation, and Δx is the spatial extent of the structure. Note that we tacitly define a 'structure' as a blob/hole perturbation of size δf, spatial extent Δx and extent in velocity Δv_T. Obviously, $\tilde{\phi}_{self} \sim \delta f \Delta v_T / \epsilon(k, ku)$ where $\epsilon(k, ku)$ is the dielectric function evaluated at the ballistic frequency of the structure centroid motion. Equation (3.23) suggests that structure growth will continue until the free energy source is depleted. Thus, we can expect an ion hole to be accelerated up the velocity profile till $(\partial f_{0i}/\partial v)(\partial f_{0,e}/\partial v) \to 0$.

As mentioned previously, $\int dv[\delta f^2/(\partial f_0/\partial v)]$ constitutes a pseudomomentum or effective dynamical pressure for the structure. From the Vlasov equation, we can immediately show that

$$\partial_t \int dv \frac{\langle \delta f_i^2 \rangle}{\partial f_0/\partial v|_u} = -\frac{q}{m_i}\langle \tilde{E}\delta n_i \rangle \,. \tag{3.24a}$$

The RHS is, of course, simply the force on the ions, so we thus arrive at a 'Charney-Drazin' theorem for Vlasov turbulence

$$\frac{d}{dt}\left\{ \int dv \frac{\langle \delta f_i^2 \rangle}{\partial f_0/\partial v|_u} + \frac{\langle p_i \rangle}{m_i} \right\} = 0 \,. \tag{3.24b}$$

We can trivially derive a corresponding relation for electrons, and then use momentum balance to re-derive Eq. (3.24b). The point of this observation is to illustrate the close correspondence between the Charney-Drazin theorem and the theory of nonlinear phase space structure growth. Both are grounded in concepts of fluctuation pseudomomentum and the conservation of an effective phase space density (q or f) along Hamiltonian trajectories. Thus, it is not surprising that one should reduce to the other.

Another topic worthy of mention at this stage is the relation of phase space structure dynamics to the Berk-Breizman (B-B) model [3–5, 31]. The B-B model is a 1D model, including resonant particles, waves (with damping), and collisions. It represents the present day extension of classic linear bump-on-tail problem [43, 44], the Laval-Pesme extension thereof [30], and the traveling wave tube system studied experimentally by Malmberg, et. al. [41]. Constructed to mimic the essential features of resonant particles in Toroidal Alfvén Eigenmodes in a simpler, more tractable context, the B-B hides a wealth of physics in the 'plain paper wrapping' of a simple model. In particular trapping, structure formation and cyclic bursts are all possible. The basic equations of the B-B model are:

$$\frac{\partial f}{\partial t} + v\frac{\partial f}{\partial x} + E\frac{\partial f}{\partial v} = C(f - f_0)\,, \tag{3.25a}$$

the collision operator

$$C(f - f_0) = \frac{\nu_f^2}{k}\frac{\partial}{\partial v}(f - f_0) + \frac{\nu_d^3}{k^2}\frac{\partial^2}{\partial v^2}(f - f_0)\,, \tag{3.25b}$$

and the effective Poisson equation (really displacement current relation)

$$\frac{\partial E}{\partial t} = -\int dvv(f - f_0) - 2\gamma_d E\,. \tag{3.25c}$$

Here γ_d is the external damping associated with all 'other' processes, including wave damping, second species, etc. Collisions can be important here, as a means to limit plateau persistence in time and to de-trap particles.

Ignoring collisions, an analysis like that given above yields

$$\partial_t \int dv\langle \delta f^2\rangle = -\left.\frac{\partial f_0}{\partial v}\right|_u \int dv\langle \tilde{E}\delta f\rangle \tag{3.26a}$$

where we take δf located at speed u localized in velocity relative to $\langle f\rangle$, i.e. $\Delta v < v_{Th}$. Then, Eq. (3.25c) gives

$$(-i\omega + 2\gamma_d)E_{k,\omega} = -\int dvv\delta f_{k,\omega}\,. \tag{3.26b}$$

So

$$\partial_t \int dv\langle \delta f^2\rangle = 2\gamma_d \left.\frac{\partial f_0}{\partial v}\right|_u \int dv' \int dv \sum_{k\omega} \frac{v'\langle \delta f(v')\delta f(v)\rangle_{k\omega}}{\omega^2 + (2\gamma_d)^2}\,. \tag{3.26c}$$

Taking $dv \sim \Delta v$, $\langle \delta f(v')\delta f(v)\rangle_{k\omega} \simeq \langle \delta f(v')\delta f(v)\rangle_k 2\pi\delta(\omega - ku)$, and noting that $\langle \delta f(v)\delta f(v')\rangle$ is sharply peaked at $v \sim v' \sim \omega/k \sim u$, by the localization of δf, we then obtain:

$$\gamma \cong (\Delta v) \left.\frac{\partial f_0}{\partial v}\right|_u \frac{4\pi\gamma_d u}{(ku)^2 + (2\gamma_d)^2} . \tag{3.26d}$$

Of course, we see there is no free lunch — i.e. there must be free energy, so $(\partial f_0/\partial v)|_u > 0$ is required for instability. However, we do note that:

(1) the growth is *nonlinear*, i.e. $\gamma \sim \Delta v$
(2) *linear* instability is *not* required, i.e. γ can be positive here even if $\gamma_{L,0} - \gamma_d < 0$, where $\gamma_{L,0} = \pi(\partial f_0/\partial v)|_{\omega/k}/2k^2$ is the usual bump-on-tail drive rate.

We also note that the sign of δf and the relation between δf and Δv must be determined by an analysis of the condition for Jeans equilibrium. We remark here that it would be interesting to explore the effects of collisions on the B-B nonlinear structure growth process.

3.4 Phase space structure dynamics in drift turbulence

In this section, we extend our discussion of phase space structure dynamics from the case of 1D to the more relevant problem of drift turbulence. This is a challenging, vast and still developing subject. Thus, we limit our treatment to: a.) the presentation and discussion of the Darmet model, a useful prototype which is both simple and relevant, b.) a discussion of the dynamics of zonal flow formation induced by relaxation of a localized structure and the relation of this process to flow momentum plus pseudomomentum conservation, c.) a calculation of drift structure growth, including both spatial and velocity scattering.

3.4.1 *Introducing the Darmet model*

The Darmet model [12], which is really an extended version of the Tagger-Pellat [37]-Diamond-Biglari [8] model descibes the dynamics of $f(x, y, E, t)$, which corresponds to a bounce averaged trapped ion distribution function. The basic equations are

$$\partial_t f + v_d \partial_y f + \{f, \phi\} = C(f) \tag{3.27a}$$

$$\alpha_e(\phi - \langle\phi\rangle_y) - \rho^2\nabla^2\phi = \frac{2}{n_{eq}\sqrt{\pi}}\int_0^\infty dE\sqrt{E}f - 1 \tag{3.27b}$$

with heat flux Q matched according to:

$$Q = -\chi_{coll}\langle T\rangle' + \int dE\sqrt{E}E\langle\tilde{v}_r\delta f\rangle \tag{3.27c}$$

given by the sum of turbulent and neoclassical processes. Note that $v_d = v_{d,0}(E/E_{th})$ is an energy dependent precession drift velocity. $\rho^2\nabla^2$ accounts for polarization charge, due to both finite Larmor radius and finite banana width. The linear waves manifested by the Darmet model are trapped ion ITG modes, and the model is easily extended to include non-Boltzmann electron response. Dissipative trapped electron dynamics are of particular relevance and simplicity. In the collisionless limit, irreversibility appears here as a consequence of trapped ion precession drift resonance. Note that the constrained nature of the bounce averaged dynamics can force long particle-spectra auto-correlation times, i.e.

$$\Delta(\omega - \omega_d) \cong \Delta k_\theta \left|\frac{d\omega}{dk_\theta} - v_{d,0}\frac{E}{E_{th}}\right|$$
$$\cong \Delta k_\theta \left|\frac{d\omega}{dk_\theta} - \frac{\omega}{k_\theta}\right| \tag{3.28a}$$

so $\tau_{ac} \sim (|d\omega/dk_\theta - \omega/k_\theta|\Delta k_\theta)^{-1}$ and the Kubo number K is

$$K = \frac{\tilde{v}}{|d\omega/dk_\theta - \omega/k_\theta||\Delta k_\theta|\Delta_r}. \tag{3.28b}$$

Given the weakly dispersive character of long wavelength trapped ion modes, it is very easy for $K \gtrsim 1$ here, even for broad spectra. Thus, quasi-coherent phase space structures are to be expected in the Darmet model.

To form a momentum theorem for the Darmet model, we exploit the connection between polarization charge and fluid vorticity. Balance of $\langle\delta f^2\rangle$ (akin to enstrophy balance in quasi-geostrophic (QG) turbulence!) implies

$$\partial_t\langle\delta f^2\rangle + \partial_r\langle\tilde{v}_r\delta f^2\rangle - \langle\delta f C(\delta f)\rangle = -\langle\tilde{v}_r\delta f\rangle\langle f\rangle' \tag{3.29a}$$

so

$$\int dE\frac{\sqrt{E}}{\langle f\rangle'}\left\{\partial_t\langle\delta f^2\rangle + \partial_r\langle\tilde{v}_r\delta f^2\rangle - \langle\delta f C(\delta f)\rangle\right\} = -\langle\tilde{v}_r\delta n_i^{GC}\rangle \tag{3.29b}$$

where δn_i^{GC} is the total ion *guiding center density*. Then the gyrokinetic Poisson equation and our friend the Taylor Identity allow

$$\delta\phi - \rho^2\nabla^2\delta\phi = \frac{2}{n_{eq}\sqrt{\pi}}\int dE\sqrt{E}\delta f_i \tag{3.29c}$$

and

$$\langle \tilde{v}_r \delta n_i^{GC} \rangle = -\langle \tilde{v}_r \rho^2 \nabla^2 \delta\phi \rangle = -\partial_r \langle \tilde{v}_r \tilde{v}_\theta \rangle \tag{3.29d}$$

so, with zonal flow momentum balance

$$\langle \tilde{v}_r \delta n_i^{GC} \rangle = \partial_t \langle v_\theta \rangle + \nu \langle v_\theta \rangle \tag{3.29e}$$

we finally arrive at the Charney-Drazin Theorem for zonal flows in the Darmet model:

$$\partial_t \left\{ \mathrm{KPD} + \langle v_\theta \rangle \right\} = -\nu \langle v_\theta \rangle - \int dE \frac{\sqrt{E}}{\langle f \rangle'} \left\{ \partial_r \langle \tilde{v}_r \delta f^2 \rangle - \langle \delta f C(\delta f) \rangle \right\} . \tag{3.29f}$$

Here

$$\mathrm{KPD} = \int dE \sqrt{E} \frac{\langle \delta f^2 \rangle}{\langle f \rangle'} \tag{3.29g}$$

is the kinetic 'phasetrophy' density. Of course, $\langle f \rangle' = \partial_r \langle f \rangle$. This strange name is motivated by the obvious resemblance of KPD to potential enstrophy density, which follows from the duality between f in Vlasov turbulence and q in quasi-geostrophic (QG) turbulence. Given the close connection between pseudomomentum (or wave activity density) and potential enstrophy density in quasi-geostrophic (QG) turbulence (as discussed in Section 3.2), it is no surprise that KPD corresponds to a kind of kinetic pseudomomentum, formulated directly in terms of δf, with no-apriori linearization or small amplitude assumption. To see this, note that in the small amplitude, non-resonant limit for waves:

$$\delta f_{\mathbf{k}} = -\frac{1}{-i\omega_{\mathbf{k}}} \tilde{v}_{r,k} \langle f \rangle' \tag{3.30a}$$

so

$$\mathrm{KPD} = \int dE \sqrt{E} \langle \tilde{v}_{r,k}^2 \rangle \frac{\langle f \rangle'}{\omega_{\mathbf{k}}^2} \sim -k_\theta \frac{E(k)}{\omega_{\mathbf{k}}} . \tag{3.30b}$$

Thus KPD reduces correctly to pseudomomentum or the negative of the linear wave momentum density in the small amplitude limit. Note that $p_{k,\theta} = k_\theta N_{\mathbf{k}}$, where $N_{\mathbf{k}} = E(k)/\omega_{\mathbf{k}}$ is the wave action density. Equation (3.29f) constitutes a non-acceleration theorem for zonal flows in the Darmet model. In particular, in the absence of KPD evolution, kinetic turbulence spreading or collisions, one cannot accelerate or maintain (against drag) a stationary zonal flow. Alternatively, apart from spreading and collisions, zonal flow growth requires decay of KPD. Thus, we again meet the idea of a law constraining the slippage of a quasi-particle gas relative to the zonal flow.

At this point, the reader — if conscious at all — may be groping for a more physical insight into the nature of KPD. Interestingly, a similar quantity appears in the Antonov Energy Principle [1] for collisionless self-gravitating matter, as discussed in the theory of stellar dynamics [9]. For that energy principle,

$$\delta W = \int d^3x \int d^3v \frac{\delta f^2}{|F_0|'} - G \int d^3x d^3x' d^3v d^3v' \frac{f(\mathbf{x},\mathbf{v}) f(\mathbf{x}',\mathbf{v}')}{|\mathbf{x}-\mathbf{x}'|} \tag{3.31}$$

consists of a fluctuation dynamic pressure term and a self-gravity term. Clearly, the former is the KPD for an unmagnetized plasma. The competition in δW is the usual and familiar one for Jeans instabilities, namely self-gravity vs. dynamical pressure. A similar KPD term appears in the Kruskal-Oberman Energy Principle [26]. Thus, we see that KPD may be thought of as a kinetically founded dynamical pressure.

To conclude this section, we compare the Charney-Drazin theorem for Hasegawa-Mima model with the corresponding relation for the Darmet model of drift wave turbulence. The relation for Hasegawa-Mima turbulence is, from Eq. (3.13)

$$\frac{\partial}{\partial t}\{\mathrm{WAD} + \langle v_\theta \rangle\} = -\nu \langle v_\theta \rangle - \frac{1}{\langle q \rangle'} \left\{ \langle \tilde{f}^2 \rangle \tau_c - \mu \langle (\nabla \tilde{q})^2 \rangle - \partial_r \langle \tilde{v}_r \tilde{q}^2 \rangle \right\} \tag{3.32a}$$

where $\mathrm{WAD} = \langle \tilde{q}^2 \rangle / \langle q \rangle'$, while for the kinetic Darmet model

$$\partial_t \{\mathrm{KPD} + \langle v_\theta \rangle\} = -\nu \langle v_\theta \rangle - \int dE \frac{\sqrt{E}}{\langle f \rangle'} \{\partial_r \langle \tilde{v}_r \delta f^2 \rangle - \langle \delta f C(\delta f) \rangle\} \tag{3.32b}$$

where $\mathrm{KPD} = \int d^3v \langle \delta f^2 \rangle / \langle f \rangle'$. The correspondence is obvious, thus confirming that conservation of momentum and phase space density really are the key common elements.

3.4.2 *Phase space structures and zonal flows*

The physical interpretation of KPD becomes problematic for resonant particles. This brings us to the case of a single phase space structure in the Darmet model or, more generally, drift wave turbulence. We consider a localized (in phase space!) ion hole or blob, with $\delta f_i = \delta f((x - x_0)/\Delta x, (E - E_0)/\Delta E)$, and proceed from phase space density conservation, as before.

Thus

$$\frac{df}{dt} = \frac{d}{dt}(f_0 + \delta f) = 0 \tag{3.33a}$$

so

$$\partial_t \int dE\sqrt{E}\langle \delta f^2 \rangle = -2\frac{d}{dt}\int dE\sqrt{E} f_0 \delta f \tag{3.33b}$$

and then

$$f_0 = f_0(x_0) + (x - x_0)\left.\frac{\partial f_0}{\partial x}\right|_{x_0} + \cdots \tag{3.33c}$$

We thus have

$$\partial_t \int dE\sqrt{E}\langle \delta f_i^2 \rangle = -2\langle \tilde{v}_r \delta n_i^{GC} \rangle \left.\frac{\partial f_0}{\partial x}\right|_{x_0} \tag{3.33d}$$

where $d(x - x_0)/dt = \tilde{v}_r$, $\int dE\sqrt{E}\delta f_i = \delta n_i^{GC}$ and assumptions similar to those in Section 3.3 apply. Now, a key point enters via the gyrokinetic Poisson equation [10, 20], which relates ion guiding center density δn_i^{GC} to electron density δn_e and polarization charge density. Thus

$$-\rho^2 \nabla^2 \tilde{\phi} = \delta n_i^{GC} - \delta n_e \,. \tag{3.33e}$$

So

$$\partial_t \int dE\sqrt{E}\frac{\langle \delta f_i^2 \rangle}{2} = -\left.\frac{\partial f_0}{\partial x}\right|_{x_0} \left\{ \langle \tilde{v}_r \delta n_e \rangle - \langle \tilde{v}_r \rho^2 \nabla^2 \tilde{\phi} \rangle \right\} . \tag{3.33f}$$

The physics of the replacement of Eq. (3.33d) with Eq. (3.33f) is a consequence of the fact that the total dipole moment of the plasma is a constant, i.e.

$$\int dx \sum_\alpha q_\alpha n_\alpha(x) x = const \tag{3.34a}$$

where $n_\alpha(x)$ is the density of charge component α. Note that the dipole moment *must* include the polarization component, which in turn guarantees a polarization flux, i.e.

$$-\rho^2 \nabla^2 \phi = \delta n_i^{GC}(\phi) - \delta n_e(\phi) \,. \tag{3.34b}$$

So

$$-\langle \tilde{v}_r \rho^2 \nabla^2 \phi \rangle = \langle \tilde{v}_r \delta n_i^{GC}(\phi) \rangle - \langle \tilde{v}_r \delta n_e(\phi) \rangle \,. \tag{3.34c}$$

Thus, ambipolarity breaking necessarily implies

$$-\rho^2 \langle \tilde{v}_r \nabla^2 \tilde{\phi} \rangle \neq 0 \,. \tag{3.34d}$$

Then, using the now all-too-familiar Taylor Identity and flow momentum balance gives

$$\partial_t \left\{ \int dE\sqrt{E} \frac{\delta f_i^2}{2\partial f_0/\partial x|_{x_0}} + \langle v_\theta \rangle \right\} = -\nu\langle v_\theta \rangle - \langle \tilde{v}_r \delta n_e \rangle \tag{3.35}$$

which states that even a *localized* phase space blob or hole cannot avoid zonal flow coupling, on account of the fact that spatial scattering of the hole produces a flux of polarization charge due to conservation of total dipole moment.

Equation (3.35) can be viewed as a kind of Charney-Drazin theorem for zonal flows produced by localized phase space structure growth. Clearly $\int dE\sqrt{E}\delta f_i^2/2\langle f\rangle'$ is the pseudomomentum, and Eq. (3.35) states that for stationary δf_i, the flow cannot grow unless $-\langle \tilde{v}_r \delta n_e \rangle < 0$ (i.e. $\langle f\rangle' < 0$ is assumed), thus requiring *electron* transport. For, say, dissipative trapped electron response [23] — relevant to trapped ion regime dynamics — we have $\langle \tilde{v}_r \delta n_e \rangle \cong -D_{DT}\partial\langle n\rangle/\partial x$, so

$$\partial_t \left\{ \int dE\sqrt{E} \frac{\delta f_i^2}{2\partial f_0/\partial x|_{x_0}} + \langle v_\theta \rangle \right\} = -\nu\langle v_\theta \rangle + D_{DT}\frac{\partial\langle n\rangle}{\partial x} \tag{3.36}$$

where D_{DT} is the dissipative trapped electron diffusivity, $\sim 1/\nu_{e,eff}$. Note that since $\langle f_i\rangle' < 0$ and $\partial\langle n\rangle/\partial x < 0$, Eq. (3.36) suggests that the electron flux will tend to drive or enhance the ion structure. Physically, this corresponds to ion structure growth by scattering off the (diffusing) electrons, so as to maintain ambipolarity. In the Bump-on-Tail or CDIA problems, momentum conservation is the key constraint, while for drift wave turbulence, ambipolarity maintenance is central. Note for stationary $\langle \delta f_i^2 \rangle$ and $\langle V_\theta \rangle$ we have

$$\langle v_\theta \rangle = -\frac{\langle \tilde{v}_r \delta n_e \rangle}{\nu} = \frac{D_{DT}}{\nu}\frac{\partial\langle n\rangle}{\partial x}. \tag{3.37}$$

Very clearly:

(1) a *localized* structure can excite a *global* (in θ) zonal flow
(2) electron transport can drive the nonlinear growth of an ion structure. Note that here, straightforward estimates give $\gamma \sim kv_d\Delta E_T/E_{th}$.

At this point we should remark that it is instructive to compare Eq. (3.36) to the Charney-Drazin Theorem for the Hasegawa-Wakatani system [13, 21], which is

$$\frac{\partial}{\partial t}\{\mathrm{WAD} + \langle v_\theta \rangle\} = -\nu\langle v_\theta \rangle - \langle \tilde{v}_r \tilde{n} \rangle - \frac{1}{\langle q\rangle'}\left\{\mu\langle(\nabla\tilde{q})^2\rangle + \partial_r\langle \tilde{v}_r \tilde{q}^2 \rangle\right\}. \tag{3.38}$$

Here, for the Hasegawa-Wakatani system $q = n - \nabla^2\phi$ and WAD $= \langle\delta q^2\rangle/\langle q\rangle'$. The correspondence is clear! It is evidently the same equation, allowing for the fact that there is no turbulence spreading or viscous dissipation in the Vlasov-Gyrokinetic structure problem. In particular, for Hasegawa-Wakatani, once again we find that a particle flux can drive the flow against drag. Finally, we note in passing that significant impurity dynamics can have a profound effect on zonal flows by altering the ambipolarity balance.

3.4.3 *Drift-hole growth*

In this section, we in essence combine the results of A.) and B.) to derive an expression for drift hole growth. Here a drift hole is a phase space structure in a slab drift wave system. For variety, we consider here an *electron* structure, which is a hole (phase space depression) so as to self-bind. The process of self-binding can be thought of as the formation of a state which is marginally stable to Jeans modes. This is discussed further in references [18, 39]. The growth calculation extends that of reference [39]. As before, we write

$$\frac{\partial}{\partial t}\int d^3v\,\delta f^2 = -2\frac{d}{dt}\int d^3v\, f_0\delta f\,. \tag{3.39a}$$

Now

$$f_0 = f_0(x_0, v_{\parallel,0}) + (x - x_0)\left.\frac{\partial f_0}{\partial x}\right|_{x_0,v_{\parallel,0}} + (v_\parallel - v_{\parallel,0})\left.\frac{\partial f_0}{\partial v_\parallel}\right|_{x_0,v_{\parallel,0}}\,. \tag{3.39b}$$

Since

$$\frac{dx}{dt} = \frac{c}{B}\tilde{E}_\theta,\ \frac{dv_\parallel}{dt} = -\frac{|e|}{m_e}\tilde{E}_\parallel \tag{3.39c}$$

we immediately find:

$$\frac{\partial}{\partial t}\int d^3v\frac{\langle\delta f^2\rangle}{2} = -\langle\tilde{v}_r\delta n_e\rangle\left.\frac{\partial f_0}{\partial x}\right|_{x_0,v_{\parallel,0}} + \frac{|e|}{m_e}\langle\tilde{E}_\parallel\delta n_e\rangle\left.\frac{\partial f_0}{\partial v_\parallel}\right|_{x_0,v_{\parallel,0}}\,. \tag{3.39d}$$

As before:

$$\delta n_e = \delta n_i^{GC} - \rho_s^2\nabla_\perp^2\tilde{\phi} \tag{3.39e}$$

we thus obtain:

$$\frac{\partial}{\partial t}\int d^3v\frac{\langle\delta f^2\rangle}{2} = -\left[\langle\tilde{v}_r\delta n_i^{GC}\rangle - \langle\tilde{v}_r\rho_s^2\nabla_\perp^2\tilde{\phi}\rangle\right]\left.\frac{\partial f_0}{\partial x}\right|_{x_0,v_{\parallel,0}}$$
$$+\frac{|e|}{m_e}\left[\langle\tilde{E}_\parallel\delta n_i^{GC}\rangle - \langle\tilde{E}_\parallel\rho_s^2\nabla_\perp^2\tilde{\phi}\rangle\right]\left.\frac{\partial f_0}{\partial v_\parallel}\right|_{x_0,v_{\parallel,0}} . \quad (3.40)$$

Here we ignore coherent (i.e. wave) non-adiabatic electron effects and hereafter drop the fourth term on RHS. Note that such correlations have been shown to potentially be significant in the context of intrinsic rotation [34]. Taking $\delta n_i^{GC} = \chi_i(k,\omega)\tilde{\phi}_{k,\omega}$, where $\chi_i(k,\omega)$ is the ion guiding center susceptibility (n.b.: this tacitly ignores any ion trapping or granulations, but should include zonal shears), we find

$$\frac{\partial}{\partial t}\int d^3v\left\{\frac{\langle\delta f^2\rangle}{2\partial f_0/\partial x|_{x_0,v_{\parallel,0}}} + \langle v_\theta\rangle\right\} = -\nu\langle v_\theta\rangle + \sum_{\mathbf{k}} k_\theta \mathrm{Im}\chi_i(\mathbf{k},k_\parallel v_{\parallel,0})|\tilde{\phi}_{\mathbf{k}}|^2$$
$$+\sum_{\mathbf{k}} k_\parallel v_{\parallel,0}\mathrm{Im}\chi_i(\mathbf{k},k_\parallel v_{\parallel,0})\frac{f_0}{\partial f_0/\partial x|_{x_0,v_{\parallel,0}}} \quad (3.41a)$$

or, collecting terms:

$$\frac{\partial}{\partial t}\int d^3v\left\{\frac{\langle\delta f^2\rangle}{2\partial f_0/\partial x|_{x_0,v_{\parallel,0}}} + \langle v_\theta\rangle\right\} = -\nu\langle v_\theta\rangle$$
$$-\sum_{\mathbf{k}}\frac{-k_\theta f_0' - k_\parallel v_{\parallel,0}f_0}{f_0'}\mathrm{Im}\chi_i(\mathbf{k},k_\parallel v_{\parallel,0})|\tilde{\phi}_{\mathbf{k}}|^2 . \quad (3.41b)$$

Here $v_{\parallel,0}$ is the hole speed, we have assumed f_0 is a Maxwellian, $\mathbf{k}$ refers to the modes excited by the structure (i.e. harmonics of the hole box size), and the hole ballistic frequency is $k_\parallel v_{\parallel,0}$. Note that Eq. (3.41b) states, absent zonal flows, that:

$$\partial_t\langle\delta f_e^2\rangle \sim -(\omega_{*e} - k_\parallel v_{\parallel,0})\mathrm{Im}\chi_i(\mathbf{k},k_\parallel v_{\parallel,0}) \quad (3.41c)$$

so that $\omega_{*e} > k_\parallel v_{\parallel,0}$ is required for free energy accessibility. This, of course, is the same as the familiar $\omega < \omega_*$ condition for instability in linear theory. That result requires that the structure motion release more energy due to radial scattering than it costs in $v_\parallel$ scattering. However, *unlike* linear theory, the dissipation which triggers growth is due to ions, i.e. $\text{Im}\chi_i(\mathbf{k}, k_\parallel v_{\parallel,0}) < 0$. This reflects the role of the ambipolarity constraint, discussed above, and implies that the regime of structure instability ($\text{Im}\chi_i(\mathbf{k}, k_\parallel v_{\parallel,0})$ large) is, in some sense, complementary to the regime of wave instability ($\text{Im}\chi_i(\mathbf{k}, k_\parallel v_{\parallel,0})$ small). This is similar to what we encountered for the case of CDIA. Also, clearly hole speed is a key parameter. $v_{\parallel,0}$ must be small enough so that $\omega_* > k_\parallel v_{\parallel,0}$, but not so small to drive $\text{Im}\chi_i(\mathbf{k}, k_\parallel v_{\parallel,0}) \to 0$. Finally, we note that via the gyrokinetic Poisson equation, even electron holes will necessarily couple to, and drive, zonal flows.

As before, hole growth is nonlinear, i.e. explosive. A straightforward estimate which ignores zonal flow effects gives $\gamma \sim k_\parallel \Delta v_{\parallel,T}(\omega_* - k_\parallel v_{\parallel,0})|\text{Im}\chi_i(\mathbf{k}, k_\parallel v_{\parallel,0})|$, where $\Delta v_{\parallel,T} \sim \phi^{1/2}$. Holes growth will surely distort $\langle f \rangle$ eventually, thus vitiating some of the assumptions of this analysis. Note that the associated zonal flow growth can be calculated using the hole fluctuation driven Reynolds stress. Obviously, coupling to the zonal flow is a major player in the saturation of hole growth. Reference [39] missed zonal flow effects, as it did not consider meso-scale fluctuation envelope variation. This omission is now rectified. Obviously, $\chi_i(k, \omega)$ must be calculated self-consistently, accounting for the zonal flow. Also, zonal flow shear will limit hole self-coherence times.

3.5 Dupree-Lenard-Balescu theory for mean evolution

In this section, we briefly sketch the essentials of the theory of $\langle f \rangle$ relaxation in a bath of turbulence which includes granulations and so has Kubo number $K \simeq 1$, rather than < 1, as nominally required for the applicability of quasilinear theory. The main novel feature of the Dupree-Lenard-Balescu (DLB) theory is the appearance of a dynamical friction or drag effect, due to Cerenkov emission by phase space granulations or eddys. This effect is physically plausible in that intuitively put, turbulence at Kubo number $K \gtrsim 1$ behaves more like a soup of blobs or structures, rather than an ensemble of waves. In a soup of structures, each blob or hole will scatter off the others, and so leave a wake as it moves. This process of wake emission is easily described by dynamical friction effects. Just as structures find new ways to tap available free energy, dynamical friction can introduce new routes to relaxation, instability and transport. This discussion is intentionally very brief, as there have been other pedagogical treatments of the DLB theory published recently [14]. The interested reader is referred to these for the gory and gruesome details.

The essence of the DLB theory is to derive an equation for the 1-time, 2-point fluctuation correlation in phase space, $\langle \delta f(x_1, v_1, t)\delta f(x_2, v_2, t)\rangle$. This equation has the generic form:

$$\partial_t \langle \delta f^2 \rangle + T_{1,2}[\langle \delta f^2 \rangle] = P_{1,2} \,. \tag{3.42}$$

Here $T_{1,2}$ refers to the two point evolution operator, including streaming, scattering and collisional dissipation. $T_{1,2}$ virtually *always* is calculated using a statistical closure of some form. The more interesting piece is the production term $P_{1,2}$. Since $df/dt = 0$ implies $d/dt\langle \delta f^2 \rangle = -\partial_t \langle f \rangle^2$, $P_{1,2}$ is obviously related to $\partial_t \langle f \rangle$ and thus to mean relaxation, transport, etc. For the 1D Vlasov prototype,

$$\partial_t \langle f \rangle = -\partial_v \langle \tilde{E} \delta f \rangle = -\partial_v [-D\partial_v \langle f \rangle + F\langle f \rangle] \,. \tag{3.43}$$

Here D is analogous to the familiar quasilinear diffusion term, and F is the drag or dynamical friction term. F arises from the fact that

$$\delta f = f^c + \tilde{f} \tag{3.44}$$

i.e. the total fluctuation is the sum of a coherent (f^c) and incoherent ($\tilde{f}$) piece. The former is proportional to (i.e. coherent with) the electric field perturbation $\tilde{E}_{k,\omega}$, the latter is not. Consideration of the x_-, $v_- \to 0$ behavior of $T_{1,2}$ forces us to confront the existence of $\tilde{f}$. Thus:

$$-D\partial_v \langle f \rangle = \langle \tilde{E} f^c \rangle \tag{3.45a}$$

$$F\langle f \rangle = \langle \tilde{E} \tilde{f} \rangle \,. \tag{3.45b}$$

The incoherent correlation $\langle \tilde{f}(1)\tilde{f}(2)\rangle$ can be obtained from the total correlation $\langle \delta f(1)\delta f(2)\rangle$. $\tilde{f}$ is related to $\tilde{\phi}$ by

$$\epsilon(k, \omega)\tilde{\phi}_{k,\omega} = \int dv \tilde{f}_{k,\omega} \,. \tag{3.46}$$

Thus, granulations resemble dressed macro-particles. It is possible for stationary turbulence to arise and persist in the absence of linearly unstable waves.

Collisionless coupling of two different species (i.e. electrons and ions) can occur via F and its dependence on ϵ. This induces dynamical friction. Just as two species interaction can lead to structure growth, dynamical friction can drive nonlinear instability, i.e. growth of δf^2 due to relaxation. Such growth has been observed in computer simulations [6]. Recently, the theory of collisionless dynamical friction was extended to include zonal flow effects [24]. The calculation of $P_{1,2}$ has also been used to estimate the efficiency of intrinsic rotation generation [25]. Several aspects of the results agree well with relevant experimental findings [36].

3.6 Conclusion

This is a fascinating and active topic, which will only grow in importance and visibility. There are no meaningful final conclusions. Many interesting topics for future research might be suggested here. However, the authors would rather do these themselves! The reader is thus invited to think for himself or herself and beat us to the punch.

Acknowledgments

The authors thank G. Dif-Pradalier, X. Garbet, O. D. Gurcan, T. S. Hahm, D. W. Hughes, C. McDevitt, M. E. McIntyre, C. Nguyen, S. M. Tobias, L. T. Neko, and Y. E. Ko for stimulating questions and discussions. The biannual Festival de Theorie has been a source of inspiration and stimulation for more than a decade. We thank the C.E.A, other supporters, and the local organizers for making this unique event possible. This work was supported by the Ministry of Education, Science and Technology of Korea via the WCI project No 2009-001 and by the US Department of Energy Contracts Nos. DE-FG02-04ER54738, DE-FC02-08ER54983 and DE-FC02-08ER54959.

References

[1] V. A. Antonov, *Vestnik Leningrad Gos. Univ.* **7** (1962), no. 135.

[2] V. I. Arnold, “Conditions for nonlinear stability of stationary plane curvilinear flows of an ideal fluid”, *Soviet Math. Dokl.* **6** (1965), p. 773–777.

[3] H. L. Berk and B. N. Breizman, “Saturation of a single mode driven by an energetic injected beam. i. plasma wave problem”, *Phys. Fluids B* **2** (1990), no. 9, p. 2226.

[4] —, “Saturation of a single mode driven by an energetic injected beam. ii. electrostatic “universal” destabilization mechanism”, *Phys. Fluids B* **2** (1990), p. 2235.

[5] —, “Saturation of a single mode driven by an energetic injected beam. iii. alfven wave problem”, *Phys. Fluids B* **2** (1990), p. 2246.

[6] R. H. Berman, D. J. Tetreault and T. H. Dupree, *Phys. Fluids* **28** (1985), p. 155.

[7] I. B. Bernstein, J. M. Greene and M. D. Kruskal, *Phys. Rev.* **108** (1957), p. 546–550.

[8] H. Biglari, P. H. Diamond and P. W. Terry, "Theory of trapped-ion-temperature-gradient-driven turbulence and transport in low-collisionality plasmas", *Phys. Fluids* **31** (1988), no. 9, p. 2644.

[9] J. Binney and S. Tremaine, *Galactic dynamics*, Princeton University Press, 2008.

[10] A. J. Brizard and T. S. Hahm, *Rev. Mod. Phys.* **79** (2007), p. 421.

[11] J. G. Charney and P. G. Drazin, "Propagation of planetary-scale disturbances from the lower into the upper atmosphere", *J. Geophys. Res.* **66** (1961), p. 83.

[12] G. Darmet, P. Ghendrih, Y. Sarazin, X. Garbet and V. Grandgirard, "Intermittency in flux driven kinetic simulations of trapped ion turbulence", *Communications in Nonlinear Science and Numerical Simulation* **13** (2008), p. 53–58.

[13] P. H. Diamond, O. D. Gurcan, T. S. Hahm, K. Miki, Y. Kosuga and X. Garbet, "Momentum theorems and the structure of atmospheric jets and zonal flows in plasmas", *Plasma Phys. Control. Fusion* **50** (2008), p. 124018.

[14] P. H. Diamond, S. I. Itoh and K. Itoh, *Modern plasma physics volume1: Physical kinetics of turbulent plasmas*, Cambridge University Press, 2011.

[15] P. H. Diamond, S. I. Itoh, K. Itoh and T. S. Hahm, *Plasma Phys. Control. Fusion* **47** (2005), p. R35.

[16] P. G. Drazin and W. H. Reid, *Hydrodynamic stability*, Cambridge University Press, 2004.

[17] T. H. Dupree, *Phys. Fluids* **9** (1966), no. 9, p. 1773.

[18] — , "Theory of phase-space density holes", *Phys. Fluids* **25** (1982), no. 2, p. 277.

[19] O. D. Gurcan, P. H. Diamond and T. S. Hahm, *Phys. Plasmas* **13** (2006), p. 052306.

[20] T. S. Hahm, *Phys. Fluids* **31** (1988), no. 7, p. 1940.

[21] A. Hasegawa and M. Wakatani, *Phys. Rev. Lett.* **50** (1983), p. 682.

[22] B. B. Kadomtsev, *Plasma turbulence*, Academic Press, 1965.

[23] B. B. Kadomtsev and O. P. Pogutse, *Reviews of plasma physics, vol. 5*, Consultants Bureau, 1970.

[24] Y. Kosuga and P. H. Diamond, *Phys. Plasmas* (2011).

[25] Y. Kosuga, P. H. Diamond and O. D. Gurcan, *Phys. Plasmas* **17** (2010), p. 102313.

[26] M. D. Kruskal and C. R. Oberman, *Phys. Fluids* **1** (1958), no. 4, p. 275.

[27] R. M. Kulsrud, *Plasma physics for astrophysics*, Princeton University Press, 2004.
[28] L. D. Landau and E. M. Lifshitz, *Fluid mechanics*, Pergamon Press, 1959.
[29] L. D. Landau, E. M. Lifshitz and L. P. Pitaevskii, *Electrodynamics of continuous media*, Pergamon Press, 1981.
[30] G. Laval and D. Pesme, *Phys. Fluids* **26** (1983), no. 1, p. 52.
[31] M. Lesur, Y. Idomura, K. Shinohara, X. Garbet and the JT-60 Team, *Phys. Plasmas* **17** (2010), p. 122311.
[32] C. C. Lin, *The theory of hydrodynamic stability*, Cambridge University Press, 1955.
[33] D. Lynden-Bell, "Statistical mechanics of violent relaxation in stellar systems", *Mon. Not. R. astr. Soc.* **136** (1967), p. 101–121.
[34] C. J. McDevitt, P. H. Diamond, O. D. Gurcan and T. S. Hahm, *Phys. Plasmas* **16** (2009), p. 052302.
[35] P. B. Rhines and W. R. Holland, *Dynamics of Atmosphere and Oceans* **3** (1979), p. 289.
[36] J. E. Rice, J. W. Hughes, P. H. Diamond, Y. Kosuga, Y. A. Podpaly, M. L. Reinke, O. D. Gurcan, T. S. Hahm, A. E. Hubbard, E. S. Marmar, C. J. McDevitt and D. G. Whyte, *Phys. Rev. Lett.* **106** (2011), p. 215001.
[37] M. Tagger, G. Laval and R. Pellat, *Nucl. Fusion* **17** (1977), p. 109.
[38] G. I. Taylor, *Phil. Trans. Roy. Soc. London.* **A215** (1915), p. 1–26.
[39] P. W. Terry, P. H. Diamond and T. S. Hahm, "The structure and dynamics of electrostatic and magnetostatic drift holes", *Phys. Fluids B* **2** (1990), p. 9.
[40] S. M. Tobias, K. Dagon and J. B. Marston, *ApJ* **727** (2011), p. 127.
[41] S. I. Tsunoda, F. Doveil and J. H. Malmberg, *Phys. Rev. Lett.* **59** (1987), no. 24, p. 2752.
[42] G. K. Vallis, *Atmospheric and oceanic fluid dynamics*, Cambridge University Press, 2006.
[43] A. Vedenov and L. I. Rudakov, *Dokl. Akad. Nauk SSSR* **159** (1964), p. 767.
[44] — , *Sov. Phys. Dokl.* **9** (1965), p. 1073.
[45] R. B. Wood and M. E. McIntyre, "A general theorem on angular-momentum changes due to potential vorticity mixing and on potential-energy changes due to buoyancy mixing", *J. Atmos. Sci.* **67** (2010), no. 4, p. 1261–1274.
[46] J. M. Ziman, *Principles of the theory of solids*, Cambridge University Press, 1979.

Chapter 4

Fast Dynamos

D. W. Hughes

Department of Applied Mathematics, University of Leeds, U.K.

4.1 Introduction

One of the most important problems in astrophysical magnetohydrodynamics (MHD) is to understand the generation and maintenance of cosmical magnetic fields — *the dynamo problem.* In an astrophysical dynamo, inductive plasma motions, such as turbulent flows in stellar convective zones, maintain a magnetic field against its tendency otherwise to decay.

For the dynamo theorist it is helpful to assign different types of dynamo action into (overlapping) categories — although nature of course has no need to follow any such categorisation. Any astrophysical MHD system is governed by a number of equations: typically, the momentum equation, the equation of conservation of mass, the magnetic induction equation, the heat equation and a gas law. The magnetic induction equation describes the evolution of a magnetic field $\boldsymbol{B}(\boldsymbol{x}, t)$ under a flow $\boldsymbol{u}(\boldsymbol{x}, t)$; it takes the form

$$\frac{\partial \boldsymbol{B}}{\partial t} = \nabla \times (\boldsymbol{u} \times \boldsymbol{B}) + \eta \nabla^2 \boldsymbol{B}\,, \tag{4.1}$$

where η is the magnetic diffusivity (here assumed constant). Equation (4.1) is formally linear in the magnetic field $\boldsymbol{B}$; the dynamo problem is though, in general, rendered nonlinear by the back-reaction of the magnetic field on the velocity via the Lorentz force. However, a theoretical simplification often employed is to consider the evolution of a magnetic field under a *prescribed* flow; in this case, only Eq. (4.1) is involved, and the problem is unambiguously linear in $\boldsymbol{B}$. A prescribed flow that, via Eq. (4.1), amplifies the magnetic field exponentially is said to act as a *kinematic* dynamo. A dynamo in which the effects of the Lorentz force on the flow are included self-consistently is said to be a *dynamic* or *magnetohydrodynamic* dynamo.

Although the kinematic dynamo problem represents a simplification of the full problem, and obviously cannot deal with issues such as the saturation of magnetic field growth, it is nonetheless instructive in focusing attention exclusively on the generation process. As such, it has received considerable attention over the past fifty years.

Within the kinematic formulation, a further categorisation can be made into *slow* and *fast* dynamos. In dimensionless form, in which velocities are scaled with a typical velocity U and lengths with a characteristic length scale L, Eq. (4.1) can be written as

$$\frac{\partial \boldsymbol{B}}{\partial t} = \nabla \times (\boldsymbol{u} \times \boldsymbol{B}) + \frac{1}{Rm}\nabla^2 \boldsymbol{B}\,, \tag{4.2}$$

where $Rm = UL/\eta$ is the magnetic Reynolds number. In astrophysics, Rm is invariably large, and often extremely so: for example, $Rm \approx 10^{17}$ in the interstellar medium. The distinction between slow and fast dynamos is illustrated most clearly for the simplest case when the velocity $\boldsymbol{u}$ is steady. The magnetic field then takes the form $\boldsymbol{B}(\boldsymbol{x}, t) = \hat{\boldsymbol{B}}\,(\boldsymbol{x}) \exp(pt)$, where p is the dynamo growth rate; $\mathrm{Re}(p) > 0$ signifies dynamo action. For a given flow, p will be a function of Rm. If

$$\lim_{Rm \to \infty} \mathrm{Re}(p) > 0 \tag{4.3}$$

then the dynamo is said to be *fast*; otherwise it is said to be *slow*. Taking the limit as $Rm \to \infty$ may be viewed as the mathematical abstraction of the case of astrophysical interest, in which Rm is finite though extremely large.

For more general, time-dependent, flows, the dynamo growth rate may be defined in terms of the magnetic energy $E_m(t)$ as

$$\gamma(Rm) = \limsup_{t \to \infty} \frac{1}{2t} \ln E_m(t)\,. \tag{4.4}$$

The dynamo is then said to be *fast* if γ remains positive and bounded away from zero in the limit $Rm \to \infty$. Physically, the defining characteristic of a fast dynamo is one in which the dynamo process proceeds on an advective time scale, independent of molecular diffusion processes.

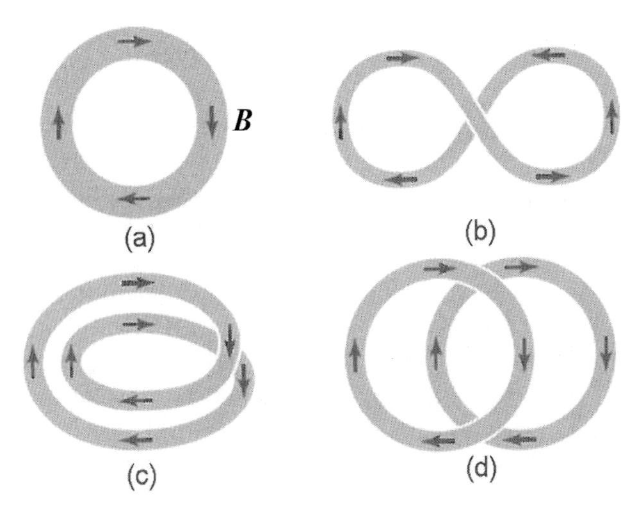

Fig. 4.1 Schematic diagram of the stretch-twist-fold dynamo. A loop of magnetic flux (a) is stretched and twisted (b), and then folded back onto itself (c). Diffusive processes (d) are assumed to smooth out the field, leaving two loops, both with the same flux (and half the cross-sectional area) as in (a). (After [1].)

The notion of a fast dynamo was first introduced by Vainshtein and Zeldovich [1], who described the process illustrated in Fig. 4.1. A torus of magnetic flux is first stretched to twice its length; it is then twisted and finally folded back onto itself, forming a torus of approximately the original shape, but with twice the original flux. Repetition of this 'stretch-twist-fold' process with a cycle period T leads to amplification of the magnetic field with a growth rate

$$p = \frac{1}{T} \ln 2 \,. \tag{4.5}$$

It is assumed that the role of magnetic diffusion is simply to smooth out the internal structure of the field, and that this is not critical to the flux doubling process. Such a dynamo is of course highly idealised, and prompts the crucial question of whether fast dynamo action can actually be realised in a fluid flow.

The investigation of fast dynamo action has proceeded in a number of complementary theoretical and computational directions. The aim of this review is to provide a brief introduction to these various ideas. An excellent comprehensive review of the development of the subject is provided by the monograph [2].

The layout of the paper is as follows. Section 4.2 considers the case of a perfectly conducting fluid (Rm strictly infinite) and addresses how the evolution of a magnetic field in such a fluid may be related to a dynamo in a fluid of finite conductivity as $Rm \to \infty$. Section 4.3 looks at an idealisation of the fast dynamo problem, in which smooth fluid flows are replaced by discrete maps, and shows how some of the ideas of nonlinear dynamical systems theory are important in understanding and describing fast dynamo action. Obtaining rigorous results for fast dynamo action is extremely difficult: Sec. 4.4 discusses an important necessary condition. In Sec. 4.5 we consider, through numerical simulations, some candidates for fast dynamo flows, and discuss the difficulties involved in any computational approach to the problem. Fast dynamo action is essentially a kinematic property; the idea can though be extended into the nonlinear regime, as discussed in Sec. 4.6. The review concludes with Sec. 4.7, which discusses briefly one of the major challenges of astrophysical dynamo theory — how to bring together the ideas of fast dynamo theory and mean field dynamo theory so as to explain the generation of large-scale astrophysical magnetic fields at very high Rm.

4.2 Perfectly conducting fluids

The idea of a fast dynamo is one in which the growth rate remains positive *in the limit* as $Rm \to \infty$. This prompts the interesting question of whether there is any relation between this limit and the case when Rm is formally infinite (i.e. a perfectly conducting fluid). Consideration of the analogous problem in hydrodynamics suggests there may not be a straightforward relation between the two cases; whereas there are indeed cases where the flow in the limit of $Re \to \infty$ (where Re is the fluid Reynolds number) is similar to that of the formally inviscid ($Re \equiv \infty$) flow, there are others where the limiting flow and the ideal flow bear no resemblance to each other. This is seen most clearly in the problem of flow past a body. For a bluff body, such as a cylinder, the flow at high values of Re is characterised by a significant turbulent wake downstream of the body; the ideal inviscid flow, on the other hand, has fore-aft symmetry, with laminar streamlines skirting the body. However, for a streamlined body, such as an aerofoil, viscous effects are, by design, confined to a thin boundary layer attached to the body; outside the boundary layer, the high Reynolds number flow is well approximated by the flow at infinite Reynolds number.

There are significant mathematical differences in the nature of the induction equation (4.2) depending whether Rm is finite or infinite;

discarding the term with the highest spatial derivative (the diffusive term) completely changes the character of the equation. If Rm is finite, Eq. (4.2) has well-defined eigensolutions; all moments $\langle |\boldsymbol{B}|^n \rangle^{1/n}$ (where $\langle\ \rangle$ denotes an appropriate average) grow (or decay) at the same rate — the dynamo growth rate. If Rm is infinite, on the other hand, different moments grow at different rates. This was recognised in [3], which showed that there can be no exponentially growing smooth solution for $\boldsymbol{B}(\boldsymbol{x}, t)$ when Rm is infinite. However, that is not to say that the evolution of $\boldsymbol{B}$ for infinite Rm cannot be related to that in the limit of $Rm \to \infty$. The interesting conjecture — *the flux conjecture* — has been made in [4] that the growth rate of the flux (i.e. the first moment) for the case of infinite Rm is equal to the true dynamo growth rate in the limit as $Rm \to \infty$. The idea behind the conjecture, simply stated, is that whereas diffusion will have only a small effect on the flux through a region (cancelling oppositely-signed fields, but leaving the flux relatively unchanged), it will dissipate other moments of the field, such as the magnetic energy.

In a perfect conductor (Rm infinite), it can be shown, by combining the induction equation with the equation for the conservation of mass (see, for example, [5]), that $\boldsymbol{B}/\rho$ (where ρ is the fluid density) satisfies the same equation as a line element $d\boldsymbol{x}$; consequently, for a field element situated at $\boldsymbol{x} = \boldsymbol{a}$ at $t = 0$, the magnetic field $\boldsymbol{B}(\boldsymbol{x}, t)$ may be expressed via the formal Cauchy solution as

$$\frac{B_i(\boldsymbol{x}, t)}{\rho(\boldsymbol{x}, t)} = \frac{B_j(\boldsymbol{a}, 0)}{\rho(\boldsymbol{a}, 0)} \left(\frac{\partial x_i}{\partial a_j} \right). \tag{4.6}$$

For incompressible fluids, for which the density is a constant, the field at time t is related to the initial field entirely via the Jacobian matrix,

$$B_i(\boldsymbol{x}, t) = B_j(\boldsymbol{a}, 0) \left(\frac{\partial x_i}{\partial a_j} \right). \tag{4.7}$$

Some support for the flux conjecture is provided by [2], which compares the magnetic flux growth in a perfectly conducting fluid for particular flows, obtained via numerical solution of (4.7), with solutions of the dynamo growth rate, for the same flows, obtained from numerical solution of the induction equation (4.2) at large but finite values of Rm. At least for certain flows the agreement is good. However, as we shall discuss further in Sec. 4.5, there are computational difficulties in obtaining both the flux growth via the Cauchy solution and the dynamo growth rate via the induction equation; hence numerical verification (or indeed repudiation) of the flux conjecture remains a challenging task.

4.3 Map dynamos

Although a 'true' fast dynamo flow $\boldsymbol{u}(\boldsymbol{x}, t)$ must be smooth in space and time, it is nonetheless instructive to consider more dramatic manipulations of the magnetic field, since these can lead to important generic insights. Perhaps the most drastic of these are *map dynamos*, in which the velocity $\boldsymbol{u}(\boldsymbol{x}, t)$ is replaced by a discrete-time mapping relating the field at successive (arbitrary) time intervals; such a map may, but need not necessarily, result from a continuous velocity field. The ideas of map dynamos are reviewed in some detail by [4] and [6].

The simplest map dynamo results from the two-dimensional *baker's map*, illustrated in Fig. 4.2. The magnetic field is assumed to be solely

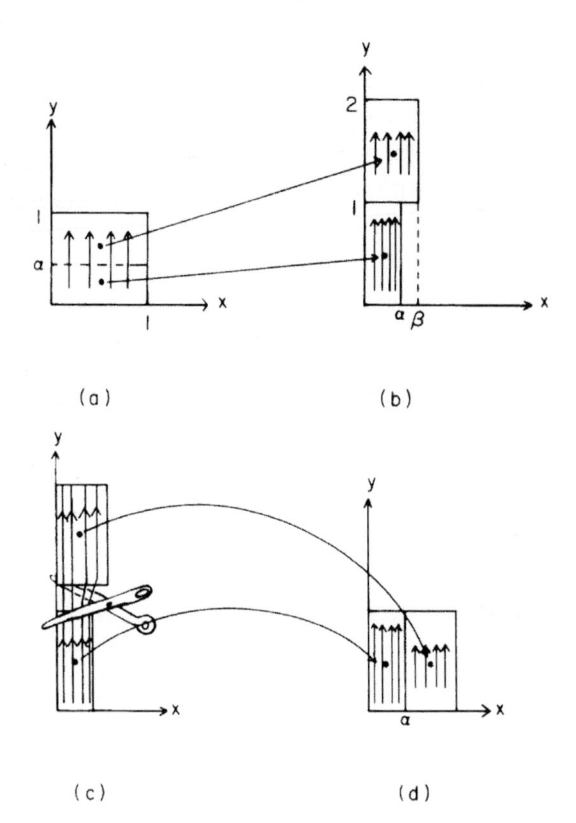

Fig. 4.2 A schematic diagram of the baker's map. (a) Magnetic field in the y-direction located in the square $0 < x, y < 1$ is (b) stretched and compressed (non-uniformly). (c) The top half of the field is then cut, and (d) placed back into the original square. The field is amplified by iteration of this process. (From [4].)

in the y-direction. It is stretched, cut and replaced, as shown, leading to an amplification of the field. A field element at position (x_n, y_n) at iteration n is transformed at the next iteration to (x_{n+1}, y_{n+1}), given by the formula

$$x_{n+1} = \begin{cases} \alpha x_n & \text{if } y_n < \alpha \\ \beta x_n + \alpha & \text{if } y_n > \alpha \end{cases}, \quad y_{n+1} = \begin{cases} y_n/\alpha & \text{if } y_n < \alpha \\ (y_n - \alpha)/\beta & \text{if } y_n > \alpha \end{cases}, \tag{4.8}$$

where $\beta = 1 - \alpha$. For the special case of $\alpha = \beta = 1/2$ the stretching is uniform and the flux doubles at each iteration. In general, however, when $\alpha \neq \beta$, the evolution of the field is more complicated. The map captures the necessary step of field stretching, but the middle step, in which the flux is cut and then displaced, glosses over all the 'tricky bits' of dynamo action — the constructive folding and reconnection of the field.

4.3.1 *Fractal dimension*

In order to obtain an idea of the distribution of magnetic field in a fast dynamo, we may inquire into the nature of the field that results from many iterations of the map (4.8). It is helpful to introduce from dynamical systems theory the idea of *fractal dimension*, which may be thought of as a non-integral measure of the 'area' of a set (see, for example, [7]). There are various different measures of fractal dimension; the simplest is what is known as the *box counting dimension* D_0, which may be determined as follows. Suppose the space in which the set under investigation lies is divided into 'cubes' of length ϵ (note that the space need not necessarily be 3-dimensional). Define $N(\epsilon)$ as the number of cubes needed to cover the set. Then D_0 is defined by

$$D_0 = \lim_{\epsilon \to 0} \frac{\ln(N(\epsilon))}{\ln(1/\epsilon)} . \tag{4.9}$$

For example, the Cantor ternary set, formed by repeatedly deleting the open middle thirds of a set of line segments, can always be covered by 2^n intervals of length $\epsilon = 3^{-n}$. Then

$$D_0 = \lim_{n \to \infty} \frac{\ln(2^n)}{\ln(3^n)} = \frac{\ln 2}{\ln 3} . \tag{4.10}$$

After n iterations of the baker's map it is straightforward to show that there are 2^n strips of 'magnetic field', with $\binom{n}{m}$ strips having width $\alpha^{n-m}\beta^m$ $(m = 0, \dots, n)$. For large n, $\binom{n}{m}$ may be approximated as a Gaussian in m, using Stirling's approximation. It can then be shown (see [4]) that most strips have values of m close to $n/2$ and, more precisely, that the strips containing a fraction $\theta < 1$ of the flux occupy a total length of order

$$2^n \theta(\alpha\beta)^{n/2} = \theta\left(2\sqrt{\alpha(1-\alpha)}\right)^n . \tag{4.11}$$

Provided that $\alpha \neq 1/2$ (i.e. provided that the stretching is non-uniform), the flux is concentrated on a set whose total length in x decreases exponentially with time. The box-counting dimension of the field is given by

$$D_0 = \frac{\ln 2}{\ln(1/\sqrt{\alpha\beta})}, \tag{4.12}$$

which is less than unity for $\alpha \neq 1/2$. Although the idea of the magnetic flux being concentrated on a fractal set has been demonstrated here for a map dynamo, it is of much more general importance. Non-uniform stretching is a generic feature of fluid flows and hence, by analogy, we may, in general, expect flux concentration on a fractal set as $Rm \to \infty$.

4.3.2 *Cancellation*

The above example demonstrates how magnetic field may be amplified by repeated stretching. However, in the simple baker's map, the 'field' is unidirectional (always in the positive y-direction), whereas in a realistic fluid flow, the field will vary with direction, leading to cancellation of fields of opposite sign. This cancellation may be incorporated into the baker's map, albeit in a very simplistic fashion, by considering a four-strip map (widths α, β, γ, δ with $\alpha+\beta+\gamma+\delta = 1$), in which, after stretching, one of the field blocks is inverted (i.e. its sign changed) before reassembling.

Each of the four strips carries the original flux, so the flux is doubled at each iteration, leading to a growth rate $\gamma = \ln 2$. Obviously, introducing cancellation reduces the growth rate (here from $\gamma = \ln 4$ without flipping one of the field blocks). In order to quantify this cancellation, the idea of a *cancellation exponent* was introduced in [8]. The idea is similar to that of the box-counting dimension. Consider, in general, the magnetic field in three-dimensional space. Suppose a planar surface S with unit normal $\boldsymbol{n}$ is divided into a grid of squares with side length ϵ. Now suppose that ϕ_i is the magnetic flux through square i, with respect to $\boldsymbol{n}$, normalised to the total flux through S. If $\chi(\epsilon)$ is defined by

$$\chi(\epsilon) = \sum_i |\phi_i| , \tag{4.13}$$

then if

$$\chi(\epsilon) \sim \epsilon^{-\kappa} \qquad \text{as} \qquad \epsilon \to 0\,, \tag{4.14}$$

κ is defined as the cancellation exponent. Note that the notion of a cancellation exponent is meaningful only for a *signed* quantity, such as magnetic field; if there is no cancellation then $\phi_i > 0$ for all i and $\sum |\phi_i| = \sum \phi_i = 1$, implying that $\kappa = 0$.

As an example, consider the four-strip baker's map with $\alpha = \beta = \gamma = \delta = 1/4$; for this two-dimensional map we consider the flux through a line $y = \text{const}$. After n iterations there are 4^n strips of flux, with each strip possessing the original flux Φ_0. Denote the number of strips with positive flux by N_+ and the number with negative flux by N_-. Then $N_+ + N_- = 4^n$. Furthermore, since the total (signed) flux doubles at each iteration, $2^n\Phi_0 = (N_+ - N_-)\Phi_0$. Hence

$$\frac{N_+}{N_+ + N_-} = \frac{1}{2} + \frac{1}{2^{n+1}}, \qquad \frac{N_-}{N_+ + N_-} = \frac{1}{2} - \frac{1}{2^{n+1}}\,. \tag{4.15}$$

Thus the fraction of strips with positive, or negative, flux tends to 1/2 exponentially in time (iteration), with the non-cancellation given by

$$\frac{N_+ - N_-}{N_+ + N_-} = \frac{1}{2^n}\,. \tag{4.16}$$

However, it is precisely this exponentially decreasing non-cancellation that leads to the growth of the total flux, since the number of strips grows faster (as 4^n) than the rate of decrease of non-cancellation (namely 2^n). For this simple example, the cancellation exponent is readily calculated by taking $\epsilon = 4^{-n}$. Then

$$\chi(\epsilon) = \sum |\phi_i| = \frac{4^n}{2^n} = 2^n = \epsilon^{-1/2}\,, \tag{4.17}$$

implying that $\kappa = 1/2$.

4.3.3 *An expression for the fast dynamo growth rate*

The interesting question of how the ideas of stretching and cancellation could be combined, quantitatively, into an expression for the fast dynamo growth rate was addressed by Du and Ott in [9]. Their analysis is based on the stretching and cancellation properties of a flow in the absence of diffusion, and so any derivation of the fast dynamo growth rate relies on the validity of the flux conjecture. The details are contained in [9]; here we shall just summarise the underlying idea.

Du and Ott considered the evolution of a magnetic field under a chaotic flow with two stretching directions; this assumption (as opposed to the other possibility of one stretching and two contracting directions) is crucial for their analysis. In the absence of magnetic diffusion, and for an incompressible flow, the magnetic field satisfies the equation

$$\frac{\mathrm{D}\boldsymbol{B}}{\mathrm{D}t} = \boldsymbol{B}\cdot\boldsymbol{u} = \boldsymbol{\mathcal{M}}\cdot\boldsymbol{B}, \quad \text{where } \boldsymbol{\mathcal{M}} = (\nabla\boldsymbol{u})^T . \tag{4.18}$$

This is equivalent to the statement made earlier that the magnetic field at time t is related to the initial field simply through the Jacobian matrix. If $J(\boldsymbol{x}_{0j};t,\tau)$ denotes the Jacobian matrix relating $\delta\boldsymbol{x}$ at time τ and location $\boldsymbol{x}_{0j}$ to $\delta\boldsymbol{x}$ at time $\tau+t$ and location $\boldsymbol{x}_j$ then

$$\delta\boldsymbol{x}_j(t) = J(\boldsymbol{x}_{0j};t,\tau)\delta\boldsymbol{x}_{0j} . \tag{4.19}$$

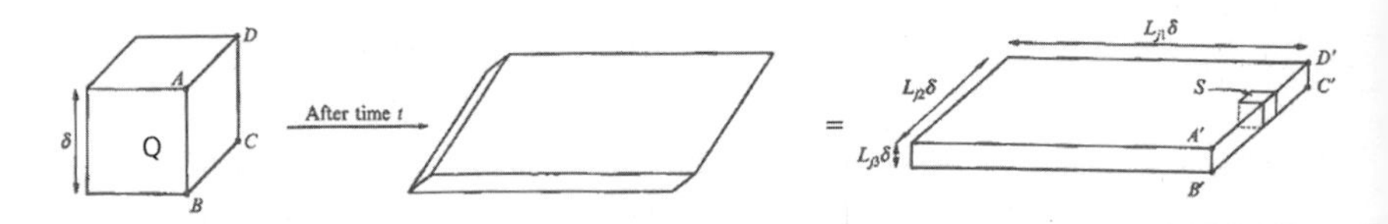

Fig. 4.3 A small cube is transformed into a parallelepiped by the flow, which is assumed to be stretching in two directions. To ease the analysis the parallelepiped is then approximated by a large aspect ratio rectangle. (From [9].)

Suppose the dynamo region is divided into small cubes (marked Q in Fig. 4.3) with side δ. Under the flow, as sketched in Fig. 4.3, the cube will be deformed into a parallelepiped of dimensions $L_{j1}\delta \times L_{j2}\delta \times L_{j3}\delta$, where the L_j (the finite-time Lyapunov numbers) are the magnitudes of the three eigenvalues of $J(\boldsymbol{x}_{0j};t,\tau)$. For definiteness, suppose $L_{j1} \geq L_{j2} \geq L_{j3}$. Also, by incompressibility, we have the relation $L_{j1}L_{j2}L_{j3} = 1$. Since, for fast dynamo action, the flow must be chaotic (see Sec. 4.4), it follows that, for a successful dynamo, $L_{j1} > 1$, and hence that $L_{j3} < 1$; by assumption, $L_{j2} > 1$. The parallelepiped can be approximated by a rectangular slab, with smallest edge length $L_{j3}\delta$. The slab contains N_j cubes of side $L_{j3}\delta$ (denoted as S in the figure), where

$$N_j = \frac{L_{j1}L_{j2}}{L_{j3}^2} = \frac{1}{L_{j3}^3} . \tag{4.20}$$

It is then possible, but by no means straightforward, to estimate the flux through a face of S to that through a face of Q, using the definition of the cancellation exponent to relate signed and unsigned flux, and noting that the diffusive cut-off of fine-scale variations in the field will vary between S and Q. This leads to the following estimate of the growth of flux involving the largest and smallest Lyapunov numbers together with the cancellation exponent:

$$\gamma = \lim_{t\to\infty} \frac{1}{t} \ln\langle L_1 L_3^{\kappa}\rangle\,, \tag{4.21}$$

where averages are taken over the fluid volume.

4.4 A necessary condition for fast dynamo action

In standard kinematic dynamo theory there are a number of rigorous results providing necessary conditions for dynamo action (reviewed in [5], for example). It has however turned out to be much harder to derive necessary conditions for *fast* dynamo action. Nonetheless, a very significant result was proved by Klapper and Young [10], building on unpublished work by Vishik.

From the discussion of map dynamos, we have seen that stretching of the magnetic field is a critical ingredient for fast dynamo action. To obtain an understanding of the Klapper and Young result, it is helpful first to discuss some measures of stretching in a fluid flow. The *Lyapunov exponents* λ_j (cf. the Lyapunov numbers introduced above) of a fluid element initially at $\boldsymbol{a}$ are the rates of exponential stretching of an infinitesimal element advected by the flow (an n-dimensional flow has n Lyapunov exponents). A chaotic flow must have at least one positive Lyapunov exponent. The largest Lyapunov exponent λ_M gives the maximum rate of stretching, and is readily calculated via

$$\lambda_M(\boldsymbol{a}) = \max_{\boldsymbol{\xi}} \limsup_{t\to\infty} \frac{1}{t} \ln|J(\boldsymbol{a},t)\boldsymbol{\xi}|\,. \tag{4.22}$$

An alternative measure is the rate of stretching of material lines, denoted by h_{line}. It should be noted that, in contrast to Lyapunov exponents, which describe the stretching of *infinitesimal* elements, h_{line} is concerned with *finite* material lines. In general, $h_{line} \geq \lambda_M$.

A related concept, which is harder to grasp physically, is the *topological entropy*, h_{top}. The topological entropy of a dynamical system is a measure of the complexity of the system, or, alternatively, of the rate at which information is revealed as time progresses; precise definitions can be found in [2] and [7]. The topological entropy is closely related to h_{line}; indeed, for two-dimensional flows they are identical, whereas for three-dimensional flows, $h_{line} \leq h_{top}$.

Klapper and Young [10] proved that for steady or time-periodic flows, the dynamo growth rate is bounded above by the topological entropy, a result conjectured in [4]. Since non-chaotic flows have zero topological entropy, a simple consequence is the following powerful anti-fast-dynamo theorem: *fast dynamo action is not possible in an integrable flow.* Finally, it is worth pointing out that λ_M is *not* an upper bound for the growth rate; as a concrete example, the dynamo resulting from flow (4.29), discussed in detail in the following section, has a growth rate greater than λ_M.

4.5 Fast dynamo flows

4.5.1 *An 'almost fast' dynamo*

Owing to the difficulties of deriving rigorous results for chaotic flows, there is no actual proof that a smooth flow can indeed act as a fast dynamo. It is instructive though to consider the rigorous analysis of a *slow* dynamo in the limit $Rm \to \infty$ [11]. Historically, dynamo theory has for the most part focussed attention on the generation of large-scale magnetic fields via mean field MHD (reviewed in [5] and, more recently, in [12]). A necessary condition for large-scale field generation is that the flow lacks reflectional symmetry, a property that is usually manifested by non-zero helicity (i.e. a non-zero correlation between velocity and vorticity). Consequently, considerable attention has been devoted to helical flows, and in particular to the flows, or modifications thereof, introduced by G.O. Roberts [13]. Incompressible, steady flows, dependent only on two Cartesian coordinates (x and y, say), may be expressed as

$$\boldsymbol{u}(x, y) = (\partial_y \psi(x, y), -\partial_x \psi(x, y), w(x, y)) \ . \tag{4.23}$$

Roberts [13] considered cellular motions, studying in detail the case of $w(x, y) = \psi(x, y) = \sin x \sin y$, for which the flow is maximally helical (i.e. the vorticity is parallel to the velocity). Steady, two-dimensional flows are integrable and hence, by the result of [10], discussed above, cannot act as fast dynamos. Nonetheless, as emphasised by Soward [11], there is some interest in seeing exactly why, and by how much, the Roberts dynamo fails to be fast.

Soward considered the modified Roberts flow given by

$$\psi = a_\epsilon \sin x \sin y, \quad w = K\psi \, , \tag{4.24}$$

where K is a constant and $a_\epsilon = 1$ everywhere except within a small radius ϵ of the stagnation point, where it is defined by

$$a_\epsilon(r) = 1 + \left[\ln\left(\frac{r}{\epsilon}\right)\right]^2 \qquad (r < \epsilon \ll 1)\,. \tag{4.25}$$

The effect of the a_ϵ term is to speed up fluid particles in the neighbourhood of the stagnation point. Soward's powerful boundary layer analysis, extending the work of [14], established a number of significant results. When $\epsilon = 0$ (the Roberts flow), the maximum growth rate as $Rm \to \infty$ is given by

$$p \sim \frac{\ln(\ln Rm)}{\ln Rm}\,, \tag{4.26}$$

which occurs when the wave number of the field in the z-direction is

$$k \sim \left(\frac{Rm}{\ln Rm}\right)^{1/2}\,. \tag{4.27}$$

Thus the dynamo is 'almost fast' (i.e. although the growth rate tends to zero as $Rm \to \infty$, it does so very slowly), and has an extremely small scale in the z-direction. Furthermore, Soward showed that the 'slowness' can be attributed to the long time spent by fluid elements in the neighbourhood of the stagnation point; for $\epsilon \neq 0$ the dynamo is indeed fast.

Although the integrability of two-dimensional steady flows rules out the possibility that they may act as fast dynamos, no such *a priori* restrictions apply either to steady (or, of course, unsteady) three-dimensional flows, or to unsteady two-dimensional flows. Since so much attention has been focused on large-scale dynamo action resulting from helical cellular flows, it is natural to consider whether such flows can act as fast dynamos. It is though worth stressing that the generation mechanisms of the two types of dynamo are very different and there is no obvious reason why, for example, helicity, which is necessary for mean field generation, should be beneficial for fast dynamo action. This issue has been explored in [15], which considered the fast dynamo growth rates of flows with identical chaotic properties but differing distributions of helicity.

Describing the nature of high Rm dynamo action driven by three-dimensional flows (steady or unsteady), or unsteady two-dimensional flows, represents an even tougher theoretical challenge than for steady two-dimensional flows. Consequently, most attacks on the problem have been computational. Of course, from a purely numerical approach it is impossible to *prove* that a dynamo is fast — the aim of such computations is simply to provide compelling evidence either for or against fast dynamo action.

4.5.2 Three-dimensional flows

Arnold and Korkina [16] considered dynamo action driven by what are known as ABC flows, with velocity

$$\boldsymbol{u} = A(0, \sin x, \cos x) + B(\cos y, 0, \sin y) + C(\sin z, \cos z, 0)\,. \tag{4.28}$$

Although, for steady flows, the induction equation may clearly be couched as an eigenvalue problem for the dynamo growth rate p, it is typically more efficient instead to solve (4.2) as an initial value problem, and then to deduce the growth rate from the exponential increase in, say, the magnetic energy. With (by today's standards) very modest computational resources, Arnold and Korkina located a window of dynamo action for $8.9 \lesssim Rm \lesssim 17.5$, for the symmetrical flow with $A = B = C = 1$. Galloway and Frisch [17] identified a new dynamo mode, of different symmetry to that in [16], with dynamo action setting in at $Rm \approx 27$; they showed that dynamo action persisted to $Rm = 550$; this was extended to $Rm = 1000$ in [18]. One major drawback of such three-dimensional simulations (which still holds today even with the great increase in computational power) is that it is hard to attain a very high Rm regime. Since diffusive structures scale as $Rm^{-1/2}$, a reasonable estimate is that the evolution of (4.2) with $Rm = 10^4$, for example, will be well resolved in a $500 \times 500 \times 500$ grid-point (or spectral mode) simulation. For $Rm = 10^6$, 5000^3 modes will be required, which represents a major computational challenge.

4.5.3 Unsteady two-dimensional flows

The problem of attaining the high Rm regime is alleviated to some extent by considering two-dimensional, necessarily unsteady flows, $\boldsymbol{u}(x, y, t)$, say. For such flows, the magnetic field satisfying the induction equation (4.2) takes the form $\boldsymbol{B}(x, y, z, t) = \hat{\boldsymbol{B}}\ (x, y, t) \exp(ikz)$. Hence, for a fixed value of k, the problem is spatially dependent on only the two coordinates x and y, thus allowing much higher Rm to be considered (for example, the computational resources that will permit exploration of $Rm = 10^6$ in three-dimensions will allow $Rm \approx 5 \times 10^9$ in two dimensions).

The most computationally efficient means of finding the growth rate is to solve Eq. (4.2) as a series of initial value problems (a different problem for each k). The hope is first to identify the optimal wave number $k_{\max}$ via a series of runs at low values of Rm, each of which is relatively inexpensive computationally; runs at higher Rm, which are more expensive, need then only consider $k = k_{\max}$.

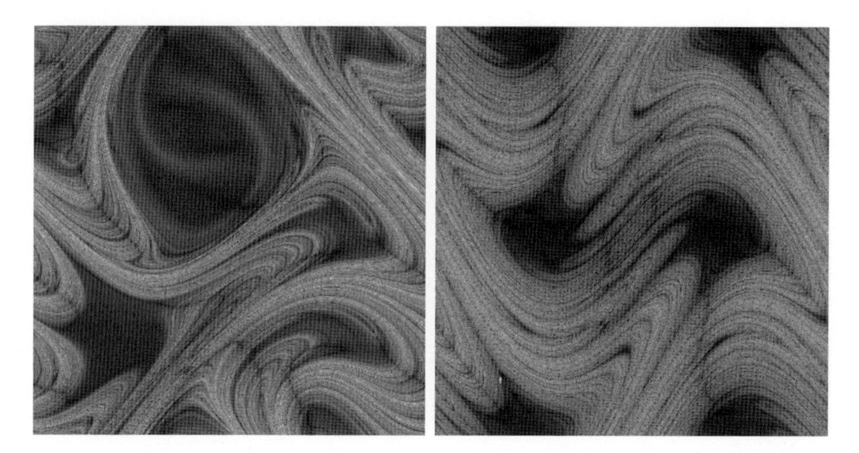

Fig. 4.4 Finite-time Lyapunov exponents (over an interval of 25 time units) for flows (4.29) (left) and (4.30) (right). For a colour version see Fig. 6.4. Light regions are chaotic (large exponential stretching), dark regions have little or no stretching.

This approach was adopted for the first time by Galloway and Proctor [19] and Otani [20]. The most successful of the flows investigated in [19] consisted of a rotation of the cellular Roberts flow pattern, with

$$\psi = \cos(x + \cos t) + \cos(y + \sin t), \qquad w = \psi \,. \tag{4.29}$$

Otani considered two similar flows, smooth versions of the pulsed flows introduced in [21], namely

$$\psi = 2\left(\cos x \cos^2 t \pm \cos y \sin^2 t\right), \quad w = \cos x \cos^2 t + \cos y \sin^2 t \,. \tag{4.30}$$

The two flows in (4.30) differ in that with the + sign, the pulses have helicity of the same sign, whereas with the − sign the two pulses have opposite helicity, leading to no overall helicity in this latter case. Flows (4.29) and (4.30) both exhibit Lagrangian chaos, a necessary condition for fast dynamo action, as illustrated in Fig. 4.4, which shows the finite-time Lyapunov exponents for the flows (for two-dimensional flows, all of the stretching is in the xy-plane).

For flow (4.29), an optimal Rm-independent wave number $k = k_{\max}$ was identified from runs over a range of moderate values of Rm. The z-wavenumber was then fixed at $k_{\max}$ and the growth rate examined for higher values of Rm [19]. For $k = k_{\max}$, the growth rate increases with Rm, levelling off at $Rm \approx 100$ and remaining essentially constant for Rm up to 10^5 (the highest value that it was possible to explore numerically at that time). The fact that the growth rate appears remarkably constant over quite

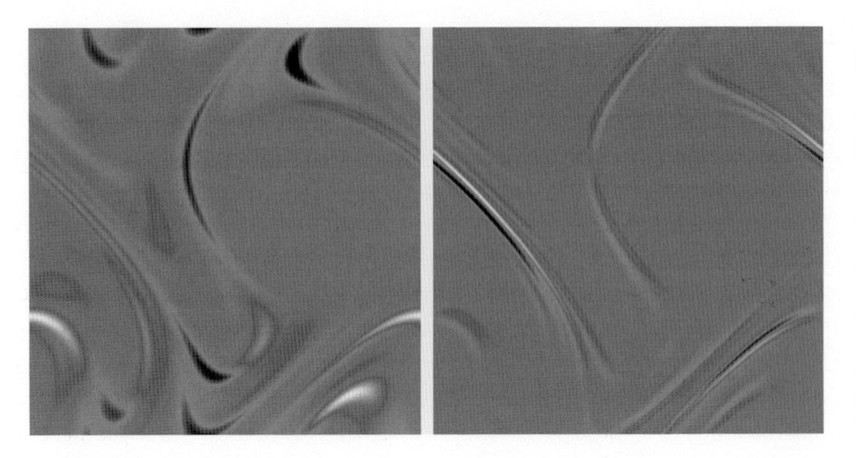

Fig. 4.5 Grey-scale plots at one instant in time of the z-components of the magnetic field (left) and the electric current (right) of the kinematic dynamo driven by flow (4.29) for $Rm = 1000$. Black denotes positive values of the field and current, white denotes negative values.

a range of Rm ($10^2 \lesssim Rm \lesssim 10^5$) provides good numerical evidence that the dynamo is indeed fast. The magnetic field develops fine-scale structure in the xy-plane (with characteristic scale $O(Rm^{-1/2})$), as can be seen from the plots of B_z and J_z ($\boldsymbol{J} = \nabla \times \boldsymbol{B}$ is the electric current) in Fig. 4.5. As Rm increases, the nature of the field remains the same, but with a decrease in the characteristic fine scale.

It is worth re-emphasising that the scale of the mode in the z-direction does *not* involve Rm — in contrast to the slow dynamo of [11], for which the scale of the optimal mode decreases according to expression (4.27) as $Rm \to \infty$. An unambiguously fast dynamo driven by a spatially two-dimensional ((x, y, t)-dependent) flow must have a z-structure that is independent of Rm; if the growth rate can be maintained at positive values only by an Rm-dependent reduction in k as $Rm \to \infty$ then the dynamo is, sometimes, referred to as an 'intermediate' dynamo [22, 23].

Rather than focusing on the growth rate, Otani showed that the nature of the 'large-scale' component of the fields (i.e. the spectral modes with small wave numbers k_x and k_y) for flows (4.30) show little change as Rm is increased, up to a value of 10^4. The inference is that an asymptotic form of the eigenfunction has been attained, and that the structure of the modes and the magnitude of the growth rate will change little for yet higher Rm, even as $Rm \to \infty$.

4.5.4 *Solving the Cauchy problem*

The previous two subsections have looked at numerical solutions of the induction equation, where the aim is to achieve as high a value of Rm as possible, in the hope that this will establish a clear asymptotic regime and thus provide information even for the limit $Rm \to \infty$. As discussed, the main computational difficulty is that as Rm increases, the diffusive layers (which must be properly resolved for a genuine solution) become thinner, as $Rm^{-1/2}$, and that the computational resources required rise accordingly.

A very different numerical approach is to attempt to solve the ideal (Rm infinite) problem via the Cauchy solution (4.7). This would then provide a growth rate for the magnetic flux and hence, if the flux conjecture holds, a value for the fast dynamo growth rate. The idea is to choose a 'typical' line (or area) and then to calculate the flux through that line at later times by relating the field to that at $t = 0$ (which is specified as an initial condition) via the Cauchy solution. There are both pros and cons for this technique in comparison to solving Eq. (4.2) for high Rm. On the plus side, it is not necessary to perform a fully three-dimensional simulation since one need consider the flux only through a fixed line in the flow (provided that this is representative of the flow as a whole). That said, it is only possible to find the flux though a line C at time t by integrating backwards in time (i.e. using the Cauchy solution) all the way to $t = 0$. Since the fluid elements lying on C at some time $t = t_2$ are, typically, different to those lying on C at all earlier times t_1, then knowledge of $\boldsymbol{B}(C, t_1)$ is of no help in determining $\boldsymbol{B}(C, t_2)$. Thus, unlike in a conventional time-stepping scheme, the workload increases with time. Furthermore, as time increases, the spatial structure of the field becomes ever more complex, with reversals on smaller and smaller scales, not limited by diffusion. To resolve these therefore requires more and more points, the number being difficult to determine *a priori*. Most numerical approaches to the fast dynamo problem have been through solving Eq. (4.2) rather than Eq. (4.7), and so it may be the case that the cons of the Cauchy solution approach outnumber the pros. Nonetheless, comparisons between the two are certainly of interest, particularly for those cases where the ideal flux growth rate and the dynamo growth rate at high Rm are not in good agreement.

4.6 Nonlinear fast dynamos

A fast dynamo, being characterised by the behaviour of its growth rate, is of course strictly defined only in the kinematic regime. However there are two ways in which the concept may be broadened so that one may speak of a *nonlinear* fast dynamo. The first is if the dynamo is oscillatory, with wholesale flux reversals on a fast (dynamic) time scale. The clearest example of such a dynamo is that of the Sun, in which both the toroidal and poloidal fields reverse (out of phase) approximately every 11 years; given that the Ohmic time for the Sun is comparable with its lifetime ($O(10^9)$ years) it is evident that the solar dynamo is operating (nonlinearly) on a fast time scale.

The second means of classifying a dynamo as a nonlinear fast dynamo is as the nonlinear evolution of a dynamo that, in its kinematic phase, is fast. To study such a dynamo it is necessary to consider a *forced* flow, with the forcing chosen such that in the absence of a magnetic field, the resulting flow acts as a fast dynamo. This idea was investigated by [24], who considered the nonlinear equilibration of a dynamo driven by a force that leads to the flow (4.29) in the kinematic regime. Two simplifying assumptions were made in [24]: one was to ignore the inertial terms, so that the role of the Lorentz force was directly to drive an induced flow; the other was to consider only the z-independent component of the Lorentz force, thus constraining the nonlinear evolution, as well as the kinematic phase, to be

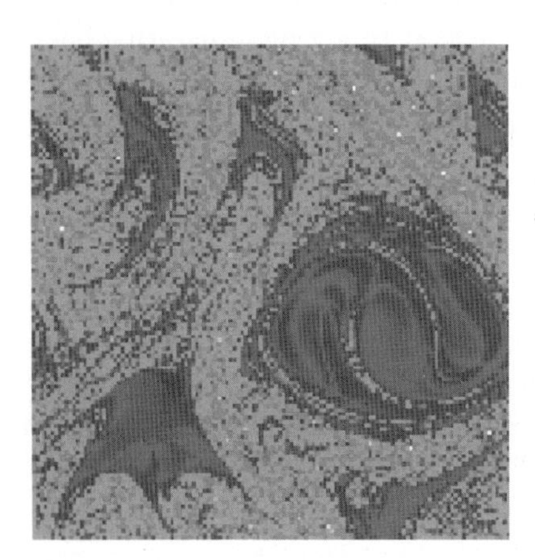

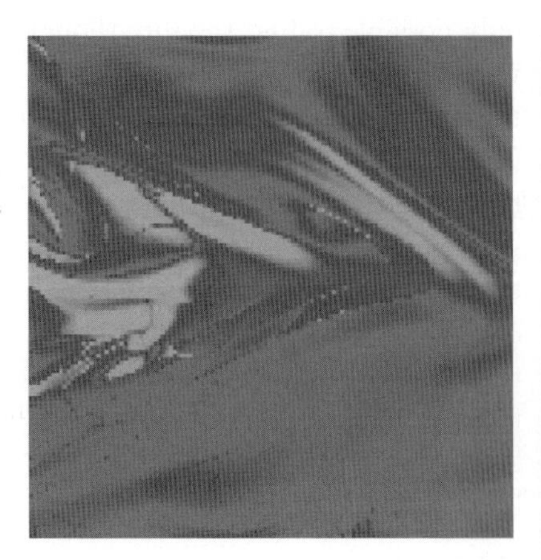

Fig. 4.6 Spatial distribution of finite-time Lyapunov exponents in the kinematic (left) and saturated (right) regimes for a forced flow, where the forcing is chosen so as to drive flow (4.29) in the absence of magnetic field. For a colour version see Fig. 6.5. The shades code the values of the exponents as a function of the initial positions. Light tones correspond to trajectories with little or no (exponential) stretching; dark tones correspond to strongly stretching trajectories. Regions of chaotic motion, which occupy a substantial fraction of the domain in the kinematic regime, are almost completely absent in the later dynamical phases.

two-dimensional. The key result was that saturation occurred not by any significant reduction in the flow amplitude, but instead by a marked reduction in the stretching properties of the flow. This is illustrated by Fig. 4.6, which shows the finite-time Lyapunov exponents (a measure of the stretching properties of the flow) in the kinematic and dynamic regimes.

The restriction on the inertial terms was dropped in the investigation of [25]; the main conclusion of [24] — a marked reduction in the chaotic properties of the flow as the dynamo saturated — was found still to hold. Zienicke et al. [26] performed a similar investigation to [24], but considered the fully three-dimensional evolution of a dynamo driven by an ABC flow forcing. They considered cases not only where the kinematic dynamo is indeed driven by the ABC flow, but also where, at higher fluid Reynolds number, the 'target' ABC flow is unstable to the forcing, leading to a more turbulent basic state flow. For both types of dynamo it was found, as in [24], that the chaos of the flow is markedly diminished in the nonlinear, saturated regime.

4.7 Future challenges

From the theoretical arguments outlined above, fast dynamo action is to be expected in a turbulent flow at high magnetic Reynolds number — precisely the conditions that pertain in most astrophysical situations. Thus, for example, it seems reasonable to ascribe the small-scale magnetic field observed at the solar surface to be the product of dynamo action by small-scale surface convection (granules and supergranules). This idea is reinforced by numerical simulations of dynamo action in turbulent convection [27].

There is though still a serious problem at the very heart of astrophysical dynamo theory. The magnetic field produced by fast dynamo action typically has the vast majority of its energy on scales comparable with, or smaller than, that of the flow. However, many astrophysical bodies possess magnetic fields that have a strong global component, on a scale that is presumably much larger than the turbulent motions responsible (at least in part) for generating the field. Traditionally, the generation of large-scale magnetic fields (i.e. fields with a strong component on scales much larger than a typical velocity scale) has been explained within the framework of mean field magnetohydrodynamics [5, 12]. A crucial step in the formulation of mean field MHD is the assumption that small-scale fields arise *only* from interactions between the mean (large-scale) field and the small-scale velocity; in other words they cannot be self-generated. But that is precisely equivalent to ignoring small-scale dynamo action!

As mentioned in Sec. 4.5, mean field dynamo theory highlights the necessity of flow helicity for the generation of large-scale magnetic fields. Fast dynamo action, on the other hand — which typically is small-scale — relies on chaotic flow trajectories. Most astrophysical flows will have a high magnetic Reynolds number Rm, and will be both turbulent (high fluids Reynolds number) *and* helical (being influenced by rotation) and so will have the necessary ingredients for both small-scale and large-scale dynamo action. Under such circumstances, which mechanism will dominate?

Cattaneo and Hughes [28] have argued that, at least in the kinematic regime, fast small-scale dynamo action, with growth rate determined by stretching and cancellation properties of the flow, will prevail over any mean field dynamo driven by small-scale flows, with growth rate proportional to the wave number (assumed small) of the large-scale field. Of course, the fact remains that nature quite happily generates large-scale magnetic fields. There are various possible mechanisms by which this process may occur. One is that it is nonlinear equilibration that leads to a large-scale dynamo, thus circumventing the kinematic problems discussed above. Another is that any mean field mechanism might be intrinsically nonlinear — for example, an electromotive force arising from the instability of a strong field, reminiscent of what occurs in RFP devices. This idea has been explored to some extent for magnetic buoyancy instabilities [29–32], but a fully self-consistent dynamo has yet to be demonstrated. Alternatively, a large-scale shear flow, as occur in many astrophysical bodies, may be able to 'rescue' an otherwise feeble large-scale dynamo. This could be via a variety of essentially mean field ideas (see, for example, [33]) or it may be that a large-scale shear flow is able, through flow interactions, to generate a large-scale velocity field that can act as a dynamo. This would then be a 'small-scale' dynamo (i.e. flow and field on the same scale), but on the large scales! Understanding the interaction between small and large-scale dynamos at high magnetic Reynolds numbers remains one of the greatest challenges in astrophysical magnetohydrodynamics.

References

[1] S. I. Vainshtein and Ya. B. Zeldovich (1972). Origin of magnetic fields in astrophysics, *Usp. Fiz. Nauk* **106**, 431–457 (English translation: *Sov. Phys. Usp.* **15** (1972), 159–172).

[2] S. Childress and A. D. Gilbert (1995). *Stretch, Twist, Fold: The Fast Dynamo* (Springer-Verlag).
[3] H. K. Moffatt and M. R. E. Proctor (1985). Topological constraints associated with fast dynamo action, *J. Fluid Mech.* **54**, 493–507.
[4] J. M. Finn and E. Ott (1988). Chaotic flows and fast magnetic dynamos, *Phys. Fluids* **31**, 2992–3011.
[5] H. K. Moffatt (1978). *Magnetic Field Generation in Electrically Conducting Fluids* (Cambridge Univ. Press, Cambridge).
[6] B. J. Bailey (1994). Maps and dynamos, in *Lectures on Solar and Planetary Dynamos* (eds. M. R. E. Proctor and A. D. Gilbert) (Cambridge Univ. Press) 305–329.
[7] E. Ott (1993). *Chaos in Dynamical Systems* (Cambridge Univ. Press, Cambridge).
[8] E. Ott, Y. Du, K. R. Sreenivasan, A. Juneja and A. K. Suri (1992). Sign-singular measures: fast magnetic dynamos, and high-Reynolds-number fluid turbulence, *Phys. Rev. Lett.* **69**, 2654–2657.
[9] Y. Du and E. Ott (1993). Growth rates for fast kinematic dynamo instabilities of chaotic fluid flows, *J. Fluid Mech.* **257**, 265–288.
[10] I. Klapper and L. S. Young (1995). Rigorous bounds on the fast dynamo growth rate involving topological entropy, *Commun. Math. Phys.* **173**, 623–646.
[11] A. M. Soward (1987). Fast dynamo action in a steady flow, *J. Fluid Mech.* **180**, 267–295.
[12] D. W. Hughes and S. M. Tobias (2009). An introduction to mean field dynamo theory, in *Relaxation Dynamics in Laboratory and Astrophysical Plasmas* (eds. P. H. Diamond, X. Garbet, P. Ghendrih and Y. Sarazin), World Scientific, 15–48.
[13] G. O. Roberts (1972). Dynamo action of fluid motions with two-dimensional periodicity, *Phil. Trans. R. Soc. A* **271**, 411–454.
[14] S. Childress (1979). Alpha-effect in flux ropes and sheets, *Phys. Earth Planet. Int.* **20**, 172–180.
[15] D. W. Hughes, F. Cattaneo and E. Kim (1996). Kinetic helicity, magnetic helicity and fast dynamo action, *Phys. Lett. A* **223**, 167–172.
[16] V. I. Arnold and E. I. Korkina (1983). The growth of a magnetic field in a three-dimensional incompressible steady flow, *Vest. Mosk. Un. Ta. Ser. 1 Matem. Mekh.* **3**, 43–46.
[17] D. J. Galloway and U. Frisch (1986). Dynamo action in a family of flows with chaotic streamlines, *Geophys. Astrophys. Fluid Dyn.* **36**, 53–83.

[18] Y.-T. Lau and J. M. Finn (1993). Fast dynamos with finite resistivity in steady flows with stagnation points, *Phys. Fluids B* **5**, 365–375.

[19] D. J. Galloway and M. R. E. Proctor (1992). Numerical calculations of fast dynamos in smooth velocity fields with realistic diffusion, *Nature* **356**, 691–693.

[20] N. F. Otani (1993). A fast kinematic dynamo in two-dimensional time-dependent flows, *J. Fluid Mech.* **253**, 327–340.

[21] B. J. Bayly and S. Childress (1987). Fast-dynamo action in unsteady flows and maps in three dimensions, *Phys. Rev. Lett.* **59**, 1573–1576.

[22] S. A. Molchanov, A. A. Ruzmaikin and D. D. Sokoloff (1985). Kinematic dynamo action in random flow, *Sov. Phys. Usp.* **28**, 307–326.

[23] P. H. Roberts and A. M. Soward (1992). Dynamo theory, *Ann. Rev. Fluid Mech.* **24**, 459–512.

[24] F. Cattaneo, D. W. Hughes and E. Kim (1996). Suppression of chaos in a simplified nonlinear dynamo model, *Phys. Rev. Lett.* **76**, 2057–2060.

[25] S. E. M. Tanner and D. W. Hughes (2003). Fast dynamo action in a family of parameterized flows, *Astrophys. J.* **586**, 685–691.

[26] E. Zienicke, H. Politano and A. Pouquet (1998). Variable intensity of Lagrangian chaos in the nonlinear dynamo problem, *Phys. Rev. Lett.* **81**, 4640–4643.

[27] F. Cattaneo (1999). On the origin of magnetic fields in the quiet photosphere, *Astrophys. J.*, **515**, L39–L42.

[28] F. Cattaneo and D. W. Hughes (2009). Problems with kinematic mean field electrodynamics at high magnetic Reynolds numbers, *Mon. Not. R. Astron. Soc.* **395**, L48–L51.

[29] D. Schmitt (1984). Dynamo action of magnetostrophic waves, in *The Hydromagnetics of the Sun* (eds. T. D. Guyenne and J. J. Hunt) (Noordwijk: ESA), 223–224, ESA SP-220.

[30] D. Schmitt (2003). Dynamo action of magnetostrophic waves, in *Advances in Nonlinear Dynamos* (eds. A. Ferriz-Mas and M. Núñez) (London: Taylor and Francis), 83–122.

[31] J.-C. Thelen (2000). A mean electromotive force induced by magnetic buoyancy instabilities, *Mon. Not. R. Astron. Soc.* **315**, 155–164.

[32] C. R. Davies and D. W. Hughes (2011). The mean electromotive force resulting from magnetic buoyancy instability, *Astrophys. J.* **727**, 112: 1–12.

[33] D. W. Hughes and M .R .E. Proctor (2009). Large-scale dynamo action driven by velocity shear and rotating convection, *Phys. Rev. Lett.* **102**, 044501: 1–4.

Chapter 5

The Effect of Flow on Ideal Magnetohydrodynamic Ballooning Instabilities: a Tutorial

H. R. Wilson[1], P. B. Buxton[1], J. W. Connor[2]

[1] *York Plasma Institute, Department of Physics, University of York, Heslington, York YO10 5DD, UK*

[2] *EURATOM/CCFE Fusion Association, Culham Science Centre, Abingdon, Oxon OX14 3DB, UK*

5.1 Introduction

Ballooning instabilities can exist whenever there is a combination of a high plasma pressure gradient and either magnetic field curvature or gravity. In terrestrial plasmas, the curvature is usually the dominant effect while in the Sun, for example, gravity plays an important role. Much of the physics is generic across a diverse range of systems. In this paper, however, we focus on tokamak plasmas. The toroidal geometry introduces periodicity into the system. This has particularly interesting consequences which, in the absence of flow, can be modelled with an elegant theoretical framework called ballooning theory [1–4]. In the presence of flows, however, the conventional ballooning theory breaks down, and a new theory is required. There have been numerous attempts at this over the last two decades. While substantial progress has been made, there are some key issues that remain to be understood. This paper reviews the progress that has been made, and discusses the challenges that remain.

In Section 5.2 we describe the toroidal magnetic geometry of a tokamak plasma and introduce a convenient orthogonal coordinate system. Section 5.3 provides an overview of the essential physics of ballooning modes, in preparation for the development of the formal theory in Section 5.4. Section

5.5 introduces the effect of a toroidal flow, describing how the physics of the ballooning mode is modified, and how to deal with it in a modified ballooning theory. We close with a summary and conclusions in Section 5.6.

5.2 Geometry

To provide a definite model on which to build the physics and theoretical framework of ballooning modes, we consider a tokamak plasma. The confining magnetic field lines lie on a set of nested toroidally symmetric flux surfaces. There are two components to the magnetic field: a toroidal component about the symmetry axis, B_ϕ, and a poloidal component, B_p (see Fig. 5.1). Thus, a magnetic field line on a particular flux surface will traverse poloidally and toroidally, tracing out a helical path on that surface. The ratio of the toroidal and poloidal components of the magnetic field varies from one flux surface to another. This ratio, which is a measure of the pitch of the field line, is characterised by a parameter called the safety factor, and denoted by q. The precise definition of q is the number of toroidal revolutions a field line must make to traverse once around the poloidal direction. It is a constant on a given flux surface, but varies from one surface to the next. In general, the field lines are effectively infinitely long and in making many toroidal and poloidal revolutions, an individual field line will cover the complete toroidal flux surface. There are, however, special flux surfaces where a magnetic field line will map back onto itself after a finite number of toroidal revolutions. These are called *rational surfaces.* It is clear that on these surfaces q will be a rational number. For example, if we write $q = m/n$ on such a rational surface, then any magnetic field line on that surface will map back onto itself after m toroidal and n poloidal revolutions. As we shall see, the concept of rational surfaces is crucial for understanding ballooning modes in tokamaks.

We need to define a convenient coordinate system. Note that a tokamak has symmetry in the toroidal direction, so it is natural to choose the toroidal angle, ϕ as one of the coordinates. A second convenient coordinate is the poloidal flux, ψ, contained within a flux surface (or more precisely, the poloidal flux per unit radian). As ψ is constant on a given flux surface, $\nabla\psi$ is a vector that points normal to the flux surface. Thus $\nabla\phi$ and $\nabla\psi$ are orthogonal. Finally, we define a poloidal angle, χ, such that $\nabla\chi$ is tangential to the flux surface and increases by 2π for one poloidal revolution.

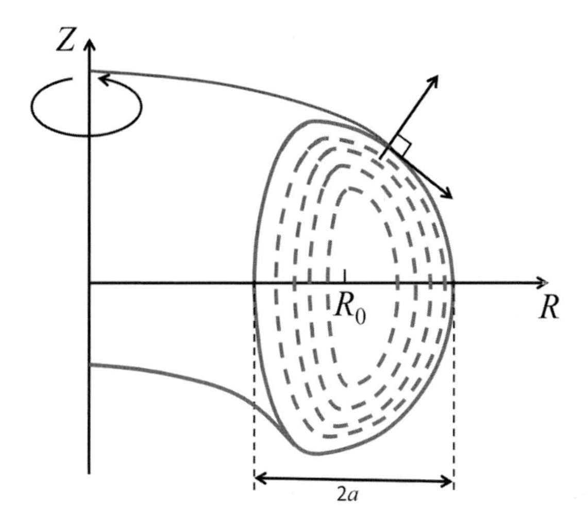

Fig. 5.1 Magnetic field geometry of a tokamak plasma illustrating the set of toroidal, nested flux surfaces (dashed curves), on which ψ is a constant. The toroidal coordinates used (ψ, χ ϕ) are shown. The toroidal component of the magnetic field is in the ϕ direction, while the poloidal component is in the χ direction. This geometry is defined by two radii: the major radius, R_0 and the minor radius a. The ratio of these is called the aspect ratio, $A = R_0/a$. A cylindrical coordinate system (R, ϕ, Z) can also be defined, as shown.

The Jacobian for this orthogonal system, $J = [\nabla\psi\cdot(\nabla\chi\times\nabla\phi)]^{-1}$. One can also define a cylindrical coordinate system, using the axis of symmetry of the toroidal plasma as the Z-axis, the azimuthal angle is simply the toroidal angle, ϕ and the radial variable is the major radius, R (note, it is conventional in tokamak physics to use upper case letters to denote the cylindrical coordinate system). These coordinates are illustrated in Fig. 5.1. With this in mind, it is clear that $|\nabla\phi| = 1/R$. The definition of ψ yields $|\nabla\psi| = RB_p$. To deduce $\nabla\chi$ requires information about the shape of the poloidal cross section of the flux surface (i.e. a cut through the torus at a fixed toroidal angle). For example, for a large aspect ratio, circular cross section flux surface, of radius r, $|\nabla\chi| = 1/r$. More generally, it can be expressed in terms of the Jacobian, $|\nabla\chi| = JB_p$. Clearly, J embodies information about the flux surface shape.

The magnetic field in a tokamak can be expressed in a compact form in terms of ϕ and ψ:

$$\boldsymbol{B} = f(\psi)\nabla\phi + \nabla\phi\times\nabla\psi \tag{5.1}$$

The first term represents the toroidal component of magnetic field, B_ϕ.

Note that $RB_\phi = f(\psi)$ is constant on a flux surface as a consequence of constraints due to Ampère's law and force balance. We will assume that plasma flows are slow compared to the sound speed, in which case force balance also constrains the pressure $p(\psi)$ to be constant on a flux surface. The precise form of $p(\psi)$ depends on the heating, fuelling and transport processes, which we shall not be involved with in this chapter. We therefore take the profile of $p(\psi)$ as an input. Similarly, the profile of $f(\psi)$ characterises the distribution of current density in the plasma. Its form therefore depends on the technique used to drive the current and the plasma resistivity distribution. Again we do not consider this physics, and assume $f(\psi)$ to be a given.

5.3 Ballooning mode physics

In the following section, we shall develop the theory of the ballooning mode, but it is first instructive to develop a physical picture of the instability. The first concept that we introduce is field line bending. Bending magnetic field lines in a plasma requires energy, and so is generally stabilising. The most unstable modes therefore tend to minimise field line bending. For a ballooning mode, we are interested in a high toroidal mode number, n. Thus, the wavelength of the mode in the toroidal direction $\sim 2\pi R/n$ is very short. However, to avoid excessive bending of field lines, we require that the wavelength along the magnetic field direction be very long. To achieve this, we shall see that we require a large poloidal mode number m. Let us consider a large aspect ratio, circular cross-section tokamak plasma. Then χ is simply a geometric angle, which we denote by θ. We can also use the minor radius, r, to label flux surfaces instead of ψ (see Fig. 5.2). To help visualise the geometry it is convenient to cut the torus at some toroidal angle and straighten it out to form a cylinder. Now cut the cylinder at some fixed poloidal angle and open that out to form a slab of plasma. The process is shown in Fig. 5.2. Note that the θ and ϕ directions of the slab must be periodic, so as a field line passes out of the side of the box on one of these faces, it re-enters at the opposite side. The upper flux surface of the slab of plasma in Fig. 5.2 is a rational surface, having $q = 2$. That is, the field line traverses the toroidal (ϕ) direction twice in order to traverse the poloidal (θ) direction once.

Let us now introduce a plasma wave to the system. As the poloidal and toroidal directions are periodic, we must fit a whole number of wavelengths in these directions. Furthermore, we ideally want the crests of the wave

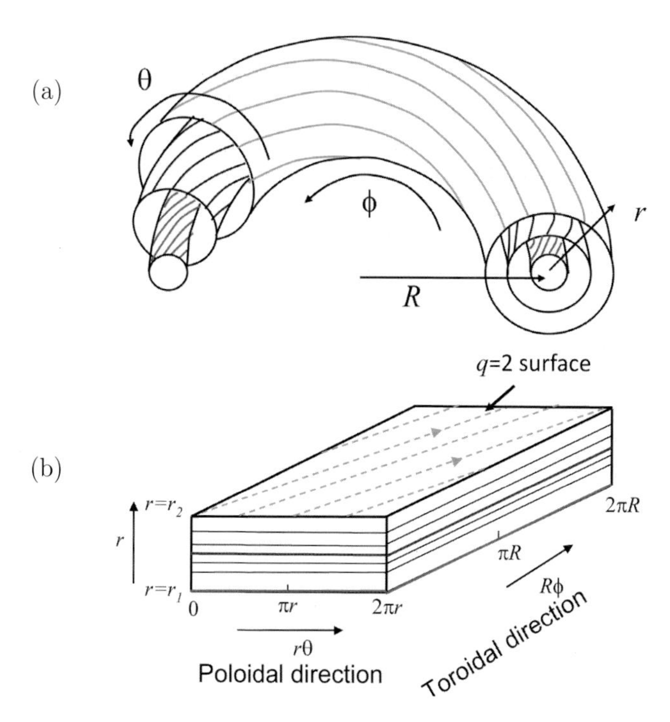

Fig. 5.2 (a) The torus of plasma is cut and straightened out to form a cylinder; (b) this is then cut at fixed poloidal angle and opened to form a slab of plasma. The lines indicate magnetic field lines.

to align with the magnetic field lines. Then, as the wave amplitude grows, it will just raise the field line as a whole, without bending it. Taking a single Fourier mode, with m wavelengths in the poloidal direction and n in the toroidal direction, this requires $m/n = q$. Figure 5.3 illustrates the situation for an $m/n = 10/5$ mode at the $q = 2$ surface at the top of the slab. At this surface, the crests of the wave are aligned with the magnetic field lines, so that as the wave rises and falls, the field line rises and falls without being bent. As one moves away from the rational surface, q will vary and the magnetic field lines will lie at a different angle to those at the rational surface. For this $m/n = 10/5$ mode, therefore, magnetic field lines away from the rational surface will not be aligned with the crests of the wave and therefore will be bent. The main conclusion of this is that in order to minimise the strong stabilisation that field line bending causes, it is necessary for this type of mode to be highly localised in the vicinity of the rational surface.

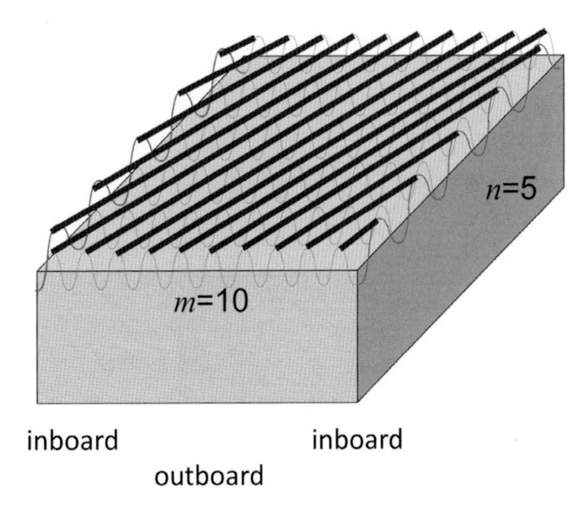

Fig. 5.3 The slab of plasma introduced in Fig. 5.2 with a $m/n = 10/5$ wave introduced onto the surface. The straight lines join the points of the crests of the wave. Note that these are parallel to the magnetic field lines of the slab introduced in Fig. 5.2(b). The inboard (minimum R) and outboard (maximum R) sides of the slab are indicated.

For a toroidal plasma, however, there is more to the story. The ballooning mode is driven by the pressure gradient in the plasma, but the conditions for the mode to be unstable depend on the concept of curvature. For a magnetic field line that traces out a circle, the curvature is the vector from the centre of the circle, pointing out to the magnetic field line. Thus, for a toroidal plasma, the curvature vector points radially outwards from the axis of symmetry (in the direction of ∇R). The pressure is peaked in the core of the plasma, at $r = 0$. Therefore the pressure gradient points in the $-\nabla\psi$ direction. On the outboard side of the plasma (where the major radius is largest) the pressure gradient and curvature vectors are opposed, and this is a destablising situation. On the inboard side, however, the pressure gradient and curvature vector are aligned with each other; this has a stabilising effect. The outboard side is therefore termed the "bad curvature" region and the inboard side is the "good curvature" region. As is evident from the cartoon in Fig. 5.3 a wave consisting of a single Fourier harmonic has the same amplitude on the inboard and outboard side and, therefore, this type of mode experiences both the good and bad curvature in equal proportions. For a tokamak, the good curvature dominates, this average curvature is "good" (partly a consequence of the Shafranov shift of flux surfaces caused by toroidal effects) and the wave is damped.

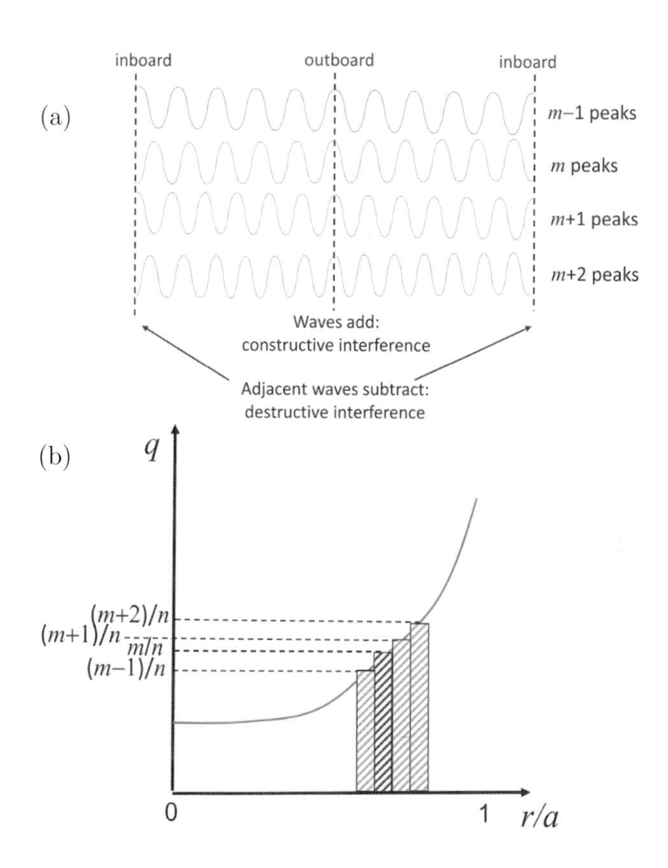

Fig. 5.4 (a) Cartoon of four poloidal Fourier harmonics in the poloidal plane with $m = 11$, arranged so that they constructively interfere on the outboard side and destructively interfere on the inboard side. (b) Sketch showing how these four harmonics are arranged radially, each centred on its own rational surface, but extending across to neighbouring rational surfaces to provide the necessary interference.

One way to beat the stabilising effect of average curvature is to construct a mode that has a maximum amplitude in the bad curvature region on the outboard side. To achieve this, we need to combine several Fourier harmonics. The toroidal symmetry of the tokamak means that we cannot couple different toroidal harmonics (at least within the linear theory we are considering here). So let us fix n and consider the consequences of combining a number of poloidal Fourier modes, with poloidal mode numbers m, $m \pm 1$, $m \pm 2$, etc. We align the waves so that they constructively interfere on the outboard side, and destructively interfere on the inboard side. Figure 5.4(a) shows how to achieve this. In order to minimise the field line bending, each poloidal Fourier mode must be localised around its rational surface, i.e. the $m+1$ mode at the position where $q(r) = (m+1)/n$, etc. Thus the different Fourier modes are localised about different positions in the plasma, as shown in Fig. 5.4(b). Clearly in order to interfere with each other, however, the mode localised around one rational surface must extend radially as far as the next rational surface. It is straightforward to show that this is a short distance $\sim 1/(nq')$, where the prime indicates a differentiation with respect to r. The result of this small, but finite, radial extent is that field lines are actually bent a little, by a moderate amount $\sim 1/n^0$. This provides a stabilising effect. The pressure gradient drive for the instability arises from the equilibrium and is also independent of n. Thus, we arrive at our main conclusion: the growth of a ballooning mode arises from the competition between field line bending and the pressure gradient. This balance results in a critical pressure gradient that must be achieved for instability. Furthermore, the ballooning mode is relatively extended in the radial direction, spanning many rational surfaces.

5.4 Ballooning theory

Before proceeding to consider how flow modifies this picture, we now describe the mathematical formalism that has been developed to analyse ballooning modes in MHD theory. We focus on the limit of large toroidal mode number, n. The displacement of a fluid element of plasma is denoted by $\boldsymbol{\xi}$, and decomposed along the directions of $\boldsymbol{\nabla}\psi$, $\boldsymbol{B}_0$ and $\boldsymbol{B}_0 \times \boldsymbol{\nabla}\psi$. For simplicity we consider incompressible perturbations, which is an accurate approximation close to marginal stability. The change in energy, δW associated with the plasma displacement can then be expressed in terms of the $\boldsymbol{\nabla}\psi$ component, ξ_ψ, only. The procedure is described in [1], and here we simply quote the result:

$$\delta W = \pi \int d\psi \oint d\chi \left\{ \frac{JB^2}{R^2 B_p^2} |k_\parallel \xi_\psi|^2 + \frac{R^2 B_p^2}{JB^2} \left| \frac{1}{n} \frac{\partial}{\partial \psi} (JB k_\parallel \xi_\psi) \right|^2 \right.$$
$$- \frac{2J}{B^2} \frac{dp}{d\psi} \left[|\xi_\psi|^2 \frac{\partial}{\partial \psi} \left(p + \frac{B^2}{2} \right) - \frac{i}{2} \frac{f}{JB^2} \frac{\partial B^2}{\partial \chi} \frac{\xi_\psi^*}{n} \frac{\partial \xi_\psi}{\partial \psi} \right]$$
$$\left. - \frac{\xi_\psi^*}{n} JB k_\parallel \left(\frac{\partial \sigma}{\partial \psi} \xi_\psi \right) \right\} . \tag{5.2}$$

Here we have written $k_{\|} = (i/B)(B \cdot \nabla)$ and σ is proportional to the component of current density along the magnetic field lines. We can recognise the physics discussed in Section 5.3 in Eq. (5.2). The first line is positive-definite, increasing the energy of the system and therefore contributing a stabilising effect. Note that each term contains a derivative of ξ_ψ along the magnetic field lines, corresponding to field line bending. Unstable perturbations must arrange that these terms are $O(n^0)$, as we discussed above.

The second line is the instability drive. It is clearly proportional to the pressure gradient, which provides the free energy to drive the ballooning instability. The terms in square brackets describe the effect of curvature. The second of these terms clearly takes opposite signs on the inboard and outboard sides (through the $\partial B^2/\partial\chi$ term; let $\chi \to \chi + \pi$). In fact, the first term also has opposite signs inboard and outboard, although it is less obvious in this case as it relies on an equilibrium relation between pressure and magnetic field. Thus, the second row contributes a stabilising (positive) contribution on the inboard side and a destabilising (negative) contribution on the outboard side. As δW involves an integration over all space, if ξ_ψ has a larger amplitude on the outboard side, then this weights the integrand towards the outboard, bad curvature region, and enables an instability to exist if the pressure gradient is sufficiently large (such that the terms in the second row make a larger contribution to δW than the field line bending terms in the first row). Finally, the third row provides an instability drive from a gradient in the current density. This is the drive for the kink mode. Note that as $n \to \infty$, this term tends to zero, so we shall not treat it here. However, it can be important for intermediate $n \sim 10$, especially when there is a large pressure gradient, driving a large bootstrap current, for example. If the mode interacts with the plasma boundary, the current density there also drives a peeling mode [5] that can couple to the ballooning mode [6, 7]. We have not included that physics here.

Equation (5.2) just has those terms that influence the instability drive. One can add the terms related to inertia (i.e. the kinetic energy) and then adopt a variational approach to derive the eigenmode equation. The procedure is carried out in [1] and is not reproduced here. We simply quote the result for a large aspect ratio, circular cross section tokamak plasma:

$$\left(\frac{\partial}{\partial\theta} + inq\right)\left[1 - \frac{r^2}{n^2q^2}\frac{\partial^2}{\partial r^2}\right]\left(\frac{\partial}{\partial\theta} + inq\right)\xi_\psi + \alpha\left[\cos\theta + \frac{ir}{nq}\sin\theta\frac{\partial}{\partial r}\right]\xi_\psi$$
$$= -\frac{\gamma^2}{\omega_A^2}\left[1 - \frac{r^2}{n^2q^2}\frac{\partial^2}{\partial r^2}\right]\xi_\psi\,. \qquad (5.3)$$

Here, $\alpha = -(2\mu_0 R q^2/B^2)(dp/dr)$ is the normalised pressure gradient and r is the minor radius labelling the flux surfaces. We have assumed a single Fourier mode in the toroidal direction $\sim e^{in\phi}$. The first term arises from field line bending, the second is the curvature and the term on the right hand side is the inertia, where γ is the growth rate of the mode, and we have introduced the Alfvén frequency $\omega_A = V_A/R$, with V_A the Alfvén velocity.

It is evident from Eq. (5.3) that at large n the field line bending term dominates unless there is a fast $\theta-$variation such that to leading order in n

$$\left(\frac{\partial}{\partial\theta} - inq\right)\xi_\psi = 0\,.$$

Thus it is tempting to write ξ_ψ in the eikonal form $\xi_\psi = \exp(inq\theta)F(r,\theta)$, where $F(r,\theta)$ has a slow spatial variation compared to the exponential factor. However, as any physical solution must be periodic in θ, $\xi_\psi(\theta + 2\pi)/\xi_\psi(\theta) = 1$, thus requiring $\exp(2\pi inq) = 1$. This is only true when nq is an integer; i.e. on the rational surfaces. Therefore this form for the solution is unphysical except at rational surfaces, and this approach to identifying an eigenfunction that minimises field line bending is flawed.

To address the problem of periodicity, let us recall the physics of the ballooning mode, discussed in the previous Section. There, we argued that the mode must couple many Fourier modes, so let us begin by representing the displacement as a Fourier series:

$$\xi_\psi = \sum_m u_m(r)e^{-im\theta}\,.$$

This is certainly periodic in θ, but because of the θ-dependence of the curvature term, the equations for each of the $u_m(r)$ depend on $u_{m\pm1}(r)$, and the result is a rather complicated set of coupled differential equations. Specifically, we find that the $u_m(r)$ satisfy the following infinite set of coupled equations:

$$\frac{r^2}{n^2q^2}\frac{d}{dr}\left[\left((m-nq)^2 + \frac{\gamma^2}{\omega_A^2}\right)\frac{du_m}{dr}\right] - \left[(m-nq)^2 + \frac{\gamma^2}{\omega_A^2}\right]u_m$$
$$+\alpha\left\{\frac{r}{2nq}\frac{d}{dr}\left[\left((m-nq)^2 + \frac{\gamma^2}{\omega_A^2}\right)(u_{m+1} - u_{m-1})\right]\right.$$
$$+\ \frac{r}{2nq}\left[(m-nq)^2 + 1 + \frac{\gamma^2}{\omega_A^2}\right]\frac{d}{dr}(u_{m+1} - u_{m-1})$$
$$\left.+\ \frac{r}{nq}(m-nq)\frac{d}{dr}(u_{m+1} + u_{m-1}) + \frac{1}{2}(u_{m+1} + u_{m-1})\right\}$$

$$-\frac{\alpha^2}{2}\left\{\left[(m-nq)^2+1+\frac{\gamma^2}{\omega_A^2}\right]\left[u_m-\frac{1}{2}(u_{m+2}+u_{m-2})\right]\right.$$

$$\left.-\left[(m-nq)\left(u_{m+2}-u_{m-2}\right)\right]\right\}=0\,. \tag{5.4}$$

Note that if the ballooning mode is radially localised (which we assume for the present, and demonstrate later), then we can neglect radial variations in all quantities except those that are multiplied by the large toroidal mode number, n. Thus, the radial variable only appears in Eq. (5.17) in the combination $m-nq$, and so we can write the solution in the form [4]

$$u_m(r)=u_0(m-nq) \tag{5.5}$$

where u_0 is a function that is not known as yet. This form shows that all of the $u_m(r)$ have the same radial profile, but they are shifted relative to each other so that each is centred on its own rational surface, where $m=nq(r)$. This so-called "ballooning symmetry" in transforming between rational surfaces is the key for ballooning theory: we shall see later that it is the loss of this symmetry that complicates the treatment of flows. To solve Eq. (5.4) in the limit that we neglect all radial variations except in the $m-nq$ factors, we define a Fourier transformed variable, $f(\eta)$ such that

$$u_m(r)=\int_{-\infty}^{\infty}f(\eta)e^{i(m-nq)\eta}e^{inq'x\eta_0}\,d\eta \tag{5.6}$$

where x is the distance from an arbitrary reference rational surface in the vicinity of the mode, say $nq(r_0)=m_0$ (then $x=r-r_0$), and a prime denotes a derivative with respect to r. While η_0 is related to a shift in the Fourier transformed variable η. This appears to be somewhat arbitrary at this stage, but we shall find that it plays a key role in the physics of ballooning modes. Indeed, one can see from Eq. (5.6) that it has an interpretation as a radial wavenumber, $k_r=nq'\eta_0$. Then it is reasonable to suggest that the reason it is undetermined is that we have neglected radial variations in equilibrium quantities that break the ballooning symmetry (this symmetry-breaking needs to be taken into account to determine the radial mode structure). We shall return to this in a moment, but for now let us substitute the form Eq. (5.6) into Eq. (5.4). The result is a single, second order differential equation:

$$\frac{d}{d\eta}\left[\left(1+P^2\right)\frac{df}{d\eta}\right]+\alpha\left[\cos\eta+P\sin\eta\right]f=\frac{\gamma_0^2}{\omega_A^2}\left(1+P^2\right)f \tag{5.7}$$

where $P = s(\eta - \eta_0) - \alpha \sin \eta$. Here $s = (r/q)dq/dr$ is called the magnetic shear, and provides a measure of the radial variation of q.

Equation (5.7) is the celebrated ballooning equation. It must be solved on each flux surface subject to the boundary condition that $f \to 0$ as $|\eta| \to \infty$ so that the Fourier transform Eq. (5.6) exists. This yields the growth rate γ_0 as an eigenvalue. Although we have so far neglected equilibrium variations with radius, note that α and s both depend weakly on r, so the resulting solution for γ_0 will also depend on r as well as η_0 (it is periodic in the latter). Thus, while we shall find that γ_0 is related to the actual growth rate γ, they cannot be identical (γ is independent of space and, as η_0 was introduced apparently arbitrarily, it cannot appear in the final answer). The effect of a radial variation of α, s, etc. is to break the symmetry between rational surfaces that led to the form of solution given in Eq. (5.5). This has two consequences. First, it provides a prescription for determining η_0. We interpreted η_0 above as a radial wavenumber. Combining the above results, it is straightforward to show that η_0 can equally be considered as a relative phase between the u_m on two adjacent rational surfaces (i.e. $u_{m+1}/u_m \sim e^{i\eta_0}$). This relative phase can only be deduced by taking account of the variations across rational surfaces, however small. The second consequence of the breaking of symmetry between rational surfaces is that it provides a prescription for calculating the relative amplitudes of the different u_m: the so-called radial envelope. The relationship between γ and $\gamma_0(r)$ is determined from an eigenvalue condition arising from the equation describing this radial mode structure. To summarise, the leading order ballooning equation describes the shape of all of the individual Fourier modes $u_m(r)$, that we must then combine to form the full ballooning mode. Taking account of the radial variations as a higher order effect (in an expansion in toroidal mode number) we learn (i) the relative phase of the u_m in this combination, and (ii) their relative amplitude.

Ballooning theory provides a rigorous theoretical framework based on these ideas to build up the full eigenmode structure and stability characteristics of high n ballooning modes. Full details are described in [1], so here we just describe the basic steps, and link them to the physical picture we have developed above. Our starting point is the 2D eigenmode equation (5.3). Guided by the mathematical development above, we introduce the so-called ballooning transform, and write:

$$\xi_\psi(r, \theta) = \sum_m e^{im\theta} \int_{-\infty}^{\infty} e^{-im\eta} \hat{\xi}(r, \eta) d\eta \,. \tag{5.8}$$

This employs the same Fourier expansion in poloidal angle that we used earlier to derive Eq. (5.7). The Fourier coefficients are then Fourier transformed, but in a subtly different way to the procedure we used to Fourier transform the $u_m(r)$ earlier. Here, we treat m as the continuous variable, and Fourier transform with respect to m, retaining the r variation in the Fourier transformed variable, $\hat{\xi}(r, \eta)$. Actually the two procedures are more similar than they might at first appear because m does play the role of a radial coordinate, as follows. Recall that m labels the different rational surfaces, and that u_m is localised about its rational surface where $q(r) = m/n$. Therefore, there are two effects that determine the radial mode structure: (i) the actual shape of each individual $u_m(r)$, and we have seen that this is described by the ballooning equation (5.7), and (ii) the relative amplitudes of the different $u_m(r)$ at their rational surfaces, which provide a radial envelope. For the latter, the value of m characterises the radial position at points along the envelope. It is in this sense that m plays the role of a radial coordinate.

The ballooning transformation has some useful properties. First, a differential operator in θ, $\partial/\partial\theta$, maps to $\partial/\partial\eta$. Second, any trigonometric function $e^{ik\theta}$ maps to $e^{ik\eta}$. As all of the coefficients in the 2-D ballooning equation (of which Eq. (5.3) is an example) are periodic in θ, they can be represented as a Fourier series. Therefore a consequence of this second property is that any periodic function of θ will map to the *same* periodic function of η. Thus, applying the ballooning transform to the 2-D ballooning equation in any arbitrary tokamak geometry leaves the equation unchanged, and simply maps the poloidal angle, θ onto the infinite "ballooning coordinate", η. While this appears to be trivial, it is extremely important. The ballooning transform ensures that ξ_ψ is periodic in θ, provided the transform exists. The existence requires that the transformed variable, $\hat{\xi}(r, \eta)$ decays sufficiently fast at large $|\eta|$. Thus, our equation in ballooning space is free of the periodicity constraint that hindered our attempts to minimise the field line bending terms in Eq. (5.3) by adopting the eikonal form $\xi_\psi \sim e^{inq\theta}$. In ballooning space we *can* adopt such an eikonal form. This then enables a formal WKB expansion procedure to be developed in ballooning space, which was not possible in real (θ) space.

In summary, we seek to solve Eq. (5.3) with $\theta \to \eta$ and $\xi_\psi(r, \theta) \to \hat{\xi}(r, \eta)$ on the infinite η domain, with boundary conditions that $\hat{\xi} \to 0$ for $|\eta| \to \infty$ so that the ballooning transform, Eq. (5.8) exists. As $\hat{\xi}(r, \eta)$ is free from any periodicity condition, we are now free to express it in the eikonal form:

$$\hat{\xi}(r,\eta) = F(r,\eta)e^{inq(\eta - S(r))} \,. \tag{5.9}$$

Note that we have introduced an additional factor $e^{-inqS(r)}$ here that embodies a fast variation with $r \sim nr$. We assume that S itself varies with r on a scale that is comparable with variations in other equilibrium quantities, while $F(r,\eta)$ is assumed to vary with an intermediate length scale $\sim n^{1/2}r$. This motivates us to expand our solution in powers of $n^{-1/2}$. Defining $\eta_0 = dS/dr$, we then find that our leading order solution $F_0(r,\eta)$ satisfies the standard ballooning equation (5.7). Thus, a general solution is $F_0(r,\eta) = A(r)f(\eta)$, with $f(\eta)$ the solution of Eq. (5.7) normalised in some appropriate way (e.g. $f(\eta = 0) = 1$). As discussed above, the growth rate, γ_0 that is determined as an eigenvalue of this equation will depend on η_0 and r, and we still have to determine how to relate it to the true growth rate, γ. We also need to evaluate $A(r)$, which is the radial envelope function referred to above (i.e. the envelope of the $u_m(r)$). These require us to develop the expansion in $n^{-1/2}$ to two more orders.

The procedure is outlined in detail in [1]. We do not reproduce the algebra here, but limit ourselves to a description of the procedure and the main results. At order $n^{-1/2}$ one derives a solubility condition by multiplying the equation by $f(r,\eta)$ and integrating over all η. That condition reduces to

$$\frac{d\gamma_0^2}{d\eta_0} = 0 \,. \tag{5.10}$$

This result makes sense physically: it says that the mode will choose the value of η_0 which is the most unstable. However, we did assume that $A(r)$ varied on a length scale $\sim n^{1/2}r$. If that is not the case, then the expansion procedure, and therefore Eq. (5.10), is invalid. It is therefore crucial to determine $A(r)$ and check this. An equation for $A(r)$ is provided by proceeding to the $\mathrm{O}(1/n)$ equation and again deriving a constraint equation by integrating over η. One finds that $A(r)$ is localised around the radial position where $\gamma_0^2(r)$ is a maximum. The full growth rate, γ^2, also appears at this order, and the eigenvalue condition that follows from requiring a localised solution is:

$$\gamma - \gamma_{00} = O(1/n) \tag{5.11}$$

where γ_{00} is the value of γ_0 at the η_0 and r values that maximise it. With this eigenvalue condition, $A(r)$ is a Gaussian with a width $\Delta r \sim r/\sqrt{n}$. Thus this solution does indeed satisfy the assumptions made about the radial length scales. Recalling that the distance between rational surfaces is $1/nq'$, this solution for $A(r)$ tells us that a large number $\sim \sqrt{n}$ rational surfaces (or poloidal harmonics, $u_m(r)$) are coupled to create the ballooning mode.

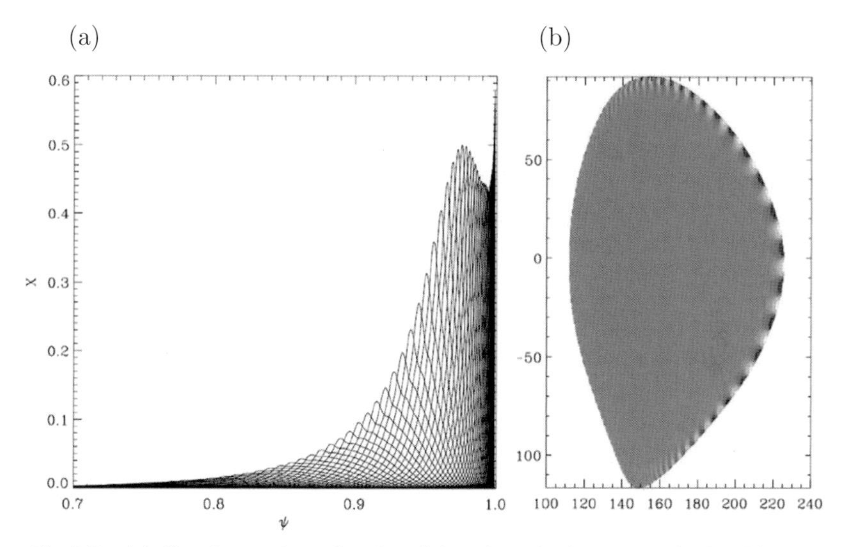

Fig. 5.5 A ballooning mode at the edge of the tokamak plasma, as calculated by ELITE. (a) The radial variation of the $u_m(r)$ and (b) the mode reconstructed in the poloidal plane from the Fourier harmonics.

The ELITE code [8, 9] is an ideal MHD code, optimised to study ballooning modes, but without resorting to the ballooning representation. Figure 5.5 shows an example of a 2D ballooning mode eigenfunction calculated by ELITE at the edge of a tokamak plasma. In Fig. 5.5(a), we show the calculation of the radial variation of each of the Fourier modes $u_m(r)$. Note: (i) each of the $u_m(r)$ have similar radial variations; (ii) each $u_m(r)$ is peaked at its rational surface, and (iii) the envelope of $u_m(r)$ varies on a longer scale than the variation of each individual $u_m(r)$. These are all consistent with the assumptions made in developing the ballooning approximations that reduce the 2D system to the 1D ballooning equation. Note that the envelope is not a Gaussian, however. This is because ELITE calculates ballooning modes at the plasma edge, and the edge has cut off the right-hand tail of the ballooning mode in this case. In Fig. 5.5(b) we show the solution for ξ_ψ that has been reconstructed from the Fourier modes. The colour contours denote the amplitude of the displacement.

This concludes our introduction to ballooning modes, and the theory that is used to analyse them. In the next section we explore the impact of plasma flow, and how our picture is changed as a consequence.

5.5 Ballooning modes: the effect of flow

In a toroidal plasma, the flow is usually dominated by the component in the toroidal direction. This is partly because that is typically the direction of momentum injection (e.g. from neutral beam injection), but also a consequence of the strong damping of poloidal flows. We therefore restrict consideration to flow in the toroidal direction, which has three consequences: a centrifugal force, a Coriolis force and a Doppler shift of the mode frequency. We restrict consideration to low flows, much less than the speed of sound in the plasma, and then the dominant effect is the Doppler shift. For a rigid body rotation, this would just shift the frequency; the growth rate, and therefore stability, would be unaffected. When there is a shear in the flow, however, this fundamentally changes the form of the equation, and stability can be significantly affected.

To understand why even a relatively low toroidal flow shear has a large influence on ballooning mode stability, consider the effect of a toroidal angular rotation, $\Omega(\psi)$. Note that Ω is constrained to be constant on flux surfaces if the flow is divergence free. This form means that each toroidal flux surface rotates as a rigid body, but that the flux surfaces can rotate relative to each other. If we consider a ballooning mode with a large toroidal mode number, n, then the Doppler shift effect is incorporated into the new eigenmode equations by the shift $\partial/\partial t \to \partial/\partial t - in\Omega$. It is the factor of n that greatly enhances the effect of a flow variation compared to other equilibrium variations. The difference in Doppler shift between two adjacent mode rational surfaces that are a distance $1/(nq')$ apart is $(n\Omega')(1/(nq')) = d\Omega/dq$. This is an $O(1)$ number, whereas for other equilibrium quantities the difference between two adjacent rational surfaces would be $O(1/n)$. Because of its important role in the impact of flow shear on ballooning mode stability, we introduce the parameter $\dot{\Omega} = d\Omega/dq$.

To illustrate the effect of flow shear, let us return to our physics description of ballooning modes and, in particular, Fig. 5.4(a). The waves shown there were waves on neighbouring rational surfaces, arranged so that they constructively interfere on the outboard side of the tokamak. Now consider what happens with rotation, shown in Fig. 5.6. Because different rational

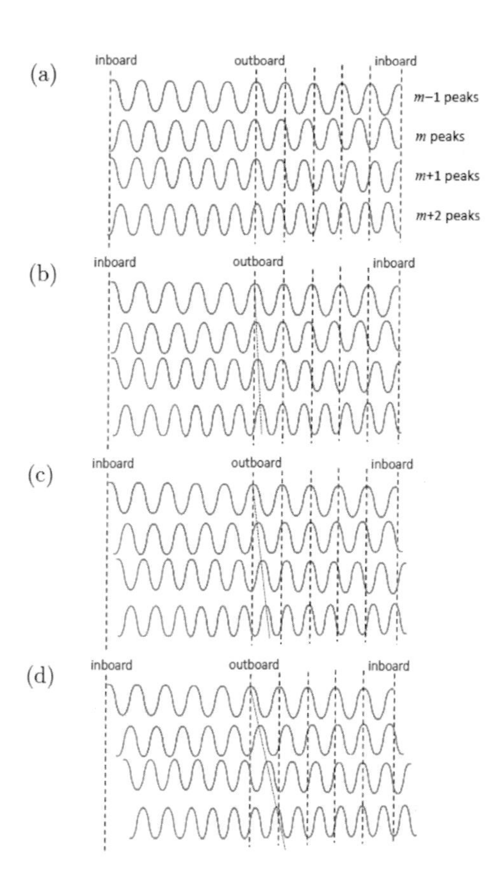

Fig. 5.6 Cartoon of wave amplitudes of the different rational surfaces as a function of time, with a flow to the right increasing from the upper rational surface $(m-1)$ to the lower one $(m+2)$. (a) shows the state at time $t=0$ with waves in phase at the outboard mid-plane, and (b)–(d) show the situation at successive later times, in the rest frame of the rational surface $m-1$. Vertical dashed lines show the positions where the wave at $m-1$ is a maximum. The dotted line indicates how the crests of the waves that were initially aligned at the outboard mid-plane at $t=0$ move in time. The consequence is that the poloidal angle where the waves constructively interfere increases with time (in each successive frame, the poloidal angle where constructive interference occurs moves from the position of one vertical dashed line to the next).

surfaces move with different speeds, the waves on those rational surfaces move relative to each other. Figures 5.6(a)–(d) show how waves that initially constructively interfere at the outboard mid-plane move relative to each other as time progresses which, in turn, moves the poloidal angle where the constructive interference occurs. Thus, the poloidal location where the ballooning mode is maximum rotates poloidally from the outboard side to the inboard side, and back to the outboard side as time progresses. This is equivalent to evolving the parameter η_0 in time. Thus we expect that the growth rate experienced by a ballooning mode in the presence of low flow shear can be obtained by averaging the growth rate obtained from the ballooning equation (5.7) over η_0. It is therefore more stable than the case with no flow, where one selects the maximally unstable value of η_0. Our goal in this Section is to modify the ballooning theory to describe this physics.

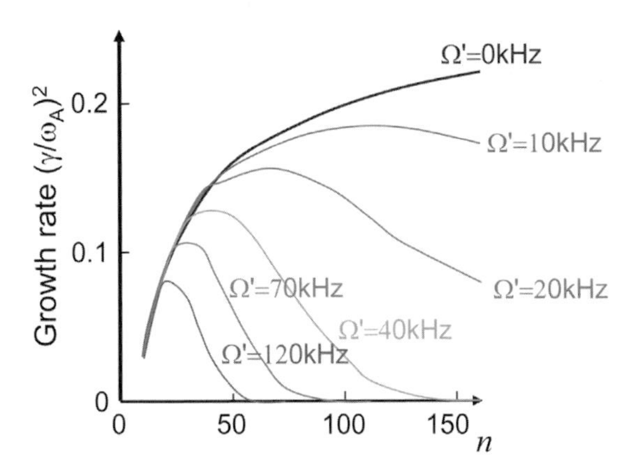

Fig. 5.7 ELITE calculations of the growth rate of ideal ballooning modes as a function of n for different levels of flow shear [10].

The ELITE code has been modified to calculate the influence of flow shear on ballooning modes. The variation of the growth rate with n at different levels of flow shear is shown in Fig. 5.7 [10]. Note that even a very small flow shear can stabilise the high n modes, whereas an increasingly large flow shear is required to stabilise the lower n modes. This is consistent with $n\dot{\Omega}$ being the relevant measure of flow shear. Our goal is to develop a modified ballooning theory that can treat the limit of large n and reduce the 2-D eigenmode equation into the 1-D ballooning equation (as for the case with zero flow).

Consider Eq. (5.3). The terms in growth rate arise because we considered a time-dependence $\sim e^{\gamma t}$. While it is possible to derive eigenmodes of the system, we shall first explore a different type of solution, called a Floquet mode. To do this, we map $\gamma \to \partial/\partial t$, and then Doppler-shift the time-derivative (working in a frame of reference where the flow at $x = 0$ is zero).

The result is our model equation for ballooning modes in the presence of toroidal flow.

$$\left(\frac{\partial}{\partial\theta}+inq\right)\left[1-\frac{r^2}{n^2q^2}\frac{\partial^2}{\partial r^2}\right]\left(\frac{\partial}{\partial\theta}+inq\right)\xi_\psi+\alpha\left[\cos\theta+\frac{ir}{nq}\sin\theta\frac{\partial}{\partial r}\right]\xi_\psi$$
$$=-\frac{1}{\omega_A^2}\left(\frac{\partial}{\partial t}+in\Omega'x\right)\left[1-\frac{r^2}{n^2q^2}\frac{\partial^2}{\partial r^2}\right]\left(\frac{\partial}{\partial t}+in\Omega'x\right)\xi_\psi\,. \quad (5.12)$$

To solve this equation, we employ the same ballooning transform, Eq. (5.8), that we used in the absence of flow shear. We can then employ the eikonal form Eq. (5.9) to remove the large terms inq. However, there are now additional large terms $in\Omega'x$ that this eikonal form does not remove. One way to eliminate these from our equation is to modify the eikonal in the presence of flow shear, and write [11–13]

$$\hat{\xi}(r,\eta)=F(r,\eta)e^{in(q(\eta-S(r)-\Omega' xt)}\,. \quad (5.13)$$

This form removes all of the terms dependent on n, but it does so at the expense of introducing a time-dependent eikonal. The result of this is that $F(r,\eta)$ satisfies an equation that is very similar to the ballooning equation without flow, i.e.

$$\frac{d}{d\eta}\left[\left(1+P^2\right)\frac{dF}{d\eta}\right]+\alpha\left[\cos\eta+P\sin\eta\right]F=\frac{1}{\omega_A^2}\frac{\partial}{\partial t}\left[\left(1+P^2\right)\frac{\partial F}{\partial t}\right]\,. \quad (5.14)$$

There is, however, one important difference: the function P now depends on time:

$$P=s(\eta-\eta_0-\dot{\Omega}t)-\alpha\sin\eta \quad (5.15)$$

so we can no longer seek eigenmodes of the system with a time dependence $\sim e^{\gamma t}$. If we define $\hat{\eta}=\eta-\eta_0-\dot{\Omega}t$ and work with $\hat{\eta}$ instead of η, then all of the coefficients of Eq. (5.14) are periodic in time. The solution of such an equation is a Floquet mode. That is $F(\hat{\eta},t)=\hat{F}(\hat{\eta},[\dot{\Omega}t])e^{\gamma t}$, where the square brackets imply that $\hat{F}$ is a periodic function of $\dot{\Omega}t$, with period 2π. Another important point to note is that flow shear and η_0 are always combined in the form $\eta_0+\dot{\Omega}t$. Thus, the effect of flow is to evolve η_0 in time. As η_0 defines the poloidal angle where the mode peaks, this is equivalent to a poloidal rotation of the peak in mode amplitude (or equivalently the position where the Fourier modes constructively interfere) as illustrated in the cartoon of Fig. 5.6.

We can write an alternative form of Eq. (5.14) by defining the slow time variable $\tau = \dot{\Omega}t + \eta_0$. We then have:

$$\frac{d}{d\hat{\eta}}\left[\left(1+P^2\right)\frac{d\hat{F}}{d\hat{\eta}}\right] + \alpha\left[\cos\left(\hat{\eta}+\tau\right) + P\sin\left(\hat{\eta}+\tau\right)\right]\hat{F}$$
$$= \frac{1}{\omega_A^2}\left(\dot{\Omega}\frac{\partial}{\partial\tau}+\gamma\right)\left[\left(1+P^2\right)\left(\dot{\Omega}\frac{\partial}{\partial\tau}+\gamma\right)\hat{F}\right] \quad (5.16)$$

where $P = s\hat{\eta} + \alpha\sin(\hat{\eta}+\tau)$. This suggests that the η_0 parameter can be absorbed into the origin of time, and is irrelevant. In fact, that is not quite true and when one considers the higher order theory (in n), a proper treatment of η_0 is important. We shall not consider this here, but rather refer the interested reader to reference [14].

Note that our ballooning theory has not managed to reduce the system to a 1-D equation, as was the case in the absence of flow. While it does involve differentials in only one spatial dimension, the flow has forced us to retain time differentials primarily because we introduced the time-dependent eikonal, Eq. (5.13). Eigenmodes of the full 2D Eq. (5.12) with a time-dependence $\sim e^{\gamma t}$ must still exist, but the conventional ballooning transformation is not a useful approach to identify them. Let us now explore how to build the eigenmodes and see if we can reduce the eigenmode equation to a single ordinary differential equation.

Our starting point is Eq. (5.12) and, as we are seeking eigenmodes, we assume a time-dependence $\sim e^{\gamma t}$ and replace $\partial/\partial t \to \gamma$. This then avoids the time-derivatives that we found in Eq. (5.16), but then how do we deal with the remaining terms that involve the large factor, n? To proceed, we introduce a Fourier transform, expressed in a convenient form:

$$\xi_\psi(r,\theta) = \int_{-\infty}^{\infty} F(k,\theta)e^{-inq_0\theta}\exp\left[-in\left(q'x(\theta-k)+Y(\theta,r)-\phi\right)\right]dk \quad (5.17)$$

where $q_0 = q(r_s)$. This special form of Fourier transform is constructed so that the argument of the exponential does not vary along a field line (the geometric function Y is related to the local magnetic shear; its precise form is not important). This captures the physics of the long wavelength along field lines, but short wavelength across them. Unlike the ballooning transform, this Fourier representation does not guarantee that ξ_ψ is periodic in θ. Examining Eq. (5.17), it is straightforward to show that the periodicity condition requires

$$F(k+2\pi,\theta+2\pi) = F(k,\theta) \quad (5.18)$$

while the condition that the Fourier transform exists requires

$$\lim_{|k|\to\infty} F(k,\theta) = 0\,. \tag{5.19}$$

Substituting this Fourier transform representation into Eq. (5.12) and writing $\eta = \theta - k$, we find (in the limit of low flow shear):

$$\frac{d}{d\eta}\left[\left(1+P^2\right)\frac{dF}{d\eta}\right] + \alpha\left[\cos(\eta+k) + P\sin(\eta+k)\right]F$$
$$= \frac{1}{\omega_A^2}\left(\dot{\Omega}\frac{\partial}{\partial k} - \gamma\right)\left[\left(1+P^2\right)\left(\dot{\Omega}\frac{\partial}{\partial k} - \gamma\right)F\right] \tag{5.20}$$

where now we have $P = s\eta - \alpha\sin(\eta+k)$. The boundary conditions (following from Eqs. (5.18) and (5.19) are that F is periodic in k at fixed η, and decays to zero at large $|\eta|$. Clearly the problem is still 2D, even though we have removed time from the system. Remarkably, the eigenmode equation that we need to solve is exactly the same as the Floquet mode equation (5.16) with τ and k interchanged! The Floquet and eigenmode approaches are equivalent, yielding the same growth rate γ (for the Floquet approach, γ is interpreted as a time-averaged growth rate, as the function $\hat{F}(\hat{\eta}, [\dot{\Omega}t])$ is periodic in time and thus contributes zero to the time-averaged growth rate).

Our analysis so far suggests that in the presence of flow shear, the ballooning mode equations are inherently 2D, and a reduction into a sequence of 1D equations (as for the case with no flow) is not possible. It is appropriate to ask whether the system reduces to 1D when the flow is very small. Let us consider this limit, and adopt the Floquet form of the ballooning mode equation (5.14). A convenient form of the equation is derived by transforming to the new function, G, where

$$F = \frac{G}{\sqrt{1+P^2}}\,. \tag{5.21}$$

The resulting equation for G can then be expressed in the form of a modified wave equation:

$$\frac{\partial^2 G}{\partial \eta^2} + V(\eta,t)G = \frac{1}{\omega_A^2}\frac{\partial^2 G}{\partial t^2} \tag{5.22}$$

where the "potential", $V(\eta,t)$ is defined as:

$$V(\eta,t) = \alpha\frac{\cos\eta + P\sin\eta}{1+P^2} - \frac{1}{\sqrt{1+P^2}}\frac{\partial}{\partial\eta}\left[\frac{P}{\sqrt{1+P^2}}\frac{\partial P}{\partial \eta}\right] + \frac{s^2}{(1+P^2)^2} \tag{5.23}$$

with P given by Eq. (5.15). The important feature of the potential is that it decays algebraically at large η. Thus the large η limit of Eq. (5.22) is the standard wave equation, and therefore has solutions:

$$\lim_{|\eta|\to\infty} G = G_+(|\eta| + \omega_A t) + G_-(|\eta| - \omega_A t)$$

representing "incoming" (ie from large $|\eta|$ towards $\eta = 0$) and "'outgoing" waves, respectively. We are interested in instabilities, in which case the solution will increase in time; on the other hand we require that G be bounded in η, and therefore must decay at large $|\eta|$. Thus, we have the boundary condition that at large $|\eta|$ we must match to the outgoing wave solution, $G_-(|\eta| - \omega_A t)$.

Now let us return to consider the form of the solution to Eq. (5.22) for $|\eta| = O(1)$, where we cannot neglect $V(\eta, t)$. We note that there are two timescales in this region: one associated with the growth rate $\sim \omega_A$, and one associated with the sheared flow $\dot{\Omega}$. We distinguish between these by defining a long time-scale variable, $\tau = \dot{\Omega}t$, noting then that the potential V depends on τ and not on the faster growth rate time-scale, labelled by t. Equation (5.22) can then be written in the form:

$$\frac{\partial^2 G}{\partial \eta^2} + V(\eta, \tau)G = \frac{1}{\omega_A^2}\left[\frac{\partial}{\partial t} + \dot{\Omega}\frac{\partial}{\partial \tau}\right]^2 G\,. \tag{5.24}$$

To solve Eq. (5.24), we seek a leading order separable solution of the form $G(\eta, \tau, t) = A(t, \tau)H(\eta, \tau)$. To leading order in $\dot{\Omega}$, we then find:

$$\frac{\omega_A^2}{H}\left[\frac{\partial^2 H}{\partial \eta^2} + V(\eta, \tau)H\right] = \frac{1}{A}\frac{\partial^2 A}{\partial t^2} = \gamma_0^2(\tau) \tag{5.25}$$

where we have used the fact that both sides of the equation must balance for all t and η, and therefore each side must be a function of τ only; we have denoted this function by $\gamma_0^2(\tau)$. As our notation suggests, γ_0 can be interpreted as the instantaneous growth rate as the poloidal position of the ballooning mode rotates with time, as demonstrated in Fig. 5.6. To see this, consider first the equation for $H(\eta, \tau)$:

$$\frac{\partial^2 H}{\partial \eta^2} + V(\eta, \tau)H = \frac{\gamma_0^2(\tau)}{\omega_A^2}H\,. \tag{5.26}$$

This is identical to Eq. (5.7), with $f = (1 + P^2)^{-1/2}H$ and $\eta_0 \to \tau$. Recall that η_0 has the interpretation of the poloidal angle where the ballooning mode has maximum amplitude; as time (τ) progresses then γ_0 adjusts to correspond to the growth rate one would obtain for a stationary plasma with the ballooning mode positioned at $\eta_0 = \tau$. As in the case with no flow, γ_0 is determined as an eigenvalue, applying appropriate boundary conditions at large $|\eta|$, which we address shortly.

Let us now consider the equation for A:

$$\frac{\partial^2 A}{\partial t^2} = \gamma_0^2 A\,. \tag{5.27}$$

Assuming a positive growth rate, the solution for A is

$$A = \exp\left(\int_0^t \gamma_0(\dot{\Omega}t')dt'\right) = \exp\left(\frac{1}{\dot{\Omega}}\int_0^\tau \gamma_0(\tau')d\tau'\right). \tag{5.28}$$

This then yields our final result for the leading order solution for G in the low flow limit:

$$G(\eta,t) = \exp\left(\int_0^t \gamma_0\left(\dot{\Omega}t'\right)dt'\right)H(\eta,\dot{\Omega}t)\,. \tag{5.29}$$

We now return to the solution for H, and address the boundary conditions at large $|\eta|$. Noting that $V \to 0$ in this limit, Eq. (5.26) has exponential solutions $\sim \exp(\pm\gamma_0\eta/\omega_A)$. We must choose the solution that matches to an outgoing wave as $|\eta| \to \infty$, but nevertheless $\dot{\Omega}|\eta|/\omega_A \ll 1$. In this limit, the solution to Eq. (5.29) can be written in the form

$$G(\eta,t) = \exp\left(\int_0^t \gamma_0\left(\dot{\Omega}t'\right)dt'\right)\left[a_-\exp\left(-\frac{\gamma_0\eta}{\omega_A}\right) + a_+\exp\left(\frac{\gamma_0\eta}{\omega_A}\right)\right]$$

which, in turn, can be written as

$$\begin{aligned} G(\eta,t) = \lim_{\dot{\Omega}|\eta|/\omega_A\to 0}\Bigg[a_-\exp\left(\frac{1}{\dot{\Omega}}\int_0^{\tau-\dot{\Omega}|\eta|/\omega_A}\gamma_0\left(\tau'\right)d\tau'\right) \\ + a_+\exp\left(\frac{1}{\dot{\Omega}}\int_0^{\tau+\dot{\Omega}|\eta|/\omega_A}\gamma_0\left(\tau'\right)d\tau'\right)\Bigg]. \end{aligned}$$

Only the outgoing wave solution is allowed, and so we must set $a_+ = 0$. This corresponds to selecting the exponentially decaying solutions for $H(\eta,\tau)$ as $|\eta| \to \infty$, which in turn provides the necessary boundary conditions to determine the eigenvalue $\gamma_0(\tau)$ from Eq. (5.26). This now completely determines our solution (at least in the situation when γ_0^2 is positive for all τ, an issue which we shall return to below). Indeed, we see that we have achieved our goal: we have reduced the 2D partial differential equation to two 1D equations (5.26) and (5.27), which are much more readily solved. In particular, we find that one of those equations is the standard ballooning equation that exists in the absence of flow shear. A complication is that now we must solve this equation for all η_0 (i.e. τ), rather than just the value of η_0 which maximises the growth rate.

Though challenging, it is possible to solve the full 2D Eq. (5.14) numerically [13]. This becomes increasingly difficult at low flow shear because of the need to resolve the two disparate time-scales ($\tau = \dot{\Omega}t$ and t) and in practice one cannot explore extremely low flow shear. For our numerical study, we select parameters $s = 0.75$, $\alpha = 2.0$. This choice ensures that the eigenvalue of Eq. (5.26), $\gamma_0^2(\tau)$, is positive for all τ; the significance of this will be discussed shortly. Rather than solve Eq. (5.14) as an initial value problem, we instead choose to solve Eq. (5.16) as a Floquet-eigenvalue problem. Thus, we impose the boundary condition that $\hat{F}$ is periodic in τ, which then provides the actual growth rate, γ, as an eigenvalue. The local growth rate is derived from the 2-D solution by evaluating $\gamma_0 = (1/\hat{F})(\partial\hat{F})/\partial t)$ at $\eta = 0$ for $0 \leq \tau \leq 2\pi$. This provides the full curves shown in Fig. 5.8(a) for γ_0 as a function of τ for a range of normalised flow shear, $s_v = \dot{\Omega}/\omega_A$. The dashed curves show the results for the eigenvalue of the 1-D equation for $H(\eta, \tau)$, from Eq. (5.26). Note the increasingly good agreement as s_v is reduced. Figure 5.8(b) shows how the full growth rate, γ, varies with flow shear, calculated using the full 2D solution. Recall that γ has two interpretations: it can either be considered as the Floquet eigenvalue in the time-dependent eikonal approach, or it can be interpreted as the growth rate of the eigenmode. Flow shear is found to be stabilising, with a linear dependence on s_v. For the chosen values of s and α, we find that a good linear fit to the numerical solutions is provided by $\gamma = 0.283 - 0.35s_v$. In the limit $s_v = 0$, we can deduce γ from our 1-D solutions for $\gamma_0(\tau)$; in particular $\gamma = (1/2\pi)\oint \gamma_0(\tau)d\tau = 0.282$. This is in excellent agreement with the zero flow-shear limit of the linear fit to the 2D solution, which yields $\gamma = 0.283$.

We close this section with a discussion of the regime of validity of our technique to reduce the 2D system describing the effect of flow on ballooning modes to a 1D system in the limit of low flow shear. Recall that our approach required us to be able to separate two timescales: one associated with the instability growth rate, which we called t, and one associated with the sheared flow, which we called τ. This separation is only possible provided $\gamma_0/\dot{\Omega} \gg 1$ for all τ. This therefore requires γ_0^2 should not go through zero as τ is varied. There is actually quite a narrow range of s and α where this holds for all τ, so the result shown in Fig. 5.8 is actually quite a special case. More generally γ_0^2 will pass through a zero as τ is varied, and the

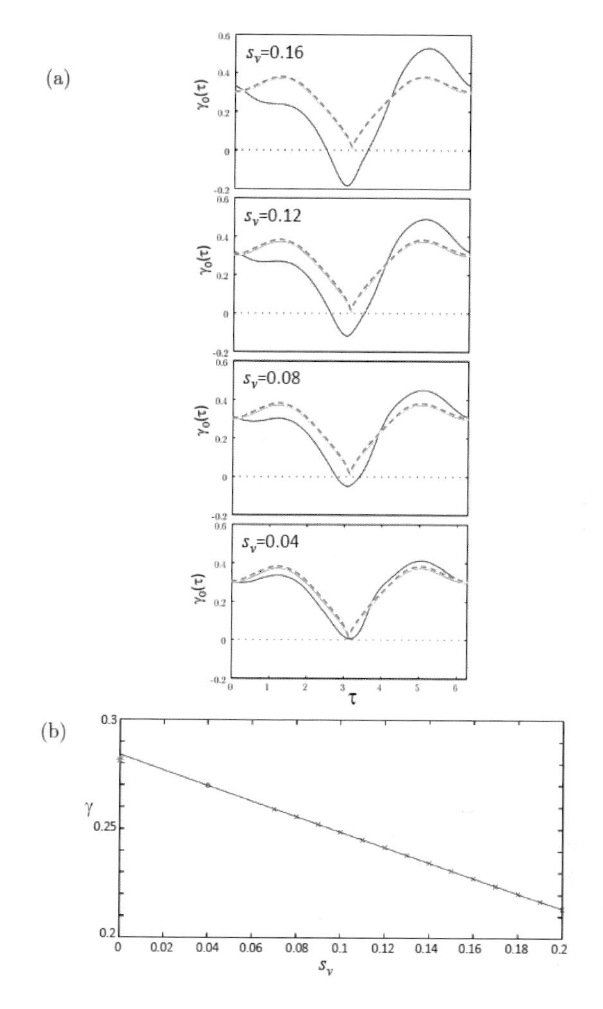

Fig. 5.8 (a) The local growth rate, γ_0 (normalised to ω_A here) calculated from the full 2D solution of Eq. (5.16) (full curves) compared with the 1D solution provided by the eigenvalue of Eq. (5.26) (dashed curves) for four values of flow shear, $s_v = \dot{\Omega}/\omega_A$. This equilibrium has $s = 0.75$, $\alpha = 2.0$. (b) The Floquet growth rate, calculated as the eigenvalue of Eq. (5.16) as a function of flow shear. The line is a linear fit to the 2D numerical solutions (crosses) described by the equation $\gamma = 0.28 - 0.35 s_v$; the star at $s_v = 0$ is the average of γ_0 over τ, derived from the 1D results.

separation into two timescales is not a valid procedure in the vicinity of $\gamma_0^2 = 0$. Another interesting effect of γ_0^2 passing through zero is that there will be a region of τ where $\gamma_0^2 < 0$. Equation (5.26) then has a continuum of eigenvalues: the so-called stable continuum of ideal MHD. How the matching to the wave solution resolves this continuum is a rather complicated process that we leave to future work.

5.6 Conclusions

When there is no flow shear in a tokamak plasma, we have shown how one can exploit the symmetry that exists between different rational surfaces at large toroidal mode number n to reduce the two-dimensional system to a series of one-dimensional equations. The requirement for periodicity in poloidal angle is a complication that can be addressed by implementing the ballooning transform. This enables the periodicity boundary condition to be replaced by one that ensures the ballooning transform exists (i.e. that the eigenfunction decays sufficiently fast in the ballooning variable, η). This allows one to describe the fastest spatial variations through an eikonal, and hence reduce the system to a single second order differential equation in η: the ballooning equation.

The introduction of toroidal flow shear causes the rational surfaces to rotate relative to each other, so that they are no longer equivalent. We then find that the system is two-dimensional in general. We showed how one can still transform into ballooning space, but now one must introduce a time-dependent eikonal to describe the fastest spatial variations. This introduces a time-dependence into the coefficients of the equation, so that the solutions are not eigenmodes of the original 2D system. Actually the system remains 2D, simply translating from radial coordinate and poloidal angle to ballooning angle and time. The resulting solutions are Floquet modes. If one attempts to seek eigenmodes with a time dependence $\sim e^{\gamma t}$, then one finds that the fast radial variation associated with the flow shear cannot be eliminated by a simple WKB form. We showed how one can introduce a Fourier transform, mapping x to k, providing a 2D partial differential equation in poloidal angle and radial wavenumber, k. The eigenmode and Floquet approaches are shown to be equivalent, with the Floquet eigenvalue found to be identical to the growth rate of the eigenmode.

In the zero flow shear limit (for $n \to \infty$), we demonstrated how one can reduce the 2D system to a pair of 1D equations, using the Floquet form as an example. We showed that an arbitrarily small flow shear has a stabilising influence compared to the zero flow case. Specifically, we showed that in the absence of flow shear, the actual growth rate of the 2D system is equal to the maximum of the 1D growth rate as η_0 is varied. With flow shear, the growth rate of the 2D system is the average of the 1D growth rate over η_0 ($0 \leq \eta_0 \leq 2\pi$; $\eta_0 = \tau$ maps the poloidal location where the mode amplitude is a maximum as time varies through τ). We provided a physical picture of why we expect this result in Fig. 5.6. Our analysis fails if the 1D growth rate becomes less than the sheared flow, $\dot{\Omega}$ for any value of η_0, or if any of the range of η_0 provides a stable 1D system. Interpreting the influence of the stable continuum of the 1D system on the 2D growth rate is then rather complicated and subtle, and left for future work.

Acknowledgments

The authors gratefully acknowledge many extremely fruitful discussions with J. B. Taylor in preparing this work.

References

[1] J. W. Connor, R. J. Hastie and J. B. Taylor, *Proc. R. Soc. London Ser. A* **365** (1979) 1.
[2] A. H. Glasser in *Finite Beta Theory, Proceedings of the Workshop*, Varenna, 1977, edited by B. Coppi and W. L. Sadowski (U.S. Dept. of Energy, Washington D.C.) (1977) CONF-7709167, p. 55.
[3] R. L. Dewar and A. H. Glasser, *Phys. Fluids* **26** (1983) 3038.
[4] Y. C. Lee and J. W. Van Dam in *Finite Beta Theory, Proceedings of the Workshop*, Varenna, 1977, edited by B. Coppi and W. L. Sadowski (U.S. Dept. of Energy, Washington D.C.) (1977) CONF-7709167, p. 93.
[5] D. Lortz, *Nucl. Fusion* **15** (1975) 49.
[6] C. C. Hegna, J. W. Connor, R. J. Hastie and H. R. Wilson, *Phys. Plasmas* **3** (1996) 584.
[7] J. W. Connor, R. J. Hastie, H. R. Wilson and R. L. Miler, *Phys. Plasmas* **5** (1998) 2687.
[8] H. R. Wilson, P. B. Snyder, G. T. A. Huysmans and R. L. Miller, *Phys. Plasmas* **9** (2002) 1277.
[9] P. B. Snyder et al., *Phys. Plasmas* **9** (2002) 2037.
[10] P. B. Snyder et al., *Nucl. Fusion* **47** (2007) 961.
[11] W. A. Cooper, *Plasma Phys. Control. Fusion* **30** (1988) 1805.
[12] F. L. Waelbroeck and L. Chen, *Phys. Plasmas B* **3** (1991) 601.
[13] R. L. Miller, F. L. Waelbroeck, A. B. Hassam and R. E. Waltz, *Phys. Plasmas* **2** (1995) 3676.
[14] J. B. Taylor and H. R. Wilson, *Plasma Phys. Control. Fusion* **38** (1996) 1999.

Chapter 6

Color Figures

Fig. 6.1 Figure 1.1 from **Chapter 1**, *The atmospheric wave–turbulence jigsaw*. From the laboratory study of Sommeria, J. and Meyers, S. D. and Swinney, H. L. 1989, *Laboratory model of a planetary eastward jet* in Nature, **337**, page 58. Courtesy Dr Joël Sommeria.

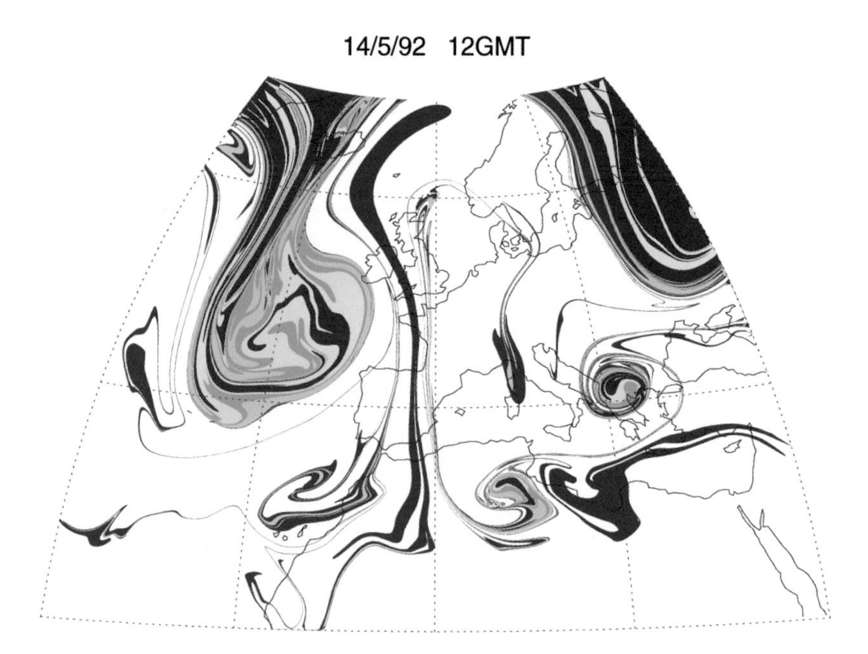

Fig. 6.2 Figure 1.4 from **Chapter 1**, *The atmospheric wave–turbulence jigsaw*. Estimated map of Q, the exact PV defined by Eq. (1.3), on a stratification surface near 10 km altitude. From Appenzeller, C. and Davies, H. C. 1992 in *Structure of stratospheric intrusions into the troposphere*, Nature **358**, page 570. The computation assumes that material invariance of Q is a good approximation over a 4-day time interval, and uses a state-of-the-art advection algorithm and weather-forecasting data to trace the flow of high-Q stratospheric air (coloured) and low-Q tropospheric air (clear). The main boundary between stratospheric and tropospheric air marks a jet core showing large-amplitude meandering, from Greenland toward Spain and then back to northern Norway. The leakage of stratospheric air into the troposphere signals intermittent attrition of the eddy-transport barrier at the jet core. The different chemical signatures of the stratospheric and tropospheric air are easily detectable and have been demonstrated in measurement campaigns, even for fine filamentary structures like those shown by Waugh, D. W. and Plumb, R. A. 1994 in *Contour advection with surgery: a technique for investigating finescale structure in tracer transport*, J. Atmos. Sci., **51**, page 530. The high-Q anomaly over the Balkans illustrates the "cutoff" or self-wrapping-up process that occurs when sufficiently large masses of stratospheric air overcome the barrier. The wrapping-up produces structures like that in Fig. 1.3.

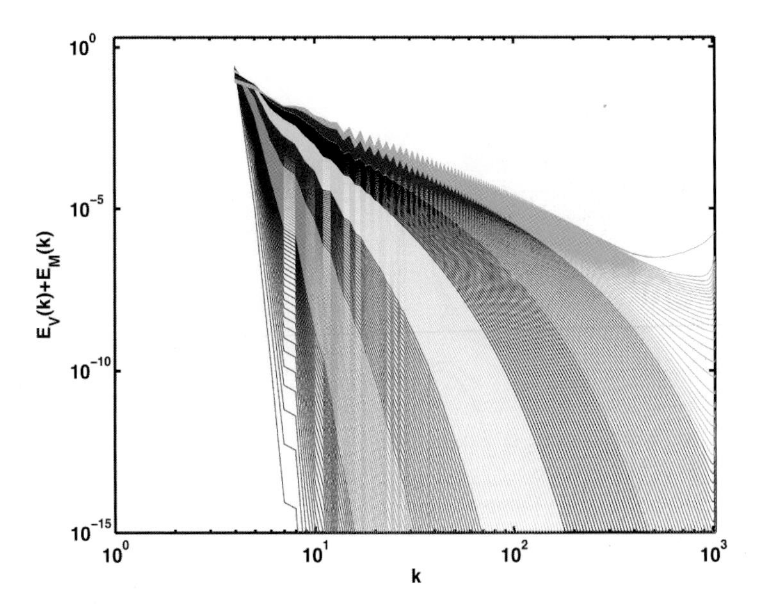

Fig. 6.3 Figure 2.1 from **Chapter 2**, *A review of the possible role of constraints in MHD turbulence.* Temporal evolution of the total energy spectrum $E_T(k,t)$ for the same flow as in Figs. 2.1 and 2.2, in logarithmic coordinates. Colors shift from blue to green, red, yellow, black, purple and cyan with increments of roughly $\Delta T = 0.4$. The first spectrum is at $t = 0$, and the last one is at $t = 2.815$, time at which pile-up of energy in the smallest scale is clearly visible; it would lead at later times to an energy spectrum $\sim k^2$, corresponding to the non-helical ideal MHD.

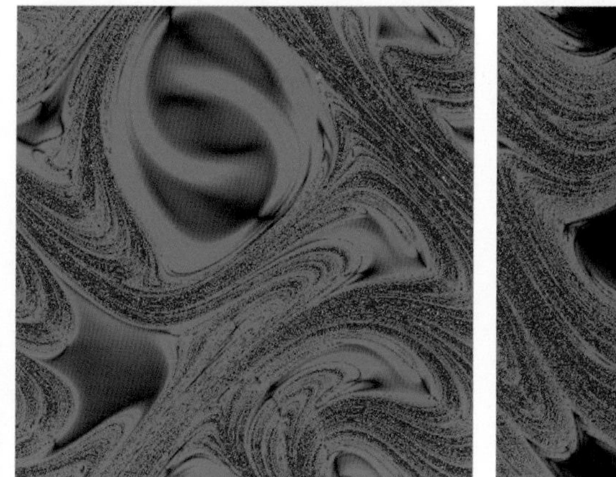

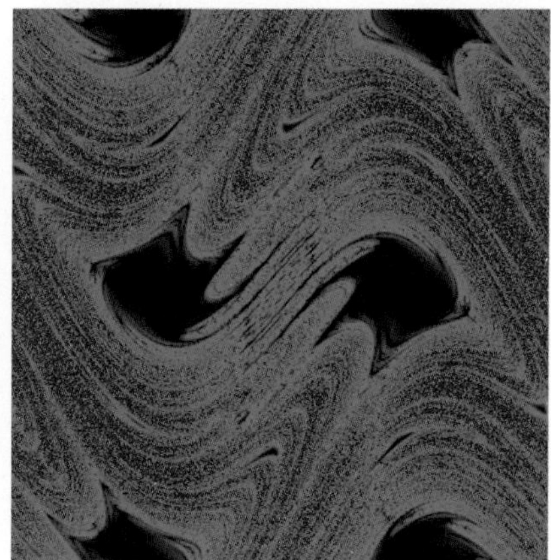

Fig. 6.4 Figure 4.4 from **Chapter 4**, *Fast Dynamos.* Finite-time Lyapunov exponents (over an interval of 25 time units) for flows (4.29) (left) and (4.30) (right). Yellow and red regions are chaotic (large exponential stretching), green and blue regions have little or no stretching.

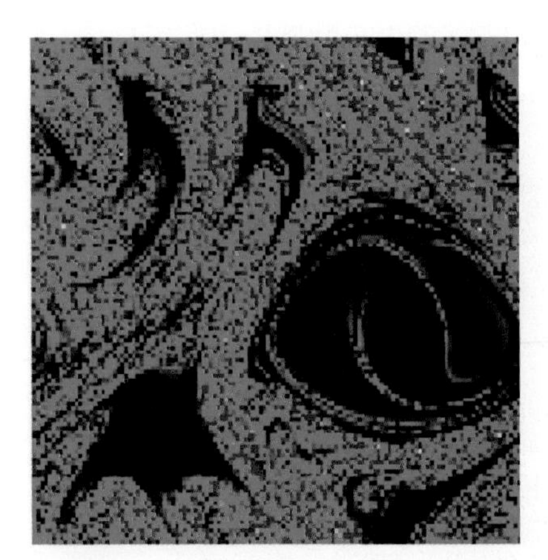

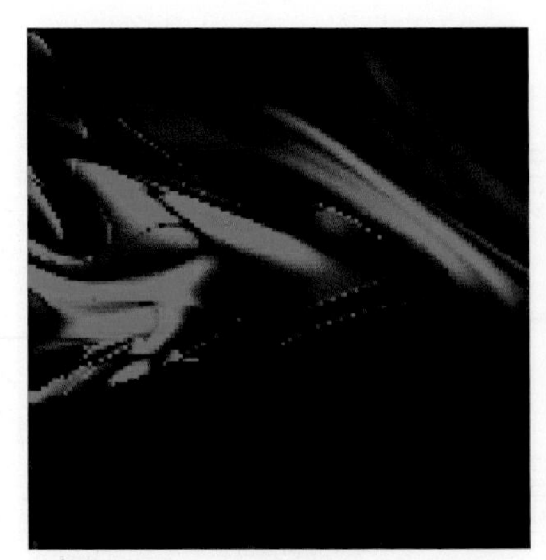

Fig. 6.5 Figure 4.6 from **Chapter 4**, *Fast Dynamos*. Spatial distribution of finite-time Lyapunov exponents in the kinematic (left) and saturated (right) regimes for a forced flow, where the forcing is chosen so as to drive flow (4.29) in the absence of magnetic field. The colours code the values of the exponents as a function of the initial positions. Yellow and red correspond to strongly stretching trajectories; green and blue to trajectories with little or no stretching. Regions of chaotic motion, which occupy a substantial fraction of the domain in the kinematic regime, are almost completely absent in the later dynamical phases.

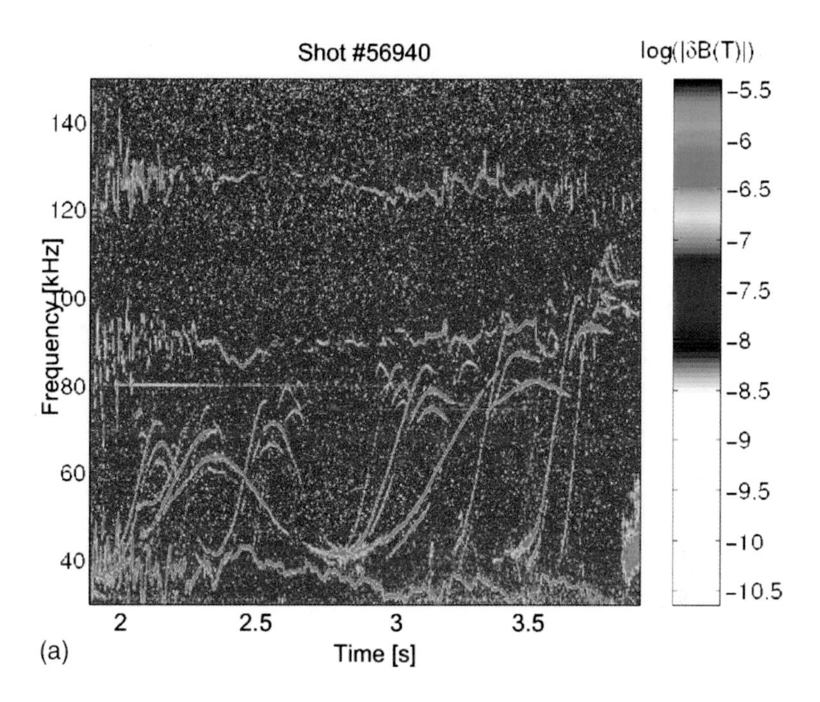

Fig. 6.6 Figure 8.1 from **Chapter 8**, *The general fishbone like dispersion relation.* Alfvén Cascades.

Chapter 7

Elements of Neoclassical Theory and Plasma Rotation in a Tokamak

A. Smolyakov

Department of Physics & Engineering Physics, University of Saskatchewan, 116 Science Place, Saskatoon, SK S7N 5E2, Canada

7.1 Introduction

Plasma rotation is the fundamental part of tokamak physics. It is critically important for many phenomena at the plasma edge and plays a central role in improved confinement regimes. Many phenomena such as error field penetration, locked modes, external Resonant Magnetic Perturbations control etc. are strongly affected by plasma rotation. Despite the importance and many years of research, plasma rotation in a tokamak is not fully understood.

Plasma rotation is related to the global electric field and pressure gradients via the force (momentum) balance. At lowest order, it is the radial electric field that is important, however, toroidal effects also generate electric field variations in poloidal (or even toroidal, for non-axisymmetric cases) directions. Since the electric field is fundamentally the result of charge separation, the problem of the determination of the electric field becomes a transport problem, in particular, transport of charged particles. The electric field generated by charge separation induces electric forces on charged plasma species. Thus, the problem of the electric field is coupled to the force (momentum) balance and momentum transport. In magnetized plasmas, the momentum balance also involves viscous forces which, in turn, are related to the gradients of plasma flow velocities and gradients of heat fluxes. As a result, the determination of the electric field becomes a problem of coupled dynamics of charges (density), momentum, and energy balance. This coupling is a cornerstone of transport (neoclassical) theory in a tokamak. Plasma turbulence provide additional important contributions and brings in new phenomena. However, the framework of neoclassical theory remains valid. Here, we attempt to overview basic elements of neoclassical transport theory in a tokamak with emphasis on plasma rotation and electric field dynamics. The presentation aims to present key physical mechanisms affecting electric field and plasma rotation and outline logical steps rather than provide detailed derivations. An auxiliary material is given in the Appendices.

7.1.1 *Quasineutrality condition*

Finding an electric field from a given distribution of electric charges is a standard problem of basic electromagnetism. Generally, the Poisson equation is to be used. For a single ion species plasma, it has the form

$$\nabla \cdot \mathbf{E} = -\nabla^2 \phi = \rho/\varepsilon_0 = e(n_i - n_e)/\varepsilon_0 \,, \tag{7.1}$$

where $E = -\nabla\phi$ is the electric field, and n_i, n_e are the ion and electron densities. In typical plasmas, however, the total number of charged particles is very large and, as a result, the difference between the electron and ion densities is very small. The charge separation occurs at a small distance of the order of the Debye length λ_D, where $\lambda_D^2 = \varepsilon_0 T/\left(n_0 e^2\right)$, resulting in quasineutrality

$$n_i = n_e \,, \tag{7.2}$$

which is valid on scales larger than the Debye length, $L > \lambda_D$. Therefore, it becomes very impractical to solve Eq. (7.1) for the electric field: "... one should never use Poisson's equation to obtain E unless it is unavoidable [8]."

The electric field determines the motion and resulting distribution of charged particles via the momentum balance (equations of motion) and continuity equations. Therefore, the standard procedure becomes to determine the electric current and charge density from the equations of motion in terms of the electric field and then find the electric field from quasineutrality condition (7.2). This condition is often used in a different, but equivalent form. From Maxwell equations, it follows

$$\nabla \cdot (\nabla \times \mathbf{B}) = 0 = \mu_0 \nabla \cdot \mathbf{J} + \varepsilon_0^{-1} \left(\partial/\partial t\right) \nabla \cdot \mathbf{E} \,. \tag{7.3}$$

Together with (7.1), this becomes the charge conservation equation

$$\nabla \cdot \mathbf{J} + (\varepsilon_0 \mu_0)^{-1} (\partial/\partial t)\, \rho = 0 \,. \tag{7.4}$$

The last term is related to the displacement current that can be neglected for low frequencies. Therefore, the quasineutrality condition is often written as

$$\nabla \cdot \mathbf{J} = 0 \,, \tag{7.5}$$

and the problem of the electric field is reduced to finding the particle velocities (and respectively the electric current) in terms of the electric field from the equations of motion and then resolving Eq. (7.5) for the electric field.

7.1.1.1 *Ambipolar diffusion*

The above arguments can be most simply illustrated for a case of ambipolar diffusion in a weakly ionized plasma, where the electron and ion densities are small compared to the density of neutrals, so the electron-neutral and ion-neutral collisions are dominant. Then, from standard equations of motion

$$m_\alpha n_\alpha \frac{d_\alpha \mathbf{V}_\alpha}{dt} = e_\alpha n_\alpha \mathbf{E} - \nabla p_\alpha - m_\alpha n_\alpha \nu_\alpha \mathbf{V}_\alpha \,, \tag{7.6}$$

one finds the particle density flow

$$\mathbf{\Gamma}_\alpha \equiv n_\alpha \mathbf{V}_\alpha = \frac{e_\alpha}{m_\alpha \nu_\alpha} \mathbf{E} - \frac{T_\alpha}{m_\alpha \nu_\alpha} \nabla n_\alpha \,, \tag{7.7}$$

where the first term describes the mobility and the second corresponds to the diffusion. The constant temperature case $\nabla T_\alpha = 0$ and steady state, $d_\alpha/dt = 0$, was assumed, where $d_\alpha/dt = \partial/\partial t + \mathbf{V}_\alpha \cdot \nabla$ is fluid (convective) derivative.

The diffusion coefficient in Eq. (7.7), $D_\alpha = T_\alpha/m_\alpha \nu_\alpha = v_T^2/\nu$, is consistent with the random walk estimate of Brownian motion diffusion $D = (\Delta r)^2/\tau$, where Δr is the random displacement and $\tau = 1/\nu$ is the characteristic time between collisions, the random displacement $\Delta r = v_T/\nu$, and characteristic particle velocity $v_T = \sqrt{T/m}$.

In general, the mobility and diffusion coefficients for electrons and ions are different, thus producing different fluxes and net electric current. The quasineutrality condition (7.5) imposes a constraint on the current. Equation (7.5) is often replaced by a stronger condition $\Gamma_e = \Gamma_i$, which defines the ambipolar electric field as

$$\mathbf{E} = \mathbf{E}_a \equiv \frac{D_i - D_e}{b_i + b_e} \,, \tag{7.8}$$

where $b_\alpha = q_\alpha / m_\alpha \nu_\alpha$ is the mobility coefficient, $\alpha = (e, i)$. Thus, the ambipolar electric field is established to equalize the (otherwise different) fluid velocities of charged plasma particles. The resulting value of the fluid velocity is expressed in the form

$$\mathbf{\Gamma}_e = \mathbf{\Gamma}_i = -D_a \nabla n\,, \tag{7.9}$$

where $D_a = (D_e b_i + D_i b_e) / (b_e + b_i)$ is the ambipolar diffusion coefficient.

7.1.2 *Diffusion in fully ionized magnetized plasma and automatic ambipolarity*

The picture of the ambipolar diffusion becomes different in fully ionized magnetized plasmas, when the friction forces are due to mutual electron-ion collisions. An estimate for the diffusion coefficient across the magnetic field in this limit can be made by taking the Larmor radius as a characteristic displacement, $\Delta r \simeq \rho$. We assume that the magnetic field is sufficiently strong so that the cyclotron frequency is larger than the collision frequency, $\omega_{c\alpha} \gg \nu_\alpha$. Then, the random walk formula gives for the diffusion coefficient

$$D_\perp = (\Delta r)^2 / \tau = \rho^2 \nu = \nu v_T^2 / \omega_c^2\,. \tag{7.10}$$

Thus, the transverse diffusion becomes suppressed by strong magnetic field, $D_\perp / D = \nu v_T^2 / \omega_c^2 / \left(v_T^2 / \nu\right) = \nu^2 / \omega_c^2 \ll 1$. Another important effect of the equilibrium magnetic field is a different role of the electric field in the diffusion process. Consider the steady state equations of motion for electrons and ions

$$e_a n_a \left(\mathbf{E} + \frac{1}{c}\mathbf{V}_\alpha \times \mathbf{B}\right) - \nabla p_a - m_a n_a \nu_{a\beta} \left(\mathbf{V}_a - \mathbf{V}_\beta\right) = 0\,, \tag{7.11}$$

where $\alpha, \beta = (i, e)$ denote plasma components. In fully ionized plasmas, the friction force $\mathbf{R}_{\alpha\beta} = -m_a n_a \nu_{a\beta} \left(\mathbf{V}_a - \mathbf{V}_\beta\right)$ is exclusively due to the electron-ion collisions so that $\mathbf{R}_{ei} = -\mathbf{R}_{ie}$. For simplicity we assume a uniform temperature so that the thermal force does not appear in Eq. (7.11). Equation (7.11) can formally be solved for each component

$$\mathbf{V} = \mathbf{V}_{E\times B} + \mathbf{V}_p + \mathbf{V}_R\,, \tag{7.12}$$

where $\mathbf{V}_{E\times B} = c\mathbf{E} \times \mathbf{B}/B^2$ is the electric drift velocity, $\mathbf{V}_p = (mn\omega_c)^{-1}\, \mathbf{b} \times \nabla p$ is a diamagnetic drift, and $\mathbf{V}_R = c\,(en)^{-1}\, \mathbf{R} \times \mathbf{b}/B$ is a friction force drift; $\mathbf{b} = \mathbf{B}/B$ is a unit vector along the equilibrium magnetic field. In the limit of strong magnetic field, $\omega_{ci} \gg \nu_{ie}$, Eq. (7.12) can be written explicitly giving

$$\mathbf{V}_e = c\frac{\mathbf{E}\times\mathbf{B}}{B^2} - \frac{1}{m_e n \omega_{ce}}\mathbf{b}\times\nabla p_e - \frac{c^2 m_e \nu_{ei}\left(T_e + T_i\right)}{e^2 B^2}\frac{\nabla n}{n}, \tag{7.13}$$

and

$$\mathbf{V}_i = c\frac{\mathbf{E}\times\mathbf{B}}{B^2} + \frac{1}{m_i n \omega_{ci}}\mathbf{b}\times\nabla p_i - \frac{c^2 m_e \nu_{ei}\left(T_e + T_i\right)}{e^2 B^2}\frac{\nabla n}{n}, \tag{7.14}$$

where $\omega_{ce} = eB/m_e c > 0$.

A notable feature of this equation, first commented upon by Tamm [70], is that the collisional diffusion, given by the last term in Eqs. (7.13) and (7.14), is independent of the electric field and is identical for electrons and ions (note that $m_i \nu_{ie} = m_e \nu_{ei}$). This feature is often referred to as an automatic ambipolarity of the diffusion in the magnetized plasma. The automatic ambipolarity (independence of the electric field) is often expressed as a statement that the ambipolarity condition $\Gamma_e = \Gamma_i$ does not produce any additional constraints on plasma flow, i.e. conditions additional to those already obtained from the equations of motion.

Though the electric field does not affect the diffusion, it does not mean that the electric field can be fully arbitrary. Equations (7.13) and (7.14) are consequences of the stationary momentum balance equations (7.11). The same stationary momentum balance equations impose some constraints on the electric field and plasma velocity. The actual form of these additional conditions depend on the symmetry of the problem. As an example, consider axially and azimuthally symmetric plasma cylinder immersed in the axial uniform magnetic field $\mathbf{B}_0 = B_0\widehat{\mathbf{z}}$ and assume that the plasma density is a function of radius, $n = n\left(r\right)$ and the only component of the electric field is in the radial direction, $E_r = -\partial\phi/\partial r$. In this case, the radial momentum balance from Eq. (7.11), or equivalently from (7.13) and (7.14), defines the relations between the radial electric field and the poloidal electron and ion velocities

$$E_r + \frac{1}{c}V_{\theta e}B_0 + \frac{T_e}{en}\frac{\partial n}{\partial r} = 0, \tag{7.15}$$

$$E_r + \frac{1}{c}V_{\theta i}B_0 - \frac{T_i}{en}\frac{\partial n}{\partial r} = 0. \tag{7.16}$$

The radial components of the ion and electron velocities from (7.13) and (7.14) are equal

$$V_{ri} = V_{re} = -\frac{c^2 m_e \nu_{ei}\left(T_e + T_i\right)}{e^2 B^2}\frac{\nabla n}{n}. \tag{7.17}$$

Thus, in the geometry of axially and azimuthally symmetric plasma cylinder in the axial magnetic field, the ambipolarity and momentum balance constraints do not determine the radial electric field (and poloidal velocities) in the main order of the classical collisional diffusion, i.e. to the order of $1/B^2$ terms. While there exist relations (7.15) and (7.16) between the radial electric field and poloidal velocity, independently they remain undetermined because of automatic ambipolarity of plasma diffusion in the main order. As a matter of fact, the axial velocity also remains a free parameter in presence of the axial symmetry, so there are two free variables that are undetermined in cylindrical geometry. As we will see in Section 7.3, the situation is different for axisymmetric toroidal geometry which has only one symmetry direction (toroidal). To determine plasma velocity in the direction of symmetry requires higher order (in $1/B$) effects, which correspond to plasma viscosity. These higher order terms affect the momentum transport and will be considered in the next sections.

7.2 Toroidal geometry and neoclassical diffusion

Magnetic field in the axisymmetric tokamak can be described by simple (quasi-) toroidal coordinates r, θ, ζ, where r is the minor radius, θ and ζ are poloidal and toroidal angles [35]. The poloidal magnetic field $B_\theta\left(r\right)$ is characterized by the safety factor parameter $q\left(r\right) \equiv r/R\left(B_\zeta/B_\theta\left(r\right)\right), r$ is the minor radius, and R is the major radius, and B_ζ constant is the amplitude of the toroidal magnetic field at the magnetic axis, $r = 0$. In general, the poloidal magnetic field (or equivalently $q = q\left(r\right)$) profile is found as a solution of Grad-Shafranov equation for the magnetic equilibrium [57]. In the low pressure plasma, the magnetic surfaces can be assumed concentric and magnetic field is represented in the form

$$\mathbf{B} = \frac{B_\zeta \mathbf{e}_\zeta + B_\theta \mathbf{e}_\theta}{1 + \varepsilon \cos\theta}, \tag{7.18}$$

where $\epsilon = r/R$ is the inverse aspect ratio, $\mathbf{e}_\theta$ and $\mathbf{e}_\zeta$ are poloidal and toroidal unit vectors.

As soon as the concept of the tokamak was suggested [70], the question was posed [2] whether the collisional diffusion in the inhomogeneous magnetic field might be higher that in the case of a uniform magnetic field. Subsequent series of works in USSR and USA revealed a new mechanism of particle transport due to effects of particles drift in inhomogeneous magnetic field (for a historical overview of the development of transport theory in toroidal systems see [37]). It was shown that due to magnetic drifts, the diffusion in toroidal magnetic field is significantly larger than classical collisional transport given by (7.10). Respectively, the transport theory in toroidal geometry was called neoclassical theory. The phenomenon of trapped particles plays a central role in neoclassical theory; see Appendix 7.10 for a brief description of particle trapping and characteristic parameters of trapped particles in a tokamak.

In this Section, we will give intuitive estimates for neoclassical diffusion coefficients. Classical references on neoclassical theory [17, 25, 32] can be consulted for technical details. A recent book [24] provides an accessible treatment of most derivations of neoclassical theory. Another review paper [54] devoted to the problem of neoclassical rotation, emphasizes effects of biasing, strong rotation, and turbulence.

The elementary estimates for neoclassical transport coefficients can be obtained by using the random walk arguments and particle trajectories characteristics as given in Appendix 7.10. In the low collisionality (banana) regime, the main contribution is due to banana particles. The number density of such particles is $n_{tr} = \sqrt{\varepsilon} n_0$, the characteristic particle displacement (banana width is $\Delta = q\rho/\sqrt{\varepsilon}$ (7.97), and the characteristic banana life-time is $\tau_{eff} = (\nu/\varepsilon)^{-1}$ (7.101). Then, the random walk diffusion coefficient in the banana regime is

$$D_B = \sqrt{\varepsilon}\frac{\Delta^2}{\tau_{eff}} = \frac{\nu q^2 \rho^2}{\varepsilon^{3/2}} \,. \tag{7.19}$$

The neoclassical diffusion in the banana regime is a factor of $q^2/\varepsilon^{3/2}$ larger than the classical diffusion.

In the plateau regime, most of thermal particles are collisionless and passing. The characteristic particle displacement in this regime is $q\rho$ (7.98), and the characteristic time is $\tau = qR/v_T$. Then, the plateau diffusion coefficient is

$$D_P = \frac{q^2\rho^2}{\tau} = q\rho^2\frac{v_T}{R} \,. \tag{7.20}$$

Note that the banana (7.19) and plateau (7.20) diffusion coefficients smoothly match at the banana-plateau boundary $\nu^* = 1$.

In highly collisional, Pfirsch-Schlüter (PS) regime, the particles experience collisions before completing the full connection length path along the magnetic field. The resulting particle motion along the magnetic field is diffusive with a characteristic diffusion coefficient $D_\parallel = v_T^2/\nu$, and the characteristic time scale over the connection length $\tau_\parallel = q^2R^2/D_\parallel$. Note, that the transverse magnetic drift is still dominant since we assume that $\omega_c > \nu$. In toroidal geometry, the direction of the magnetic drift v_D is always in vertical direction. As a result, the direction of the particle displacement with respect to the magnetic surface (in radial direction) alternates as particle diffuses between the inner and outer regions of the torus. The characteristic amplitude of the radial displacement is $v_D \tau_\parallel$. This process is random so the net effective diffusion in radial direction is

$$D_{PS} = \frac{\left(v_D \tau_\parallel\right)^2}{\tau_\parallel} = \nu \rho^2 q^2 \,. \tag{7.21}$$

This is the Pfirsch-Schlüter (PS) diffusion coefficient, which is applicable for $\nu > v_T/qR$. Note, that at the boundary of the plateau and Pfirsch-Schlüter regimes $\nu = v_T/qR$ (or $\nu^* = \varepsilon^{-3/2}$), the plateau (7.20) and PS (7.21) coefficients are smoothly matched. In all three collisionality regimes, neoclassical diffusion greatly exceeds the classical diffusion (7.10) (as least by order of magnitude).

7.3 Diffusion and ambipolarity in toroidal plasmas

The role of the electric field in neoclassical transport can be illustrated from a perspective of transport fluxes and neoclassical transport coefficients reviewed in Section 7.2. It is easy to see that in all regimes, the ion transport coefficients are much larger than those of electrons, because of the condition $\rho_i \gg \rho_e$. Such difference is acceptable for heat fluxes, but not for fluxes of charged particles. Different fluxes of the ions and electrons would create radial current and plasma polarization. The resulting buildup of the electric field acts to compensate the difference in radial ion and electron fluxes so that $J_r = en\left(V_{ri} - V_{re}\right) = 0$, which provides the relation for the radial electric field. Since $D_i \gg D_e$, the condition $J_r = 0$ means that most of the ion flux is set to zero and the resulting (ambipolar) neoclassical particle flux is of the order of that for electrons with the D_e diffusion coefficient. Such ambipolar constraint does not exist for heat fluxes so that the ion heat flux exceeds that for electrons.

The above arguments can be followed quantitatively by using the expressions for particle fluxes obtained in [16, 66]

$$\Gamma_\alpha = -nD_\alpha \left(\frac{1}{n}\frac{dn}{dr} + \gamma_\alpha \frac{1}{T}\frac{dT}{dr} - \frac{e_\alpha}{T_\alpha}\left(E_r - B_\theta V_{\alpha\parallel}\right)\right), \tag{7.22}$$

where $\alpha = (e, i)$, and

$$D_\alpha = \frac{\sqrt{\pi} q}{4} \frac{v_{T\alpha}^3}{R\omega_{c\alpha}^2}, \tag{7.23}$$

in the plateau regime, and

$$D_\alpha = 0.73 \frac{\nu_\alpha q^2 \rho_\alpha^2}{\varepsilon^{3/2}}, \tag{7.24}$$

in the banana regime. The factor γ_α depends on the collisionality; in the plateau regime $\gamma_{i,e} = 3/2$, and in the banana regimes $\gamma_e = -0.17$, and $\gamma_i = -0.37$ [25].

When $D_i \gg D_e$, the ambipolarity condition approximately means $J_r \simeq J_{ri} = 0$ and the radial electric field is

$$E_r - B_\theta V_{i\parallel} = \frac{T_i}{e} \left(\frac{1}{n} \frac{dn}{dr} + \gamma_i \frac{1}{T_i} \frac{dT_i}{dr} \right). \tag{7.25}$$

Thus, this relation between the radial electric field and ion parallel flow velocity is required for the ambipolarity. Then, the ambipolar radial particle flux, obtained from (7.22) for electrons, is

$$\Gamma_e = \Gamma_i = -nD_e \left(\frac{1}{n} \left(1 + \frac{T_i}{T_e} \right) \frac{dn}{dr} + \gamma_e \frac{1}{T_e} \frac{dT_e}{dr} + \gamma_i \frac{1}{T_i} \frac{dT_i}{dr} \right). \tag{7.26}$$

In (7.26), the terms leading to Ware pinch, due to the difference of the ion and electron velocities in parallel direction, were omitted. Note that the toroidal velocity here remains undetermined. Alternatively, one can say that the ambipolarity condition defines the combination of the electric field and parallel velocity in Eq. (7.25), but not E_r and $V_{i\parallel}$ separately. Though in (7.25), E_r and $V_{i\parallel}$ are both unknown, they are not "equivalent" unknown quantities. We will see later in Sections 7.5 and 7.8 that the characteristic evolution times for E_r and $V_{i\parallel}$ are different. Because of the toroidal symmetry, the evolution of $V_{i\parallel}$ is much slower. Its determination requires that the processes responsible for toroidal momentum balance/transport be included. In standard neoclassical (tokamak) theory such processes appear when the higher order terms in the ion Larmor radius are accounted for.

7.4 Ambipolarity and equilibrium poloidal rotation

The ambipolarity condition (7.25) can also be obtained directly from the steady state momentum balance equations. In fact, the condition (7.25) defines the equilibrium state of plasma (ion) poloidal rotation. Thus, imposing the ambipolarity condition $J_{re} + J_{ri} = 0$ actually does not provide new information compared to what is available from the steady-state momentum balance. This stationary condition can be found from the parallel projections of the momentum balance equations for ions and electrons

$$\mathbf{B} \cdot m_e n \frac{d\mathbf{V}_e}{dt} = -en\mathbf{B} \cdot \mathbf{E} - \mathbf{B} \cdot \nabla p_e - \mathbf{B} \cdot \nabla \cdot \mathbf{\Pi}_e + \mathbf{B} \cdot \mathbf{R}_{ei} \,, \tag{7.27}$$

$$\mathbf{B} \cdot m_i n \frac{d\mathbf{V}_i}{dt} = en\mathbf{B} \cdot \mathbf{E} - \mathbf{B} \cdot \nabla p_i - \mathbf{B} \cdot \nabla \cdot \mathbf{\Pi}_i + \mathbf{B} \cdot \mathbf{R}_{ie} \,. \tag{7.28}$$

Summing up these equations and neglecting the inertial terms on the left hand side (for steady state), one obtains for the magnetic surface averaged equations

$$\langle \mathbf{B} \cdot \nabla \cdot \mathbf{\Pi}_i \rangle + \langle \mathbf{B} \cdot \nabla \cdot \mathbf{\Pi}_e \rangle = 0 \,. \tag{7.29}$$

The pressure gradient terms are eliminated by the magnetic surface averaging. In the main order (the order considered in standard neoclassical theory), the main contribution to (7.29) comes from the parallel viscosity defined in the Appendix 7.12, Eq. (7.128). Since the ion parallel viscosity is the factor of $(m_i/m_e)^{1/2}$ larger than that for the electrons, Eq. (7.29) approximately means $\langle \mathbf{B} \cdot \nabla \cdot \mathbf{\Pi}_i \rangle = 0$. From the expression for the ion parallel viscosity (7.152), one obtains that the ion poloidal velocity in stationary state is determined by the ion temperature gradient [23]

$$V_\theta = k \frac{c}{eB} \frac{\partial T_i}{\partial r} \,. \tag{7.30}$$

On the other hand, the ion momentum balance gives the following general expression for the poloidal velocity

$$V_\theta = \frac{E_r B_\zeta}{B^2} + V_\parallel \frac{B_\theta}{B} + \frac{1}{mn\omega_c} \frac{B_\zeta}{B} \frac{\partial p_i}{\partial r} \,, \tag{7.31}$$

which provides a link between the poloidal and toroidal velocities, radial electric field, and pressure gradient. Alternatively, this equation can also be viewed as a consequence of the stationary ion momentum balance in the radial direction

$$E_r + \frac{1}{c} \left(V_\theta B_\zeta - V_\zeta B_\theta \right) - \frac{e}{n} \frac{\partial p_i}{\partial r} = 0 \,. \tag{7.32}$$

Note that the total ion velocity is $\mathbf{V} = V_\parallel \mathbf{b} + \mathbf{V}_\perp$, and that only the velocity $\mathbf{V}_\perp$ perpendicular to the magnetic field enters the momentum balance. Combining (7.30) and (7.31) one obtains (7.25), where $\gamma_i = 1 + k$. Thus, ambipolarity condition (7.25) is equivalent to the condition of full damping of poloidal velocity to the stationary value determined by (7.30).

7.5 Ambipolarity paradox and damping of poloidal rotation

In Section 7.3, the ambipolar value of radial electric field was found from the ambipolarity condition $\Gamma_e = \Gamma_i$. On the other hand, it is easy to see that in steady-state, the ambipolarity directly follows from the momentum conservation properties of the collision operator [39, 55, 70]. Indeed, from the toroidal projection of the steady state momentum balance, one finds the following equations for the radial electron and ion currents

$$\langle \mathbf{e}_\zeta \cdot \mathbf{J}_{re} \times \mathbf{B} \rangle + \langle \mathbf{e}_\zeta \cdot \mathbf{R}_{ei} \rangle = 0\,, \tag{7.33}$$

$$\langle \mathbf{e}_\zeta \cdot \mathbf{J}_{ri} \times \mathbf{B} \rangle + \langle \mathbf{e}_\zeta \cdot \mathbf{R}_{ie} \rangle = 0\,. \tag{7.34}$$

The toroidal projections of the electric field, pressure and parallel viscosity forces disappear after averaging over the magnetic surface. From the momentum conservation during binary particles collisions, $R_{ei} + R_{ie} = 0$, one immediately gets the condition that the diffusion is automatically ambipolar, $J_{re} + J_{ri} = 0$, and that it does not depend on the electric field. These two superficially different results: the independent from the electric field automatic ambipolarity condition, which follows from (7.33) and (7.34), and the electric field that is determined from the condition $\Gamma_e = \Gamma_i$ in Section 7.3, were, at some point, a subject of the so called ambipolarity paradox [30]. It was clarified in [30], where it was shown that automatic ambipolarity occurs only in the steady state, when the poloidal rotation velocity satisfies the condition (7.29). Arbitrary value of the poloidal rotation that does not satisfy (7.30), and respectively, arbitrary values of the electric field and toroidal rotation that do not satisfy (7.25), will generate non-ambipolar radial current that will damp rotation to the value defined by (7.30). On time scales shorter than the poloidal rotation damping time (before the equilibrium rotation (7.30) is established) the radial diffusion is not ambipolar.

The equation for poloidal rotation damping that was derived in [30] can be most easily formulated from the quasineutrality condition (7.5), which can be written in the form

$$\mathbf{B} \cdot \nabla \left(\frac{J_\parallel}{B} \right) + \nabla_\perp \cdot \mathbf{J}_I + \nabla_\perp \cdot \mathbf{J}_\pi + \nabla_\perp \cdot \mathbf{J}_p = 0\,. \tag{7.35}$$

The parallel current does not give any contribution to surface average equation so it will dropped below. The perpendicular electric current, which is found from the momentum balance equation, consists of several contributions. The ion inertial (polarization) current $\mathbf{J}_I$ is

$$\mathbf{J}_I = \frac{cm_i n_0}{B}\mathbf{b} \times \frac{d}{dt}\mathbf{V}_E \simeq -\frac{cm_i n_0}{B}\widehat{\mathbf{e}}_r \frac{\partial}{\partial t}\widetilde{V}_{E\theta} \,. \tag{7.36}$$

Here, the $\mathbf{V} \cdot \nabla$ nonlinear corrections in the fluid derivative are neglected. In what follows, the tilde is used to denote the perturbed components. The neoclassical inertia current $\mathbf{J}_\pi$ is the current due to the neoclassical (parallel) viscosity $\mathbf{\Pi}$,

$$\mathbf{J}_\pi = \frac{c}{B}\mathbf{b} \times \boldsymbol{\nabla} \cdot \mathbf{\Pi} \,, \tag{7.37}$$

$$\mathbf{\Pi} = \frac{3}{2}\pi_\parallel \left(\mathbf{bb} - \frac{1}{3}\mathbf{I}\right) , \tag{7.38}$$

and $\mathbf{J}_p$ is the diamagnetic current

$$\mathbf{J}_p = \frac{c}{B}\mathbf{b} \times \nabla p \,. \tag{7.39}$$

The following identities are useful for the parallel viscosity gradients

$$\nabla \cdot \mathbf{\Pi} = \frac{3}{2}\pi_\parallel \left(\mathbf{b}\nabla \cdot \mathbf{b} + \mathbf{b} \cdot \nabla \mathbf{b}\right) + \frac{3}{2}\mathbf{b}(\mathbf{b} \cdot \nabla \pi_\parallel) - \frac{1}{2}\nabla \pi_\parallel \,, \tag{7.40}$$

$$\mathbf{B} \times \nabla \cdot \mathbf{\Pi} = \frac{3}{2}\pi_\parallel \mathbf{B} \times (\mathbf{b} \cdot \nabla)\mathbf{b} - \frac{1}{2}\mathbf{B} \times \nabla \pi_\parallel \,, \tag{7.41}$$

where $\pi_\parallel = 2(p_\parallel - p_\perp)/3$. By using these identities one can reduce the neoclassical viscous current contribution to the following form

$$\nabla \cdot \mathbf{J}_\pi \equiv \nabla \cdot \left(\frac{c}{B^2}\mathbf{B} \times \nabla \cdot \mathbf{\Pi}\right) = \frac{3}{2}\frac{c}{B^2}\nabla \pi_\parallel \cdot \mathbf{B} \times \nabla \ln B - \frac{c}{2}\nabla \cdot \left(\frac{1}{B^2}\mathbf{B} \times \nabla \pi_\parallel\right) ,$$

where we have used relations $(\mathbf{b} \cdot \nabla)\,\mathbf{b} = \nabla_\perp \ln B$ and $\nabla \times \mathbf{B} \simeq 0$ that are valid for low pressure plasma. After flux surface averaging, the parallel viscosity and diamagnetic contributions combine into the form

$$\left\langle \nabla \cdot \widetilde{\mathbf{J}}_\pi + \nabla \cdot \widetilde{\mathbf{J}}_p \right\rangle = -\frac{c}{2B_0}\left\langle \frac{\partial(4\widetilde{p} + \widetilde{\pi}_\parallel)}{\partial r}\frac{1}{r}\frac{\partial}{\partial \theta}\ln B \right\rangle . \tag{7.42}$$

It is worth noting here that the equilibrium radial current is zero. In the perturbed state, the finite values of divergence of $\mathbf{J}_\pi$, $\mathbf{J}_p$ come from the perturbed parts of the pressure and parallel viscosity $\widetilde{p}$ and $\widetilde{\pi}_\parallel$. Part of this current can be expressed by using the momentum balance

$$m_i n_0 \frac{\partial \widetilde{V}_\parallel}{\partial t} = -\frac{1}{qR}\frac{\partial}{\partial \theta}\left(\widetilde{p} + \widetilde{\pi}_\parallel\right) . \tag{7.43}$$

In the low frequency approximation, the toroidal oscillation of parallel velocity can be found from the incompressibility condition $\nabla \cdot \mathbf{V} = 0$,

$$\nabla \cdot \widetilde{\mathbf{V}}_E + \frac{1}{qR}\frac{\partial}{\partial \theta}\widetilde{V}_{\parallel} = 0\,, \tag{7.44}$$

which gives $\widetilde{V}_{\parallel} = -2q\widetilde{V}_{E\theta}\cos\theta$. Using this and (7.43) in (7.42), one obtains

$$\left\langle \nabla \cdot \widetilde{\mathbf{J}}_{\pi} + \nabla \cdot \widetilde{\mathbf{J}}_{p} \right\rangle = \frac{3c}{2B_0}\left\langle \frac{\partial \widetilde{\pi}_{\parallel}}{\partial r}\frac{1}{r}\frac{\partial}{\partial \theta}\ln B \right\rangle - \frac{2m_i n_0 q^2}{B_0}\frac{\partial}{\partial t}\frac{\partial}{\partial r}\widetilde{V}_{E\theta}\,. \tag{7.45}$$

The complete quasineutrality equation (7.35) takes the form

$$-\frac{m_i n_0}{B_0}\left(1+2q^2\right)\frac{\partial}{\partial t}\frac{\partial}{\partial r}\widetilde{V}_{E\theta} + \frac{3}{2B_0}\left\langle \frac{\partial \widetilde{\pi}_{\parallel}}{\partial r}\frac{1}{r}\frac{\partial}{\partial \theta}\ln B \right\rangle = 0\,. \tag{7.46}$$

It turns out that similar combination due the parallel viscosity also occurs in the parallel momentum balance. Indeed, the parallel component of the viscous force is

$$\mathbf{b}\cdot\nabla\cdot\mathbf{\Pi} = \frac{3}{2}\pi_{\parallel}\nabla\cdot\mathbf{b} + \mathbf{b}\cdot\nabla\pi_{\parallel} = -\frac{3}{2}\pi_{\parallel}\mathbf{b}\cdot\nabla\ln B + \mathbf{b}\cdot\nabla\pi_{\parallel}\,, \tag{7.47}$$

which produces the surface average force in the from

$$\left\langle \mathbf{b}\cdot\nabla\cdot\widetilde{\mathbf{\Pi}} \right\rangle = -\frac{3}{2}\left\langle \widetilde{\pi}_{\parallel}\frac{B_\theta}{B_0}\frac{\partial}{r\partial\theta}\ln B \right\rangle\,. \tag{7.48}$$

Equations (7.45) and (7.48) finally give the equation for damping of the poloidal velocity (and hence for the radial electric field evolution) [30]:

$$m_i n_0\left(1+2q^2\right)\frac{\partial}{\partial t}\frac{\partial}{\partial r}\widetilde{V}_{E\theta} = -\frac{B_0}{B_\theta}\frac{\partial}{\partial r}\left\langle \mathbf{b}\cdot\nabla\cdot\widetilde{\mathbf{\Pi}} \right\rangle\,. \tag{7.49}$$

Using the closures for the parallel viscosity from Section 7.13, one can get an estimate for the characteristic time scale of the poloidal rotation damping, e.g. in the plateau regimes, one has

$$\tau_\theta \simeq \left(\frac{v_{Ti}}{qR}\right)^{-1}\left(2+\frac{1}{q^2}\right)\,. \tag{7.50}$$

In summary, the neoclassical diffusion in a tokamak becomes ambipolar after the poloidal rotation is damped to the stationary value given by (7.30). In this state, the steady state conditions (7.33) and (7.34) are achieved, and the argument for automatic ambipolarity due to the momentum conservation by the collision operator becomes valid. On times scales shorter that the poloidal rotation damping time, the neoclassical diffusion is not automatically ambipolar and there exists an equation which defines the evolution of the radial electric field, i.e. Eq. (7.49).

7.6 Neoclassical plasma inertia

Equation (7.49), together with poloidal viscosity closures (7.146), are able to provide only qualitative estimates for the poloidal rotation damping time. It was noted already in [30] that in banana regime, Eqs. (7.49) and (7.154) produce a result that violates the initial assumption of $\tau_\theta < \nu_i^{-1}$. The problem is that the expressions (7.146) in Section 7.13 were derived for static cases, when $\partial/\partial t = 0$, and, thus, do not take into account a number of dynamical processes. Some insight on the nature of these processes can be obtained by noting that the right hand side of the Eq. (7.49) also involves the mean parallel velocity $V_\parallel$, which has its own dynamics. In general, evolutions of poloidal and parallel velocities in toroidal geometry are coupled, e.g. via Eq. (7.31). In presence of poloidal flow damping, the evolution of the poloidal velocity is constrained, e.g. full damping of V_θ would mean that the relation (7.25) remains valid at all times, even when the radial electric field evolves. This constraint exists on a (typically short) time scale of the poloidal flow damping (7.50). Thus, on the longer time scale, the evolution of the radial electric field induces changes in toroidal rather than poloidal velocity, which stays constant. This leads to an increased, or neoclassical plasma inertia (neoclassical polarization) effect [27]. Fuller description of this effect, in weakly collisional regimes, $\nu_i < \omega$, requires more accurate model for parallel viscosity, in particular, the time dependent viscosity [20, 59].

Coupling of poloidal and toroidal flows in tokamak geometry and resulting neoclassical enhancement of plasma inertia can be illustrated with a simple model. The neoclassical inertia is characterized by the contribution of the parallel viscosity

$$\left\langle \nabla \cdot \widetilde{\mathbf{J}} \right\rangle_{neo} = \frac{3c}{2B_0} \left\langle \frac{\partial \widetilde{\pi}_\parallel}{\partial r} \frac{1}{r} \frac{\partial}{\partial \theta} \ln B \right\rangle . \tag{7.51}$$

Consider the averaged neoclassical viscous force in the form (cf. with (7.146))

$$\left\langle \mathbf{B} \cdot \nabla \cdot \widetilde{\mathbf{\Pi}} \right\rangle = -\frac{3}{2} \frac{\varepsilon}{q} \left\langle \widetilde{\pi}_\parallel \frac{1}{r} \frac{\partial}{\partial \theta} \ln B \right\rangle_{\widehat{\varsigma}} = \frac{\varepsilon}{q} m_i n_0 \chi_\theta \widetilde{V}_\theta . \tag{7.52}$$

Here, for simplicity, we neglect the ion temperature gradient. We consider the banana regime when $\nu_i < \varepsilon^{3/2} v_{ti}/qR$, where ν_i is the ion-ion collision frequency. Several collisionality regimes can be distinguished in this case depending on the value of the ion-ion collision frequency ν_i compared to the characteristic frequency of the island evolution/rotation ω. We use here a simplified model when the neoclassical viscosity coefficient can be taken in the form [59]

$$\chi_\theta = \frac{q^2}{\varepsilon^{1/2}}\left(\frac{d_0}{dt} + \frac{\nu_i}{\varepsilon}\right), \tag{7.53}$$

where the numerical coefficients of the order of unity are omitted. The model expression (7.53) applies both in collisional, $\nu_i > \omega\varepsilon$, and in collisionless, $\omega > \nu_i/\varepsilon$, regimes. The poloidal component of the plasma flow velocity from (7.31) is

$$V_\theta = V_E + \frac{\varepsilon}{q} V_\parallel \,. \tag{7.54}$$

In this equation, V_E is the velocity perpendicular to the magnetic field

$$V_E \simeq \frac{c}{B_0}\frac{\partial \phi}{\partial r}, \tag{7.55}$$

and $V_\parallel$ is the velocity parallel to the magnetic field.

From (7.52) and the parallel component of the momentum balance, one can obtain the following equation for the parallel flow velocity

$$\frac{d_0}{dt} V_\parallel = -q\varepsilon^{1/2}\left(\frac{d_0}{dt} + \frac{\nu_i}{\varepsilon}\right)\left(V_E + \frac{\varepsilon}{q} V_\parallel\right). \tag{7.56}$$

This equation can be easily solved in two limiting cases. In the limit of large collisional frequency , $\nu_i/\varepsilon \gg d_0/dt$, one has

$$V_\parallel^{(0)} = -\frac{q}{\varepsilon} V_E \,, \tag{7.57}$$

and the neoclassical inertia becomes [4]

$$\langle \nabla \cdot \mathbf{J} \rangle_{neo} = \frac{c}{B_0}\frac{q}{\varepsilon}\frac{\partial}{\partial r}\frac{d}{dt} m_i n_0 V_\parallel = -\frac{c^2 m_i n_0}{B_0^2}\frac{q^2}{\varepsilon^2}\frac{d_0}{dt}\frac{\partial^2}{\partial r^2}\phi \,. \tag{7.58}$$

The contribution of the neoclassical viscosity is additional to the inertial term (7.36), which provides the following contribution to the quasineutrality equation (7.35)

$$\nabla \cdot \mathbf{J}_I = -\frac{c^2 m_i n_0}{B_0^2}\frac{d_0}{dt}\nabla_\perp^2 \phi \simeq -\frac{c^2 m_i n_0}{B_0^2}\frac{d_0}{dt}\frac{\partial^2}{\partial r^2}\phi \,, \tag{7.59}$$

for $\partial^2/\partial r^2 \gg r^{-2}\partial^2/\partial^2\theta$. In the limit of large collisionality, $\nu_i/\varepsilon \gg d_0/dt$, the neoclassical inertia term (7.58) is a factor of q^2/ε^2 larger than the standard inertial (polarization) term (7.59).

In the low collisional regime, $\nu_i/\varepsilon \ll d_0/dt$, the equation for the parallel flow velocity takes the form

$$\frac{d_0}{dt}V_{\parallel}^{(0)} = -q\varepsilon^{1/2}\frac{d_0}{dt}V_E\,, \tag{7.60}$$

so that

$$V_{\parallel}^{(0)} = -q\varepsilon^{1/2}V_E\,. \tag{7.61}$$

Respectively one obtains

$$\langle\nabla\cdot\mathbf{J}\rangle_{neo} = \frac{c}{B_0}\frac{q}{\varepsilon}\frac{\partial}{\partial r}\frac{d}{dt}m_i n_0 V_{\parallel} = -\frac{c^2 m_i n_0}{B_0^2}\frac{q^2}{\varepsilon^{1/2}}\frac{d_0}{dt}\frac{\partial^2}{\partial r^2}\phi\,, \tag{7.62}$$

Thus, in the low collisional regime, $\nu_i/\varepsilon \ll d_0/dt$, the neoclassical inertia is larger than the standard inertia (polarization) by a factor of $q^2/\sqrt{\varepsilon}$.

The neoclassical enhancement of inertia modifies the poloidal rotation damping. It also affects the dynamics of small scale fluctuations, e.g. magnetic islands [64]. Damping of poloidal rotation taking into account neoclassical inertia effects was analyzed in [34, 59, 69]. The damping is usually obtained in the exponential form, e.g. as Eqs. (7.49) and (7.50). However, it was also shown that, in general case, the damping is sensitive to the initial form of the distribution function for particles close to trapped-passing boundary and the universal expression for the decay rate appears only in the long time asymptotic regime [28, 45].

7.7 Oscillatory modes of poloidal plasma rotation

We have noted already that the expressions for viscosity coefficients in Section 7.13, and to a certain degree, the expressions used in Section 7.6, are only valid in static ($\partial/\partial t = 0$), or sufficiently low frequency limit, i.e. at low rotation frequencies. More accurate analysis of the time dependent plasma response reveals the existence of different regimes of poloidal plasma rotation, namely, the existence of oscillatory regimes of poloidal rotation with a finite frequency. Such modes were discovered long time ago and called Geodesic Acoustic Modes (GAM) [72]. However it is only recently that GAMs have attracted great deal of interest both experimentally [12, 15, 41, 42] and in theory [14, 22, 46, 56]. These rotational modes are coupled to drift-waves and can be nonlinearly driven by Reynolds stress from small-scale fluctuations, and presumably play an important role in regulation of drift wave turbulence and transport.

The basic physics of GAMs consists in the balance of the radial current due to plasma inertia (the second term in Eq. (7.35)) against the radial current due to diamagnetic and parallel viscosity drifts (the third and fourth terms, respectively, in Eq. (7.35)). Essentially, it is the same dynamical equation (7.49) that controls the poloidal flow damping. The important difference though is that the parallel viscosity has to be calculated in the high frequency regime opposite to the low frequency (incompressible) limit of the Eq. (7.44).

In the compressible (high frequency) limit, the expressions for oscillations of plasma pressure and parallel viscosity can be obtained from moment equations (7.117) and (7.119) given in the Appendices 7.11 and 7.12. Neglecting unimportant for us terms, these equations are reduced to [63]

$$\frac{3}{2}\frac{d\widetilde{p}}{dt} - 5p_{0i}\mathbf{V}_E \cdot \nabla \ln B = 0\,, \tag{7.63}$$

$$\frac{d}{dt}\widetilde{\pi}_{\|} - \frac{2}{3}p_{0i}\mathbf{V}_E \cdot \nabla \ln B = 0\,. \tag{7.64}$$

Consider poloidal plasma rotation due to the axisymmetric perturbation of plasma potential $\widetilde{\phi} = \widetilde{\phi}(r)$; with poloidal and toroidal mode numbers equal zero, $m = n = 0$. Equations (7.63) and (7.64) result in

$$-i\omega\widetilde{p} - \frac{10}{3}p_{0i}\frac{c}{B_0}\frac{\partial\widetilde{\phi}}{\partial r}\frac{\partial}{r\partial\theta}\ln B = 0\,, \tag{7.65}$$

$$-i\omega\widetilde{\pi}_{\|} - \frac{2}{3}p_{0i}\frac{c}{B_0}\frac{\partial\widetilde{\phi}}{\partial r}\frac{\partial}{r\partial\theta}\ln B = 0\,. \tag{7.66}$$

Using these equations in (7.42), together (7.35) and (7.59), one obtains the dispersion equation for the oscillating mode of the radial electric field

$$-7\frac{p_{oi}}{m_i n_0}\left(\frac{\partial}{r\partial\theta}\ln B\right)^2\frac{\partial^2\widetilde{\phi}}{\partial r^2} + \omega^2\frac{\partial^2\widetilde{\phi}}{\partial r^2} = 0\,. \tag{7.67}$$

Taking into account that $B \simeq B_0\left(1 + \varepsilon\cos\theta\right)$, $\partial \ln B/\partial\theta \simeq -\varepsilon\sin\theta$, and thus, the averaged part of $r^{-2}\left(\partial \ln B/\partial r\right)^2 = 1/\left(2R^2\right)$, one obtains the GAM eigenmode frequency for the oscillating mode of the poloidal rotation

$$\omega^2 = \omega_{GAM}^2 \equiv \frac{7}{4}\frac{v_{Ti}^2}{R^2}\,. \tag{7.68}$$

These calculations do not include dissipative part of the non-ambipolar current that is responsible for radial particle transport that was discussed in Section 7.3. The effect of the dissipative ion transport can schematically be added to the quasineutrality equation in the form

$$-\frac{cm_in_0}{B}\frac{\partial}{\partial r}\frac{\partial}{\partial t}V_{E\theta}-\frac{c}{2B_0}\left\langle\frac{\partial(4\widetilde{p}+\widetilde{\pi}_{\|})}{\partial r}\frac{1}{r_s}\frac{\partial}{\partial\theta}\ln B\right\rangle+\frac{\partial}{\partial r}D_i\frac{e}{T_i}E_r=0\,, \tag{7.69}$$

where the last term describes the dissipative part of the radial ion current from Eq. (7.22). The first two terms give the dispersion relation (7.68), and the last term provides a dissipative contribution due to neoclassical particle transport. The latter leads to GAM damping. Physically, this damping is similar to the poloidal flow damping in (7.49).

The last term in (7.69) is not qualitatively correct since it was calculated in the static limit. Indeed, the dissipative current is a part of the parallel viscosity term $\widetilde{\pi}_{\|}$ in (7.69). The ν_{ii} collisional damping can be simply evaluated by using the fluid model for the parallel viscosity in the Appendix 7.12. Taking into account the collision operator (7.132) in Eq. (7.130), the Eq. (7.64) for parallel viscosity is modified

$$\left(-i\omega+\frac{6}{5}\nu_{ii}\right)\widetilde{\pi}_{\|}-\frac{2}{3}p_{0i}\frac{c}{B_0}\frac{\partial\widetilde{\phi}}{\partial r}\frac{\partial}{r\partial\theta}\ln B=0\,. \tag{7.70}$$

Proceeding as in (7.67), one gets the dispersion relation

$$\omega^2-\frac{1}{12}\frac{v_{Ti}^2}{R^2}\frac{\omega}{\omega+6i\nu_{ii}/5}=\frac{5}{3}\frac{v_{Ti}^2}{R^2}\,. \tag{7.71}$$

This equation has three roots: two correspond to GAM with weak damping, and the third is a strongly damped (aperiodic) root, $\omega\sim-i\nu_{ii}$. For weak collisions, $\nu_{ii}\ll\omega$, the oscillating rotation modes have the frequency

$$\omega\simeq\pm\sqrt{\frac{7}{4}}\frac{v_{Ti}}{R}-\frac{2}{70}i\nu_{ii}\,. \tag{7.72}$$

The aperiodic root is $\omega\simeq-8i\nu_{ii}/7$. For strong collisions, $\nu_{ii}\gg(\omega,v_{Ti}/R)$, the rotational modes are

$$\omega\simeq\pm\sqrt{\frac{3}{5}}\frac{v_{Ti}}{R}-\frac{5}{144}i\frac{v_{Ti}^2}{R^2\nu_{ii}}\,, \tag{7.73}$$

and the aperiodic mode is $\omega\simeq-6i\nu_{ii}/5$. It is important to note that the oscillating rotating modes (GAM) "survive" the collisional damping even when ν_{ii} is large, albeit with slightly different frequency, see Eqs. (7.72) and (7.73). For these modes, the damping is maximal at $\nu_{ii}\simeq v_{Ti}/R$, but quantitatively, the damping rate always remains the order of magnitude smaller than the real part of the frequency.

The total GAM damping includes both effects of collisions and collisionless Landau damping. Kinetic calculations were reported in [18, 38, 40, 47, 67, 71, 78] . In general, there are two distinct scales in the damping rate: ν_{ii} and v_{Ti}/qR, however, interaction of various regimes and sensitivity to initial conditions [45] creates several different regimes [28].

In summary, oscillating modes of poloidal rotation (GAM) represent dynamical state away from the neoclassical equilibrium (and not automatically ambipolar). Their damping is related to the dissipative part of the non-ambipolar (ion) radial current. Hence, one can expect that GAM damping may be accompanied by finite particle transport; generally at the level corresponding to the ion non-ambipolar flux with $D \simeq D_i$. Such additional neoclassical transport will also be modulated at the GAM frequency.

Non-ambipolar radial electric field (in the sense that it is not determined by the condition (7.25)) can also be generated by nonlinear Reynolds stresses from small scale fluctuations [14]. In toroidal geometry and presence of poloidal flow damping, nonlinear Reynolds stresses also generate toroidal velocity leading to several different regimes of poloidal and toroidal flow excitation [43]. Relaxation of these flows to stationary values will also be accompanied by additional non-ambipolar diffusion.

7.8 Dynamics of the toroidal momentum

Even within the linear neoclassical theory, evolution of parallel (or toroidal velocity) is one of the most involved part of the whole problem of plasma rotation in a tokamak. There are two basic reasons for this. In general, the higher order expansion (in Larmor radius) is required and, as a result, calculations are more difficult. The second reason is due to the fact that there exist several different (and interacting) mechanisms that affect toroidal rotation.

7.8.1 *Momentum diffusion in strongly collisional, short mean free path regime*

When the poloidal rotation is damped and steady state (7.29) has been established, the toroidal rotation velocity still remains undetermined. The evolution of the toroidal momentum occurs on the slower time scale compared to the poloidal rotation damping. It is instructive to consider the toroidal momentum dynamics from the perspective of contributions of different viscosity components presented in Section 7.12. One can use the toroidal momentum conservation equation, which has the form (after summing up the electron and ion equations):

$$\frac{\partial}{\partial t}\langle m_i n V_\zeta\rangle + m_i\langle n\mathbf{e}_\zeta\cdot(\mathbf{V}\cdot\nabla)\mathbf{V}\rangle = \langle \mathbf{e}_\zeta\cdot\mathbf{J}\times\mathbf{B}\rangle - \langle \mathbf{e}_\zeta\cdot\nabla\cdot\mathbf{\Pi}\rangle + S_\zeta\,. \tag{7.74}$$

Here V_ζ is the toroidal projection of the ion flow velocity, $\langle \mathbf{e}_\zeta\cdot\mathbf{J}\times\mathbf{B}\rangle$ is the Maxwell stress, which describes the exchange of the momentum between plasma and electromagnetic field, $\langle \mathbf{e}_\zeta\cdot(\mathbf{V}\cdot\nabla)\mathbf{V}\rangle$ is the Reynolds stress, and S_ζ is the input of toroidal momentum such as that due to neutral beam heating. The electromagnetic term $\langle \mathbf{e}_\zeta\cdot\mathbf{J}\times\mathbf{B}\rangle$ term is important for external biasing when $J_r\neq 0$ [54], as well as for the interaction with external magnetic fields. Here we only describe possible contributions from the $\langle \mathbf{e}_\zeta\cdot\nabla\cdot\mathbf{\Pi}\rangle$ term.

The toroidal momentum balance equation in the form of (7.74) is convenient because a number of symmetries can be exploited. The pressure contribution disappears from the surface averaged equation (7.74) due to toroidal symmetry. The flux surface average of the toroidal projection of parallel viscosity is also identically zero in an axisymmetric tokamak, $\left\langle \mathbf{e}_\zeta\cdot\nabla\cdot\mathbf{\Pi}_{\|}\right\rangle = 0$, when $\Pi_{\|}$ has a structure as given by (7.128).

The remaining contributions to the toroidal momentum balance (7.74) such as nonstationary term with time evolution $\partial V_\zeta/\partial t$, Reynolds stress, Maxwell stress, higher order viscosity (gyro- andq perpendicular viscosity) are not ambipolar. These are contributions that define the evolution and final state of the toroidal velocity V_ζ. The evolution equation for V_ζ is obtained by imposing the quasineutrality condition $J_r = 0$ (in absence of biasing)

$$\frac{\partial}{\partial t}\langle m_i n V_\zeta\rangle + m_i\langle n\mathbf{e}_\zeta\cdot(\mathbf{V}\cdot\nabla)\mathbf{V}\rangle$$
$$= -\langle \mathbf{e}_\zeta\cdot\nabla\cdot\mathbf{\Pi}_\wedge\rangle - \langle \mathbf{e}_\zeta\cdot\nabla\cdot\mathbf{\Pi}_\perp\rangle + S_\zeta\,. \tag{7.75}$$

From Section 7.12, we have: $\Pi_\wedge$ is the gyroviscosity (7.135), $\Pi_\wedge\sim p(\nabla V)/\omega_c$, and $\Pi_\perp$ is the perpendicular viscosity (7.139), which generally scales as $\Pi_\perp\sim \nu p(\nabla V)/\omega_c^2$, and is the next order term in the collision frequency expansion, $\nu/\omega_c\ll 1$. The gyroviscosity is independent of the collisionality regimes, and $\Pi_\perp$ should be calculated separately depending on the collisionality (banana, plateau, and Pfirsch-Schlüter regimes). Remember though that gyroviscosity may contain the collisional and parallel fluxes contributions as noted in Section 7.12. The actual expressions are cumbersome and not readily available in all regimes. Here we only give a qualitative picture of various contributions.

Conceptually, the collisional Pfirsch-Schlüter regime should have been the simplest case. The momentum diffusion terms is simply obtained from the perpendicular diffusion tensor in (7.145).

$$\langle \mathbf{e}_\zeta \cdot \nabla \cdot \mathbf{\Pi}_\perp \rangle = \frac{\partial}{\partial r} \Pi^{(1)}_{r\zeta} = -\frac{6}{5} \frac{p\nu}{\omega_c^2} \frac{\partial^2 V_\zeta}{\partial r^2} . \qquad (7.76)$$

Note that the diffusion of the angular momentum in Pfirsch-Schlüter regime corresponds to the classical diffusion coefficient $D_\perp = \rho^2 \nu$ (7.10) [11, 29]. This equation does not take into account a number of additional effects. Most important is the temperature gradient, which drives the rotation. When the temperature gradient is included, the following equation was obtained in the collisional Pfirsch-Schlüter regime [74, 76]:

$$\Pi_{r\zeta} = -mn\nu_{ii}\rho_\theta^2 \left[1.2 \frac{\varepsilon^2}{q^2} \frac{\partial}{\partial r} \frac{E_r}{B_\theta} + \frac{0.47\varepsilon^2}{1 + ZT_i/T_e} \frac{T_i}{ZeB_\theta} \left(\frac{\partial \ln T_i}{\partial r} \right)^2 \right] . \qquad (7.77)$$

The first term in this expression identically corresponds to (7.76). Note that some numerical coefficients in Eq. (B9) of [74] were corrected in [76]. Similar equations were derived in [6, 9]. The collisional result (7.77) was obtained as a result of contributions from the perpendicular tensor(7.139) and the gyroviscosity (7.135) tensor. In the latter, the effect of heat fluxes are crucial [6, 9]. The earlier work on the effects of gyroviscosity on the toroidal momentum balance in collisional regimes [51, 65] neglected the heat flux contributions to $\Pi_\wedge$ (the last terms in (7.135)) and, as a result, overestimated the effect of gyroviscosity. The gyroviscosity tensor also provides additional terms [6, 74, 75] , which can be large in up-down asymmetric configurations (such as the one with a single-null divertor).

7.8.2 *Diffusion of toroidal momentum in the weak collision (banana) regime*

In the collisionless (banana) regime earlier calculations of the toroidal momentum balance were performed by using the kinetic theory [52]. The evolution equation was obtained in the form

$$n \frac{\partial V_\zeta}{\partial t} = \frac{1}{r} \frac{\partial}{\partial r} \left[r \chi_i n \frac{\partial}{\partial r} V_\zeta + \frac{3.5q}{\varepsilon} \frac{\partial}{\partial r} r \left(\frac{R}{r} \right)^{1/2} \chi_i \frac{n}{eB_0} \frac{\partial T_i}{\partial r} \right] , \qquad (7.78)$$

where

$$\chi_i = 0.1 \nu_{ii} q^2 \rho_i^2 . \qquad (7.79)$$

Essentially, such ion viscosity coefficient corresponds to the ion perpendicular viscosity (7.139) enhanced in toroidal geometry by q^2 factor due to the magnetic drift induced deviation of particle trajectories from the magnetic surface. Trapped particles are not involved in toroidal momentum transport. As a result, angular momentum diffusion coefficient does not show any trapped particles effects typical for banana diffusion as in (7.19). Similar result for the momentum diffusion coefficient was obtained in [29, 73]. Equation (7.78) shows that the toroidal rotation may be driven by the ion temperature gradient (cf. with 7.30). The analysis in [73] produced the result similar in structure to (7.78) but with different (and opposite in sign) numerical coefficient, −3.7 instead of 3.5 in (7.78). The problem was reconsidered again in [76], where it was shown that self-consistent poloidal variations of the electrostatic potential provide an important contribution to the toroidal momentum balance. Taking these into account, the following equation was derived [76]

$$\frac{\partial}{\partial t} mnV_\zeta + \frac{1}{r}\frac{\partial}{\partial r} r\Pi_{r\zeta} = 0\,, \tag{7.80}$$

where

$$\Pi_{r\zeta} = mn\nu_{ii}\rho_\theta^2 \left[-\frac{\varepsilon^2}{10}\frac{\partial}{\partial r}\frac{E_r}{B_\theta} + \frac{0.15\varepsilon}{1+ZT_i/T_e}\frac{T_i}{ZeB_\theta}\left(\frac{\partial \ln T_i}{\partial r}\right)^2 \right]. \tag{7.81}$$

The radial electric field in this expression can be expressed by using (7.25). Then, the angular momentum diffusion coefficient becomes identical to (7.79). The second term in (7.81) is due to the poloidal variations of the electrostatic potential. It is worth noting here that driving term in (7.77) and (7.81) due to the temperature gradient occurs as a result of the symmetry breaking related to finite Larmor radius corrections to the perturbed distribution function.

7.8.3 *Toroidal momentum diffusion and momentum damping from drift-kinetic theory and fluid moment equations*

The scaling of the toroidal momentum diffusion coefficient (7.79) corresponds to the contribution of passing particles. In the banana regime, most particles are passing and have $q\rho_i$ as a characteristic deviation from the magnetic surface. They complete many connection length qR passes before they collide (when their motion becomes random and contributes to an irreversible diffusion) process. The characteristic time step for such a diffusion process is $\tau \simeq \nu_{ii}^{-1}$ and the diffusion coefficient is $(q\rho_i)^2/\tau$, which is consistent with the diffusion coefficient in (7.81).

The above arguments can be made more specific by using the drift kinetic equation in the form

$$\frac{\partial f}{\partial t} + \frac{v_\parallel}{qR}\frac{\partial f}{\partial \theta} + \mathbf{V}_d \cdot \nabla f = -\widehat{\nu} f\,, \tag{7.82}$$

where V_d is the magnetic drift frequency

$$\mathbf{V}_d = \frac{1}{\omega_c}\left(\frac{v_\perp^2}{2} + v_\parallel^2\right)\mathbf{b}\times\nabla\ln B\,, \tag{7.83}$$

and $\widehat{\nu}$ is the model collision operator for ion-ion collisions. The required toroidal momentum balance equation is obtained from the $v_\parallel$ moment of (7.82) which gives

$$\frac{\partial}{\partial t}V_\parallel - \frac{\partial}{\partial r}\left\langle \frac{v_\parallel}{\omega_c}\left(\frac{v_\perp^2}{2} + v_\parallel^2\right)\frac{\partial \ln B}{\partial \theta}\widetilde{f}\right\rangle = 0\,, \tag{7.84}$$

where $\langle \ldots \rangle$ means averaging over magnetic flux surface, $\int d\theta\,(\ldots)$, and averaging over particle distribution function, $n^{-1}\int d^3v\,(\ldots)$.

Equation (7.82) can be symbolically solved by iterations, $\widetilde{f} = f^{(1)} + f^{(2)} + \cdots$, in the weakly collisional limit $\partial/\partial t \ll \left(v_\parallel/qR, \widehat{\nu}\right)$, $v_\parallel/qR \gg \widehat{\nu}$. In the first order, one has

$$\frac{v_\parallel}{qR}\frac{\partial f^{(1)}}{\partial \theta} + \mathbf{V}_d \cdot \nabla f_T = 0\,, \tag{7.85}$$

where

$$f_T = f_0 \frac{m v_\parallel V_\zeta}{T}\,, \tag{7.86}$$

is the shifted Maxwellian describing toroidal rotation. The solution of (7.85) is

$$f^{(1)} = \left(\frac{v_\perp^2}{2} + v_\parallel^2\right)\frac{qR}{r\omega_c}\frac{m v_\parallel}{T}\frac{\partial V_\zeta}{\partial r} f_0 \int \frac{1}{B}\frac{\partial B}{\partial \theta}d\theta\,, \tag{7.87}$$

where $\int B^{-1}(\partial B/\partial\theta)\,d\theta = \widetilde{(\ln B)} = -\varepsilon\cos\theta$ is the poloidal oscillation of the magnetic field. Here $\widetilde{(\ln B)}$ denotes the poloidally oscillating part. The next order equation

$$\frac{v_\parallel}{qR}\frac{\partial f^{(2)}}{\partial \theta} = -\widehat{\nu} f^{(1)}\,, \tag{7.88}$$

gives

$$f^{(2)} = -\widehat{\nu}\left(\frac{v_\perp^2}{2} + v_\parallel^2\right)\frac{q^2R^2}{v_\parallel\omega_c}\left(\frac{\partial}{r\partial\theta}\right)^{-1}\left(\widetilde{\ln B}\right)\frac{m}{T}\frac{\partial V_\zeta}{\partial r}f_0 . \tag{7.89}$$

Then, the toroidal momentum balance equation takes the form

$$\frac{\partial}{\partial t}V_\zeta + \frac{\partial}{\partial r}\left\langle \frac{q^2R^2}{\omega_c^2}\left(\frac{v_\perp^2}{2} + v_\parallel^2\right)^2\left(\frac{\partial}{r\partial\theta}\right)^{-1}\left(\widetilde{(\ln B)}\right)\frac{\partial\widetilde{(\ln B)}}{r\partial\theta}\frac{m}{T}f_0\frac{\partial V_\zeta}{\partial r}\right\rangle = 0 , \tag{7.90}$$

where we assume $V_\zeta \simeq V_\parallel$. For a simple case of circular magnetic surfaces

$$\left\langle\left(\frac{\partial}{r\partial\theta}\right)^{-1}\left(\widetilde{\ln B}\right)\frac{\partial\widetilde{(\ln B)}}{r\partial\theta}\right\rangle = -\frac{1}{2R^2} , \tag{7.91}$$

and (7.90) gives the diffusion equation for the toroidal momentum

$$\frac{\partial}{\partial t}V_\zeta = \chi_\zeta\frac{\partial^2}{\partial r^2}V_\zeta , \tag{7.92}$$

with a diffusion coefficient

$$\chi_\zeta = \left\langle\widehat{\nu}\frac{q^2}{\omega_c^2}\left(\frac{v_\perp^2}{2} + v_\parallel^2\right)^2\frac{m}{2T}f_0\right\rangle \simeq \nu q^2\rho_i^2 . \tag{7.93}$$

For the order of magnitude estimates, in (7.92), we have neglected the effects of cylindrical geometry, which was approximated by the local Cartesian coordinate system.

The complicated structure of the collision operator $\widehat{\nu}$ and boundary conditions at the trapped-passing boundary (only passing particle have to be included in the solution) make the exact solution quite complicated [52, 76], however the above procedure does predict correct scaling of the momentum diffusivity. The poloidal variations of the electrostatic potential and temperature gradient will have to be included [76], to get the driving term $\sim (\partial T/\partial r)^2$ in (7.81). One can notice that in moment approach, as in Eq. (7.75), we operated with toroidal projection of the momentum equation which involve the toroidal projection of the viscosity tensor and the radial electric field from the Maxwell stress, while in the approach based on the drift kinetic equation, Eqs. (7.84) and (7.92) are obtained by taking the $v_\parallel$ moments of the drift kinetic equation. In the latter, the parallel (along $\mathbf{b} = \mathbf{B}/B$) projection is exploited which does not include the radial electric current: in the parallel projection, the Maxwell stress term disappears, $\mathbf{b}\cdot\mathbf{J}\times\mathbf{B} = 0$. Of course, both lead to the same results in the end.

One aspect of the relation between these two approaches, namely appearance of the J_r in the parallel projection of (7.82) was discussed in [76]. The other aspect was clarified in [62] , where it was shown that the $\left\langle v_{\|}\mathbf{V}_d\cdot\nabla f\right\rangle$ term in drift kinetic equation corresponds to the contribution of the gyroviscosity to the moment balance equation. The parallel momentum balance equation derived in [62] has the form

$$mn\left(\frac{\partial}{\partial t}+V_{\|}\mathbf{e}_0\cdot\nabla+\mathbf{V}_E\cdot\nabla+\frac{4p+5\pi_{\|}/2}{mn\omega_c}\mathbf{e}_0\times\nabla\ln B\cdot\nabla\right)V_{\|}$$
$$-mnV_{\|}\mathbf{V}_E\cdot\nabla\ln B-2V_{\|}\frac{c}{\omega_c}\mathbf{e}_0\times\nabla\left(p+\pi_{\|}\right)\cdot\nabla\ln B$$
$$=enE_{\|}-\nabla_{\|}p-\mathbf{b}\cdot\nabla\cdot\Pi_{\|}\,. \tag{7.94}$$

The fourth term on the left in this equation originates from gyroviscosity, this term corresponds to the $\left\langle v_{\|}\mathbf{V}_d\cdot\nabla f\right\rangle$ term in (7.82), which is responsible for the diffusion of the toroidal momentum. Note that the parallel heat flux term of a similar structure has to be added to (7.94) according to (7.135). It is worth noting that the equation (7.94), also contains the term responsible for fast poloidal flow damping, $\mathbf{b}\cdot\nabla\cdot\Pi_{\|}$, as it was discussed in Section 7.5.

7.8.4 *Comments on non-axisymmetric effects*

So far, the discussion of ambipolarity and toroidal momentum balance in this Section was concerned with toroidally symmetric configurations. In a non-symmetric case, such as a stellarator or rippled tokamak, the toroidal component of the parallel viscosity tensor does not average identically to zero because of absence of symmetry, $\left\langle\mathbf{e}_\zeta\cdot\nabla\cdot\mathbf{\Pi}_{\|}\right\rangle\neq 0$. Then, an additional ambipolarity condition, similar to (7.29), occurs in non-axisymmetric toroidal geometry. Resulting ambipolarity equations, $\left\langle\mathbf{B}_\theta\cdot\nabla\left(p\mathbf{I}+\mathbf{\Pi}_{\|}\right)\right\rangle=0$ and $\left\langle\mathbf{B}_\zeta\cdot\nabla\left(p\mathbf{I}+\mathbf{\Pi}_{\|}\right)\right\rangle=0$, will uniquely define the poloidal and toroidal rotation velocities, or radial electric field and toroidal velocity. These conditions generally occur in the lower order compared to the effects considered in previous Sections. On the discussion of the ambipolarity in quasi-symmetric systems see [68].

The toroidal component of parallel viscosity due to broken toroidal symmetry has been termed Neoclassical Toroidal Viscosity [58]. It is believed to play a major role in breaking the toroidal rotation and error penetration [10]. For a recent review of work on Neoclassical Toroidal Viscosity see [3].

A simple picture and classification of trapped particles regimes (as in Section 7.10) becomes more complicated in a rippled tokamak. Toroidal ripples create additional variations of the magnetic field and additional particles trapping in toroidal direction [36, 77]. At lower collisionality, the interaction of resonances and trapping in poloidal and toroidal variations of the magnetic field create a multitude of various collisionality regimes depending on plasma parameters. Previous work and recent results on rotation and electric field in a rippled tokamak problem are described in [19].

7.9 Summary

This overview described key processes that define plasma rotation in a tokamak. In an axisymmetric case, the neoclassical theory predicts two different time scales for the evolution of plasma rotation. On a short scale, of the order of the ion-ion collision time, poloidal rotation is established at the equilibrium value determined by Eq. (7.30). In this state, the radial particle diffusion becomes ambipolar and the radial electric field cannot be determined separately from toroidal rotation, as it is shown in Eq. (7.25). Depending on the initial state, the relaxation of poloidal velocity to the steady state velocity (7.30), may also occur on a faster (collisionless) time scale v_{Ti}/qR, such as for e.g. oscillating rotation due to Geodesic Acoustic modes. Characteristic feature of the poloidal rotation damping is absence of the spatial scale in the rotation damping time, $\partial V_\theta/\partial t \sim -V_\theta/\tau_p$, though the characteristic time τ_p may depend on the geometrical scale related to the fraction of trapped particles at a given radial location. We would like to note also that the poloidal rotation damping is inherently related to a non-ambipolar particle flux. In nonstationary states, such flux may strongly exceed standard (steady state) neoclassical particle flux.

In axisymmetric system, the evolution of toroidal rotation occurs on a slower time scale which is defined by the diffusion of toroidal momentum, $\partial V_\zeta/\partial t = \partial/\partial r\left(\chi_i \partial V_\zeta/\partial r + P\right)$. Additional driving terms in the diffusion equation occur as a result of radial temperature gradient. This process is not automatically ambipolar; indeed, the diffusion equation for toroidal velocity is obtained as a result of the imposed ambipolarity condition $J_r = 0$. As it follows from (7.77), (7.78), and (7.81), neoclassical theory for toroidal rotation predicts low values of diffusivity of the toroidal momentum χ_i. It is worth noting here that, in relevant banana regimes, neoclassical diffusivity of the toroidal momentum χ_i is lower than the neoclassical ion heat flux (e.g. from (7.19)), while the experimental values of the diffusivity of the toroidal momentum are of the same order as the ion heat transport even in those regimes when the heat flux is close to neoclassical values [13]. It suggests that other effects, e.g. turbulence, play important role in toroidal momentum transport [54].

Analysis of Section 7.8.3 and [62] shows that both neoclassical and turbulent effects in the toroidal momentum evolution can be accounted for in the moment approach via contributions of Reynolds stress and gyroviscosity tensors, $mn\left(\mathbf{V}\cdot\nabla\mathbf{V}\right)+\nabla\cdot\Pi_{\wedge}$, of which both poloidal and toroidal components must be included. As well, Reynolds stress and gyroviscosity must include contributions of perpendicular $V_{\perp}$ and parallel $V_{\|}$ components of the velocity (and heat fluxes $q_{\perp}, q_{\|}$). In general, parallel components of the velocity and heat flux $V_{\|}$, $q_{\|}$ require kinetic treatment. Effects of fluctuating $\mathbf{E}\times\mathbf{B}$ drift in (7.94) lead to turbulent transport of toroidal momentum (the third term in (7.94)) and other effects, e.g. such as turbulent equipartition pinch [49], (the fifth term in (7.94)). Assuming that $\mathbf{V}_E$, $V_{\|}$, and $q_{\|}$ are all of the same, first order $\sim \rho/a$, effects, one can make a general observation that neoclassical and turbulent dynamics of toroidal momentum are contained within the second order momentum conservation equation such as (7.94) (other terms in Eq. (7.94) due to gyroviscosity were described in [61]). Full understanding and accurate description of turbulent momentum transport is still lacking and is currently a subject of active research.

Acknowledgments

The author expresses his gratitude to X. Garbet for stimulating and enlightening discussions on a number of topics related to this overview. The author also would like to acknowledge many useful discussions that took place at the Festival de Theorie, Aix-en-Provence (2009) and the programme "Gyrokinetics in Laboratory and Astrophysical Plasmas' at the Isaac Newton Institute, Cambridge (2010).

This work was supported in part by Natural Sciences and Engineering Research Council of Canada.

7.10 Appendix: Trapped (banana) particles and collisionality regimes in a tokamak

The charged particle moving in the stationary inhomogeneous magnetic field has its energy conserved, $v_{\|}^2 + v_{\perp}^2 = const$, where $v_{\|}$ is the particle velocity along the magnetic field, and $v_\perp$ is the absolute value of the particle velocity in the direction perpendicular to the magnetic field. In addition, if the particle Larmor frequency is large, there is an approximate (adiabatic) invariant corresponding to the particle magnetic moment (magnetic flux inside the particle Larmor circle)

$$\mu = \frac{v_\perp^2}{B(\mathbf{r})} = const. \tag{7.95}$$

In a tokamak, the magnetic field has a maximum on the inner side of the torus. Consider the particle which starts its motion with initial velocity $(v_{\parallel 0}, v_{\perp 0})$ from the point on the outer side of the torus where the magnetic field has the lowest magnitude $B_{\min}$. As the particle moves inward into the region of the larger field, its perpendicular velocity increases according to the relation

$$v_\perp^2 = v_{\perp 0}^2 \frac{B}{B_{\min}} .$$

while the parallel velocity, $v_\parallel$, decreases. If the particle initial velocity $v_{\parallel 0}$ is not too large, $v_{\parallel 0} < \sqrt{\epsilon} v_{\perp 0}$, there exists a reflection point at which $v_\parallel = 0$ and $v_\perp^2 = v_{\perp 0}^2 + v_{\parallel 0}^2$. Such particle will be reflected and start bouncing between two conjugate points. Projection of its trajectory onto the poloidal cross section has banana shape, hence the name - banana particles or trapped particles. Number of such trapped particles is proportional to the phase space of the "trapped" cone with $v_{\parallel 0} < \sqrt{\epsilon} v_{\perp 0}$, so that the density of trapped particles is given by $n_{tr}/n_0 = \sqrt{\epsilon}$, where n_0 is the total particle density.

Particles moving in the inhomogeneous magnetic field of the tokamak experience magnetic drift which moves them off the magnetic surface

$$\mathbf{v}_D = \frac{1}{\omega_c} \frac{v_\perp^2}{2} \mathbf{b} \times \nabla \ln B + \frac{v_\parallel^2}{\omega_c} \mathbf{b} \times (\mathbf{b} \cdot \nabla) \mathbf{b} \sim \frac{v_\perp^2/2}{\omega_c} \frac{1}{R}.$$

The magnitude of such a displacement for trapped particles (banana width) can be estimated by taking into account that for trapped particles $v_\parallel / v_\perp \sim \sqrt{\epsilon}$, $v_\perp \simeq v_T$, respectively, the particle bounce time is

$$\tau_b = \frac{qR}{v_\parallel} = \frac{qR}{v_T \sqrt{\epsilon}} , \tag{7.96}$$

where qR is the characteristic length in the torus along the magnetic field line (connection length). The banana width then becomes

$$\Lambda_b = \mathbf{v}_D \tau_b = \frac{v_\perp^2}{\omega_c R} \frac{qR}{v_\perp \sqrt{\epsilon}} = \frac{\rho q}{\sqrt{\epsilon}} , \tag{7.97}$$

where $q = B_T r/(B_p R)$. Passing particles are also displaced off the magnetic surface. For a typical passing particle with $v_\perp \simeq v_\parallel \simeq v_T$, the displacement is

$$\Lambda = \mathbf{v}_D \frac{qR}{v_T} = q\rho \,. \tag{7.98}$$

Banana (trapped) particles trajectories exist if particle experiences no collisions while it makes the full bounce period. The collisions rotate the particle velocity vector in the phase space $(v_\perp, v_\parallel)$. This process takes the particles out of the trapped particle cone, $v_\parallel < \sqrt{\epsilon} v_\perp$ and, thus, defines the banana life-time. To evaluate it, we have to take into account that the Coulomb collision operator is of the diffusion nature, so that the diffusion equation in the particle phase space is

$$\frac{\partial}{\partial t} f = D_v \frac{\partial^2}{\partial v^2} f, \tag{7.99}$$

where the diffusion coefficient in the velocity space is $D_v = (\Delta v)^2 / \tau = v_T^2 \nu$. It is assumed that $(\Delta v) \simeq v_T$ and $\tau \simeq \nu^{-1}$. The effective banana collision time (or banana life-time) can be introduced as a time required to move the particle out of the trapped phase space cone, $\Delta v_\parallel = \sqrt{\varepsilon} v_T$. Then

$$\frac{\partial}{\partial t} \simeq \tau_{eff}^{-1} \,, \tag{7.100}$$

and

$$\frac{\partial^2}{\partial v^2} \simeq (\Delta v_\parallel)^{-2} \simeq (v_T^2 \epsilon)^{-1} \,.$$

From the diffusion equation in the phase space one finds the effective collision frequency

$$\nu_{eff} = \tau_{eff}^{-1} = v_T^2 \nu_c (v_T^2 \epsilon)^{-1} = \frac{\nu}{\epsilon} \,. \tag{7.101}$$

For the banana trajectory to exist, this frequency should be smaller than the banana bounce frequency (the inverse of the trapped particle bounce time) $\nu_{eff} < \tau_b^{-1}$. This inequality defines the weaklyl collisional trapped particles (or the so called banana) regime

$$\frac{\nu}{\epsilon} < \frac{\epsilon^{1/2} v_T}{q R_0} \,. \tag{7.102}$$

This criterion is often written by using the normalized frequency, ν_*,

$$\nu^\star \equiv \frac{\nu q R}{\epsilon^{3/2} v_T} \,. \tag{7.103}$$

The banana collisionality regime exists for $\nu^* < 1$.

When the particle collision frequency increases and ν^* becomes larger than unity, $\nu^* > 1$, closed banana trajectories no longer exist, but particle trajectories will remain essentially collisionless over the connection length till $\nu^* < \varepsilon^{-3/2}$. In other words, for $\nu < v_T/qR$, the particle performs the full periodic motion over qR length without experiencing a collision. This region $1 < \nu^* < \varepsilon^{-3/2}$ is called the plateau regime. When the collisions become more frequent, $\nu > v_T/qR$, the particle experiences the collision before it completes the full motion over the connection length. This is the highly collisional, short mean path, or Pfirsch-Schlü ter regime.

7.11 Appendix: Hierarchy of moment equations

Plasma viscosity in the magnetic field has several distinct contributions. It is instructive to consider the structure of the viscosity tensor by employing the moment approach due to Grad [21]. This approach is reviewed in this section.

The moment equations are obtained by taking the formal moments of the Boltzmann kinetic equation in the form

$$\frac{\partial}{\partial t}f + \nabla \cdot (\mathbf{v}f) + \nabla_v \cdot \left(\frac{e}{m}\left(\mathbf{E} + \frac{1}{c}\mathbf{v} \times \mathbf{B}\right) f\right) = C\,, \tag{7.104}$$

where C is the collision operator. The density moment $n = \int d^3 v f$ trivially results in the continuity equation,

$$\frac{\partial n}{\partial t} + \nabla \cdot (n\mathbf{V}) = 0\,, \tag{7.105}$$

where fluid velocity is defined as

$$n\mathbf{V} = \int d^3 v \mathbf{v} f\,. \tag{7.106}$$

The fluid velocity moment (momentum balance) equation becomes

$$\frac{\partial}{\partial t}\int d^3 v \mathbf{v} f + \nabla \cdot \int d^3 v\,(\mathbf{v}\mathbf{v}f) + \int d^3 v \mathbf{v} \nabla_v \cdot \left(\frac{e}{m}\left(\mathbf{E} + \frac{1}{c}\mathbf{v} \times \mathbf{B}\right) f\right)$$
$$= \int d^3 v \mathbf{v} C. \tag{7.107}$$

This can be further transformed by introducing general definitions for plasma pressure

$$p = m \int d^3 v \frac{v'^2}{3} f\,, \tag{7.108}$$

and viscosity moments

$$\mathbf{\Pi} = m \int d^3v \left(\mathbf{v}'\mathbf{v}' - \frac{v'^2}{3}\mathbf{I}f \right). \tag{7.109}$$

Here $\mathbf{I}$ is the isotropic unit tensor; the random velocity $\mathbf{v}'$, $\mathbf{v} = \mathbf{v}' + \mathbf{V}$, is introduced so that

$$0 = \int d^3v\mathbf{v}'f. \tag{7.110}$$

The second term in (7.107) can be transformed as follows

$$\begin{aligned}\int d^3v \left(\mathbf{vv}f\right) &= \int d^3v \left(\left(\mathbf{v}' + \mathbf{V}\right)\left(\mathbf{v}' + \mathbf{V}\right)f\right) = n\mathbf{VV} + \int d^3v \left(\mathbf{v}'\mathbf{v}'f\right) \\ &= n\mathbf{VV} + \int d^3v \left(\mathbf{v}'\mathbf{v}' - \frac{v'^2}{3}\mathbf{I}f\right) + \int d^3v \frac{v'^2}{3}\mathbf{I}f \\ &= n\mathbf{VV} + \mathbf{\Pi} + p\mathbf{I}. \end{aligned} \tag{7.111}$$

Here, all terms linear in $\mathbf{v}'$ disappear due to the condition (7.110). This results in the standard momentum balance equation

$$m\left(\frac{\partial}{\partial t}\left(n\mathbf{V}\right) + \nabla \cdot \left(n\mathbf{VV}\right)\right) = en\left(\mathbf{E} + \mathbf{V} \times \mathbf{B}\right) - \nabla \cdot \mathbf{\Pi} - \nabla p + \mathbf{F}, \tag{7.112}$$

where $\mathbf{F} = \int Cf d^3v$ is the total force due to the momentum exchange (which may include the friction and thermal force).

Similarly, the second order moment equation is obtained in the form

$$\begin{aligned}\frac{\partial}{\partial t}\int d^3v\mathbf{vv}f + \nabla \cdot \int d^3v \left(\mathbf{vvv}f\right) + \int d^3v\mathbf{vv}\nabla_v \cdot \left(\frac{e}{m}\left(\mathbf{E} + \frac{1}{c}\mathbf{v} \times \mathbf{B}\right)f\right) \\ = \int d^3v\mathbf{vv}fC. \end{aligned} \tag{7.113}$$

The first term in this equation is written by using (7.111). The second and third terms on the left are transformed as

$$\int d^3v \left(\mathbf{vvv}f\right) = \int d^3v \left(\mathbf{v}'\mathbf{v}'\mathbf{v}' + \mathbf{VVV} + \mathbf{v}'\mathbf{v}'\mathbf{V} + \mathbf{v}'\mathbf{V}\mathbf{v}' + \mathbf{V}\mathbf{v}'\mathbf{v}'\right), \tag{7.114}$$

and

$$\begin{aligned}\int d^3v\mathbf{vv}\nabla_v \cdot \left(\frac{e}{m}\left(\mathbf{E} + \frac{1}{c}\mathbf{v} \times \mathbf{B}\right)f\right) \\ = -\frac{e}{m}\left(\mathbf{EV} + \mathbf{VE}\right) - \omega_c\left(\mathbf{\Pi} \times \mathbf{b} - \mathbf{b} \times \mathbf{\Pi}\right). \end{aligned} \tag{7.115}$$

Equation (7.113) is a tensor (second order) equation. According to the structure in (7.111) it contains several distinct contributions from the lower order moments, which are determined by the evolution equations for plasma density, n, velocity, $\mathbf{V}$, and pressure p. In certain sense, the tensor contributions $n\mathbf{VV}$, $p\mathbf{I}$, and $\mathbf{\Pi}$ to (7.111) are orthogonal. The tensor term $n\mathbf{VV}$ is calculated using (7.112) and continuity equation. The second order scalar moment (pressure) equation is obtained as the trace of (7.113) minus the contributions from $\partial\left(n\mathbf{VV}\right)/\partial t$ terms. Equivalently, the evolution equation for the scalar second order moment can be obtained by taking the scalar moment of (7.104) with a random velocity energy, v'^2 :

$$\frac{\partial}{\partial t}\int d^3v\mathbf{v}'^2 f + \nabla\cdot\int d^3v\left(\mathbf{v}'^2 f\right) + \int d^3v\mathbf{v}'^2\nabla_v\cdot\left(\frac{e}{m}\left(\mathbf{E}+\frac{1}{c}\mathbf{v}\times\mathbf{B}\right)f\right)$$
$$= \int d^3v v'^2 fC. \tag{7.116}$$

This becomes the standard pressure evolution equation

$$\frac{3}{2}\left(\frac{\partial}{\partial t}p + \mathbf{V}\cdot\nabla p\right) + \frac{5}{2}p\nabla\cdot\mathbf{V} + \nabla\cdot\mathbf{q} + \mathbf{\Pi}:\nabla\mathbf{V} = Q\,, \tag{7.117}$$

where the heat flux $\mathbf{q}$ was introduced as

$$\mathbf{q} = \frac{m}{2}\int \mathbf{v}'\mathbf{v}'^2 d^3v f\,. \tag{7.118}$$

Finally, evolution equation for viscosity is obtained by subtracting from Eq. (7.113) the symmetric part for $3/2\mathbf{I}\left(\partial/\partial t\right)p...$ (represented by Eq. (7.117)) and the Reynolds stress part, $\left(\partial/\partial t\right)\left(mn\mathbf{VV}\right)...$, which is obtained by multiplying equation (7.112) by $\mathbf{V}$. This results in the equation

$$\frac{d\mathbf{\Pi}}{dt} + \mathbf{\Pi}\nabla\cdot\mathbf{V} + \left(\mathbf{\Pi}\cdot\nabla\mathbf{V} + \mathbf{Tr}\right) + \omega_c\left(\mathbf{b}\times\mathbf{\Pi} - \mathbf{\Pi}\times\mathbf{b}\right)$$
$$+ \left[p\nabla\mathbf{V} + p\left(\nabla\mathbf{V}\right)^T - \frac{2}{3}\mathbf{I}p\nabla\cdot\mathbf{V}\right] + \frac{2}{5}\left[\nabla\mathbf{q} + \left(\nabla\mathbf{q}\right)^T - \frac{2}{3}\mathbf{I}\nabla\cdot\mathbf{q}\right]$$
$$+ \nabla\cdot\tau = \mathbf{C}_\pi\,, \tag{7.119}$$

where

$$\tau_{ijk} = m\int\left(v_i'v_j'v_k' - \frac{v'^2}{5}\left(v_i'\delta_{jk} + v_j'\delta_{ik} + v_k'\delta_{ij}\right)\right)d^3v \tag{7.120}$$

is the third order tensor, and $\mathbf{C}_\pi = m\int d^3v\left(\mathbf{v}'\mathbf{v}' - \frac{v'^2}{3}\mathbf{I}f\right)Cf$ is the tensor moment of the collision operator.

In a similar way, one obtains the evolution equation for the third order moment, heat flux, $\mathbf{q}$. Neglecting collisional effects and some other secondary terms, one has [44]

$$\frac{d}{dt}\mathbf{q}+\frac{7}{5}(\mathbf{q}\cdot\nabla)\mathbf{V}-\omega_c\mathbf{q}\times\mathbf{b}=-\frac{5}{2}\frac{p}{m}\nabla T-\nabla\cdot\left(\frac{T}{m}\mathbf{\Pi}\right)-\frac{5}{2m}\mathbf{\Pi}\cdot\nabla T+\mathbf{\Pi}\cdot\mathbf{F}\,, \tag{7.121}$$

where $\mathbf{F}\equiv(\nabla p+\nabla\cdot\mathbf{\Pi})/mn$.

All these moment equations are most easily obtained by transforming the initial kinetic equation into the reference frame moving with the fluid velocity $\mathbf{V}$, where equation for $f(\mathbf{v}', t, \mathbf{x})$ takes the form

$$\frac{df}{dt}+\mathbf{v}'\cdot\nabla f+\frac{\partial f}{\partial\mathbf{v}'}\cdot\left(\mathbf{F}-(\mathbf{v}'\cdot\nabla)\,\mathbf{V}+\omega_c\mathbf{v}'\times\mathbf{b}\right)=C\,. \tag{7.122}$$

Such a procedure is described in [44]. It is worth noting in summary, that the trace part of the tensor equation (7.113), which determines the time derivative of the scalar second order moment (thermal pressure) does not enter the equation for the viscosity tensor 7.119. The trace part of Eq. (7.113) is simply the energy evolution equation (7.117).

7.12 Appendix: Plasma viscosity tensor in the magnetic field: parallel viscosity, gyroviscosity, and perpendicular viscosity

In strongly magnetized plasmas, the cyclotron frequency is a large parameter, $\omega_c > (kV, d/dt)$, and it offers a natural way to introduce various components of the viscosity. The central idea is to separate in (7.119) the components which have the terms with ω_c. These terms (with ω_c) will be the largest and can be used to construct the perturbation series expansion. Most conveniently, it can be done by introducing the operator $\mathbf{K}$ [33]

$$\widehat{\mathbf{K}}\mathbf{\Pi}\equiv(\mathbf{\Pi}\times\mathbf{b}-\mathbf{b}\times\mathbf{\Pi})\,. \tag{7.123}$$

This operator is a tensor counterpart of the standard vector product $(\mathbf{V}\times\mathbf{B})$ which arises in the moment equation for the fluid velocity (7.112). Inverse of $\mathbf{K}$ for any symmetric tensor $\mathbf{A}$ is

$$\mathbf{K}^{-1}\mathbf{A}=\frac{1}{4}\left\{[\mathbf{b}\times\mathbf{A}\cdot(\mathbf{I}+3\mathbf{bb})]+[\mathbf{b}\times\mathbf{A}\cdot(\mathbf{I}+3\mathbf{bb})]^T\right\}\,. \tag{7.124}$$

Important properties of $\mathbf{K}$ are

$$\mathbf{KI}=0\,, \tag{7.125}$$

$$\widehat{\mathbf{K}}^{-1}\mathbf{I} = 0\,, \tag{7.126}$$

and

$$\mathbf{K}\,(\mathbf{bb}) = \mathbf{0}\,. \tag{7.127}$$

These properties suggest that one can introduce the general "parallel" viscosity tensor (traceless)

$$\mathbf{\Pi}_{\|} = \frac{3}{2}\pi_{\|}\left(\mathbf{bb} - \frac{1}{3}\mathbf{I}\right) = \left(p_{\perp} - p_{\|}\right)\left(\mathbf{bb} - \frac{1}{3}\mathbf{I}\right), \tag{7.128}$$

so that

$$\widehat{\mathbf{K}}^{-1}\mathbf{\Pi}_{\|} = \widehat{\mathbf{K}}\mathbf{\Pi}_{\|} = 0\,. \tag{7.129}$$

The tensor $\mathbf{\Pi}_{\|}$ has to be traceless because the finite trace (symmetric) part of the total stress tensor $\mathbf{P} = p\mathbf{I} + \mathbf{\Pi}$ is separated into the isotropic pressure tensor $p\mathbf{I}$, and $\mathbf{\Pi}$ is automatically traceless. The parallel component of (7.119) can be obtained by applying to Eq. (7.119) the $\mathbf{bb}:$ operator

$$\begin{aligned}
&\frac{d\pi_{\|}}{dt} + \pi_{\|}\nabla\cdot\mathbf{V} + \mathbf{bb}\colon(\mathbf{\Pi}\cdot\nabla\mathbf{V} + \mathbf{Tr}) \\
&\quad + \mathbf{bb}\colon\left[p\nabla\mathbf{V} + p\,(\nabla\mathbf{V})^{T} - \frac{2}{3}\mathbf{I}p\nabla\cdot\mathbf{V}\right] \\
&\quad + \frac{2}{5}\,\mathbf{bb}\colon\left[\nabla\mathbf{q} + (\nabla\mathbf{q})^{T} - \frac{2}{3}\mathbf{I}\nabla\cdot\mathbf{q}\right] + \mathbf{bb}:\nabla\cdot\tau = \mathbf{bb}:\mathbf{C}_{\pi}\,,
\end{aligned} \tag{7.130}$$

where the semi-colon denotes the double contraction operation, in dyad form $\mathbf{ab}:\mathbf{cd} \equiv (\mathbf{a}\cdot\mathbf{d})\,(\mathbf{b}\cdot\mathbf{c})$.

This parallel component of the general evolution equation does not involve the magnetic field terms with ω_c but involves the higher order parallel moments (due to the third rank tensor τ). In general, for collisionless plasmas one cannot neglect those higher order contributions and Eq. (7.130) remains unclosed. One can say that the parallel component of the viscosity tensor $\pi_{\|}$ is a tensor counterpart of the parallel velocity (or parallel heat flux), which, in general, cannot be determined from fluid equations. Similarly, the parallel viscosity tensor $\pi_{\|}$ should be calculated from kinetic theory.

In highly collisional, short mean free path (along the magnetic field), limit, $\nu > \left(d/dt, k_{\|}v_T\right)$, the collision frequency becomes the largest parameter in the parallel equation and can be used as an expansion parameter.

The collision frequency occurs from the moment of the collision operator on the right hand side of the equation (7.119). In the lowest order, $\mathbf{bb} : \mathbf{C}_\pi = -6\nu\pi_\parallel/5$, where ν is the collision frequency between like particles. Expansion of Eq. (7.119) for large ν defines the parallel viscosity in the collisional regime in the form

$$\pi_\parallel = -\frac{5}{6\nu}\mathbf{bb}\colon \left[p\nabla\mathbf{V} + p\left(\nabla\mathbf{V}\right)^T - \frac{2}{3}\,\mathbf{I}p\nabla\cdot\mathbf{V} + \nabla\mathbf{q} + \left(\nabla\mathbf{q}\right)^T - \frac{2}{3}\,\mathbf{I}\nabla\cdot\mathbf{q}\right]. \tag{7.131}$$

To get more accurate value (corresponding to the two term polynomial approximation of Braginskii [1]) the energy weighted parallel tensor $\Pi^*_\parallel$ must be included, so that

$$\mathbf{C}_\pi = -\frac{6}{5}\nu\left(\mathbf{\Pi} + \frac{4}{3}\mathbf{\Pi}^*_\parallel\right), \tag{7.132}$$

$$\mathbf{bb} : \mathbf{C}_\pi = -\frac{6}{5}\nu\left(\pi_\parallel + \frac{4}{3}\pi^*_\parallel\right), \tag{7.133}$$

and the equation for $\pi^*_\parallel$, analogous to the (7.119) has to be employed.

It was recognized also that nonlinear terms in the collisional operator also contribute to the parallel collisional viscosity [5]. These nonlinear terms also proportional to the collision frequency and quadratic in heat fluxes, schematically, $\mathbf{bb} : \left[\mathbf{C}_\pi\right]_{NL} \sim \nu\left[\mathbf{q}\right]^2$. Balancing linear and nonlinear terms in the moment of the collisional operator one can find the parallel nonlinear viscosity which is independent of collisional frequency [5, 43].

Contrary to the parallel component (7.130), which does not allow any asymptotic expansion in general case (except the case of the short mean free path regime), the transverse components of (7.119) involve large parameter ω_c which can be used for natural expansion series for the viscosity. Such perturbation series can be obtained by applying the operator $\widehat{\mathbf{K}}^{-1}$ to the Eq. (7.119) giving

$$\begin{aligned}\mathbf{\Pi} = {} & \frac{1}{\omega_c}\widehat{\mathbf{K}}^{-1}\left(\left[p\nabla\mathbf{V} + p\left(\nabla\mathbf{V}\right)^T\right] + \frac{2}{5}\nabla\mathbf{q} + \left(\nabla\mathbf{q}\right)^T\right) \\ & + \frac{1}{\omega_c}\widehat{\mathbf{K}}^{-1}\left(\frac{d\pi}{dt} + \pi\nabla\cdot\mathbf{V} + \left(\pi\cdot\nabla\mathbf{V} + \mathbf{Tr}\right)\right) \\ & + \frac{1}{\omega_c}\widehat{\mathbf{K}}^{-1}\left(\nabla\cdot\tau\right) + \frac{1}{\omega_c}\widehat{\mathbf{K}}^{-1}\left(\mathbf{C}_\pi\right).\end{aligned} \tag{7.134}$$

The oblique (or gyroviscosity) arises from transverse to the magnetic field components of (7.119) in the first order of the expansion. In the main order of (7.119), only the terms with gradients of flow and heat flux are retained ($\omega_c > k_\perp V$, d/dt) leading to the gyroviscosity in the form

$$\mathbf{\Pi}_\wedge = \frac{1}{\omega_c}\widehat{\mathbf{K}}^{-1}\left(\left[p\nabla\mathbf{V} + p\left(\nabla\mathbf{V}\right)^T\right] + \frac{2}{5}\left[\nabla\mathbf{q} + \left(\nabla\mathbf{q}\right)^T\right]\right). \qquad (7.135)$$

The momentum and heat balance equations, (7.112) and (7.121), can also be solved by using ω_c as an the expansion parameter. At lowest order, one finds from (7.112) and (7.121):

$$\mathbf{V} = \mathbf{V}_E + \mathbf{V}_p + \mathbf{V}_\parallel\,, \qquad (7.136)$$

$$\mathbf{q} = \frac{5}{2}\frac{cp}{eB}\mathbf{b}\times\nabla T + \mathbf{q}_\parallel\,. \qquad (7.137)$$

The lowest order gyroviscosity tensor is calculated from (7.135) by using (7.136) and (7.137). Explicit expressions are given in [44, 61].

The terms with $\nabla\mathbf{V}$ in (7.135) were obtained by Braginskii [1] by a different method. He called this viscosity oblique, and later it became known as gyroviscosity. It is collisionless in the main order when the lowest order drifts for $\mathbf{V}$ and $\mathbf{q}$ are due to $\mathbf{E}\times\mathbf{B}$ and diamagnetic effects. The third rank tensor τ gives the higher order contributions [61].

It is worth noting here that the evolution equation for total second order moment (7.113) includes the tensorial contributions due to the symmetric part $p\mathbf{I}$ and the Reynolds stress part $mn\mathbf{VV}$. The finite trace component of (7.113) (related to $p\mathbf{I}$) is independent of $\mathbf{\Pi}$, and does not appear in the gyroviscosity or perpendicular viscosity tensors, contrary to the statements in [48]; e.g. Eq. (11) in [48]. The contribution of the Reynolds stress $mn\mathbf{VV}$ is also independent of $\mathbf{\Pi}$. The Reynolds stress, $mn\mathbf{VV}$, in general, provides the same order contribution as $\mathbf{\Pi}_\wedge$, leading to the so called gyroviscous cancellation [26, 53]. It was generalized later to include temperature gradient effects [61, 62]. Independently of the collisionality regime, the Reynolds stress terms $mn\mathbf{VV}$ are additional, but not a part of the gyroviscosity tensor $\mathbf{\Pi}_\wedge$ (7.135), contrary to the Eqs. (31) and (73) in [60].

The expansion procedure (7.134) can be repeated and next order terms in $(1/B)$ parameter can be added to the viscosity (they will depend on a particular ordering). The terms due to collisions are

$$\mathbf{\Pi}_\perp = \frac{1}{\omega_c}\widehat{\mathbf{K}}^{-1}\left(\mathbf{C}_\pi\right) = -\frac{6}{5}\frac{\nu}{\omega_c}\mathbf{\Pi}_g\,, \qquad (7.138)$$

$\widehat{\mathbf{K}}^{-1}\left(\mathbf{C}_\pi\right)/\omega_c \simeq \left(\nu/\omega_c\right)\widehat{\mathbf{K}}^{-1}\mathbf{\Pi}_g$ (the last term in (7.134)) will produce the so called perpendicular (collisional) viscosity which has a scaling

$$\mathbf{\Pi}_\perp \sim \frac{\nu p}{\omega_{c^2}}\left(\nabla V, \nabla q\right). \qquad (7.139)$$

The collisionless terms $(d/dt)\,\widehat{\mathbf{K}}^{-1}\mathbf{\Pi}_g/\omega_c$ describe finite Larmor radius effects [50].

Usually gyroviscosity is referred to as a collisionless effect. However, the velocity and heat flux in (7.135) may also contain collisional terms, e.g. collisional contribution of the heat flux in (7.135) produces the collisional contribution to the gyroviscosity that is of the same order as the perpendicular viscosity [7].

It is important to note that gyroviscosity contains parallel components of the velocity and heat fluxes. The latter cannot be determined from fluid equations (except of the highly collisional short mean free path regimes) and, in general, must be calculated from kinetic equation. Note, that parallel components also appear through the third rank tensor τ in (7.134). Parallel components of the velocity (and heat fluxes in general) in the gyro- and perpendicular viscosity are important in the problem of the toroidal momentum transport (see Section 7.8). For completeness we give these expression here in the explicit form.

The evolution equations for the components of the interest have the form (only essential for our purposes terms are retained):

$$\frac{d}{dt}\Pi_{xz} + \frac{6}{5}\nu\Pi_{xz} - \omega_c\Pi_{yz} = -p\left(\frac{\partial V_x}{\partial z} + \frac{\partial V_z}{\partial x}\right) - \frac{2}{5}\left(\frac{\partial q_x}{\partial z} + \frac{\partial q_z}{\partial x}\right), \quad (7.140)$$

$$\frac{d}{dt}\Pi_{yz} + \frac{6}{5}\nu\Pi_{yz} + \omega_c\Pi_{xz} = -p\left(\frac{\partial V_y}{\partial z} + \frac{\partial V_z}{\partial y}\right) - \frac{2}{5}\left(\frac{\partial q_y}{\partial z} + \frac{\partial q_z}{\partial y}\right). \quad (7.141)$$

These equations are written in the Cartesian coordinate system, x, y, z, for the case of the uniform magnetic field $\mathbf{B} = B_0\mathbf{z}$. Such Cartesian system is a local approximation for toroidal coordinates, r, θ, and $\zeta : (x, y, z) \rightarrow (r, r\theta, R\zeta)$. In the lowest order one has

$$\Pi_{yz}^{(0)} = \frac{p}{\omega_c}\left(\frac{\partial V_x}{\partial z} + \frac{\partial V_z}{\partial x}\right) + \frac{2}{5\omega_c}\left(\frac{\partial q_x}{\partial z} + \frac{\partial q_z}{\partial x}\right), \quad (7.142)$$

$$\Pi_{xz}^{(0)} = -\frac{p}{\omega_c}\left(\frac{\partial V_y}{\partial z} + \frac{\partial V_z}{\partial y}\right) - \frac{2}{5\omega_c}\left(\frac{\partial q_y}{\partial z} + \frac{\partial q_z}{\partial y}\right). \quad (7.143)$$

In the next order, one gets

$$\begin{aligned}\Pi_{yz}^{(1)} &= \frac{6}{5}\frac{\nu}{\omega_c}\Pi_{xz}^{(0)} \\ &= -\frac{6}{5}\frac{p\nu}{\omega_c^2}\left(\frac{\partial V_y}{\partial z} + \frac{\partial V_z}{\partial y}\right) - \frac{12\nu}{25\omega_c^2}\left(\frac{\partial q_y}{\partial z} + \frac{\partial q_z}{\partial y}\right),\end{aligned} \quad (7.144)$$

$$\Pi_{xz}^{(1)} = -\frac{6}{5}\frac{\nu}{\omega_c}\Pi_{yz}^{(0)}$$
$$= -\frac{6}{5}\frac{p\nu}{\omega_c^2}\left(\frac{\partial V_x}{\partial z} + \frac{\partial V_z}{\partial x}\right) - \frac{12\nu}{25\omega_c^2}\left(\frac{\partial q_x}{\partial z} + \frac{\partial q_z}{\partial x}\right). \tag{7.145}$$

Equations (7.142)–(7.143) and (7.144)–(7.145) are, respectively, π_4 and π_2 components in notations of Braginskii [1]. Typically, (7.142)–(7.143) are referred as gyroviscosity, while (7.144)–(7.145), in some papers, are referred as perpendicular viscosity (based on the ν/ω_c^2 scaling).

7.13 Appendix: Closure relations for the flux surface averaged parallel viscosity in neoclassical (banana and plateau) regimes

As it was noted in Section 7.1.1, in the banana and plateau regimes parallel viscosity has to be calculated kinetically. The calculation of time dependent and non-averaged viscosity is a difficult problem and has not been fully accomplished [20, 59]. In neoclassical theory, most commonly, surface averaged and stationary viscosity is used, which (for ions) can be expressed in the form [31, 32]

$$\langle \mathbf{B}\cdot\nabla\cdot\mathbf{\Pi}_i\rangle = 3\left\langle \left(\nabla_\parallel B\right)^2\right\rangle\left(\mu_{i1}u_{i\theta} + \mu_{i2}\frac{2q_{i\theta}}{5p_i}\right). \tag{7.146}$$

Here, the ion coefficients are related to the collisional moments

$$\mu_{i1} = K_{11}^i\,, \tag{7.147}$$

$$\mu_{i2} = K_{12}^i - \frac{5}{2}K_{11}^i\,, \tag{7.148}$$

where $K_{\alpha\beta}^i$ are some functions related to the moments of the collisional frequency [31, 32], see also [24]. The variables $u_{i\theta}$ and $q_{i\theta}$ are normalized ion fluid flow velocity and heat flux in poloidal direction (both are the flux functions)

$$u_{i\theta} = \frac{\mathbf{V}\cdot\nabla\theta}{\mathbf{B}\cdot\nabla\theta} = \frac{V_\theta}{B_\theta}\,, \tag{7.149}$$

$$q_{i\theta} = \frac{\mathbf{q}\cdot\nabla\theta}{\mathbf{B}\cdot\nabla\theta} = \frac{q_\theta}{B_\theta}\,. \tag{7.150}$$

The fluid velocity consists of $\mathbf{E}\times\mathbf{B}$ and diamagnetic drifts. The heat flux is

$$\mathbf{q} = \frac{5}{2}\frac{cp}{eB}\mathbf{b}\times\nabla T. \tag{7.151}$$

Finally, from (7.149), (7.150), and (7.151), the parallel ion viscosity in (7.146) can be written as

$$\langle \mathbf{B} \cdot \nabla \cdot \mathbf{\Pi}_i \rangle = \frac{3\left\langle (\nabla_{\parallel} B)^2 \right\rangle}{B_\theta} \mu_{i1} \left(V_{i\theta} + k \frac{c}{eB} \frac{\partial T_i}{\partial r} \right), \tag{7.152}$$

where $k \equiv \mu_{i2}/\mu_{i1}$; in the banana regime $k = -1.17$, and $k = 1/2$ in the plateau regime. The viscosity coefficient in the plateau regime is

$$\mu_{i1} = \frac{2\sqrt{\pi} n T_i q R}{3 v_{Ti}} . \tag{7.153}$$

In banana regime, the viscosity coefficient depends on geometrical factor (fraction of trapped particles) but generally scales as

$$\mu_{i1} = c \frac{n T_i q R}{v_{Ti}} \nu^* , \tag{7.154}$$

where c is a numerical factor [31, 32].

References

[1] S. Braginskii, “Reviews of plasma physics”, Reviews of Plasma Physics (M. Leontovich, éd.), vol. 1, Consultants Bureau, New York, 1965, p. 184–272.

[2] G. Budker, “On particles drifts in toroidal magnetic thermonuclear reactor”, Plasma physics and the problem of controlled thermonuclear reactions (M. Leontovich, éd.), vol. 1, Pergamon Press, London, New York, 1959, p. 66.

[3] J. D. Callen, A. J. Cole and C. C. Hegna, “Toroidal rotation in tokamak plasmas”, *Nuclear Fusion* **49** (2009), no. 8, p. 085021.

[4] J. D. Callen and K. Shaing, “Neoclassical MHD equations for tokamaks”, Tech. Report UWPR 85-8, University of Wisconsin Plasma Report, 1985.

[5] P. J. Catto and A. N. Simakov, “A drift ordered short mean free path description for magnetized plasma allowing strong spatial anisotropy”, *Physics of Plasmas* **11** (2004), no. 1, p. 90–102.

[6] — , “Evaluation of the neoclassical radial electric field in a collisional tokamak”, *Physics of Plasmas* **12** (2005), no. 1, p. 012501.

[7] — , “A new, explicitly collisional contribution to the gyroviscosity and the radial electric field in a collisional tokamak”, *Physics of Plasmas* **12** (2005), no. 11, p. 114503.

[8] F. Chen, *Introduction to plasma physics and controlled fusion*, 2nd ed., éd., Springer, 1984.
[9] H. A. Claassen, H. Gerhauser, A. Rogister and C. Yarim, "Neoclassical theory of rotation and electric field in high collisionality plasmas with steep gradients", *Physics of Plasmas* **7** (2000), no. 9, p. 3699–3706.
[10] A. J. Cole, C. C. Hegna and J. D. Callen, "Neoclassical toroidal viscosity and error-field penetration in tokamaks", *Physics of Plasmas* **15** (2008), no. 5, p. 056102.
[11] J. W. Connor, S. C. Cowley, R. J. Hastie and L. R. Pan, "Toroidal rotation and momentum transport", *Plasma Physics and Controlled Fusion* **29** (1987), no. 7, p. 919–931.
[12] G. D. Conway, C. Troster, B. Scott and K. Hallatschek, "Frequency scaling and localization of geodesic acoustic modes in asdex upgrade", *Plasma Physics and Controlled Fusion* **50** (2008), no. 5, p. 055009.
[13] J. S. Degrassie, "Tokamak rotation sources, transport and sinks", *Plasma Physics and Controlled Fusion* **51** (2009), no. 12, p. 124047.
[14] P. H. Diamond, S. I. Itoh, K. Itoh and T. S. Hahm, "Zonal flows in plasma — a review", *Plasma Physics and Controlled Fusion* **47** (2005), no. 5, p. R35–R161.
[15] A. Fujisawa, "A review of zonal flow experiments", *Nuclear Fusion* **49** (2009), no. 1, p. 013001.
[16] A. A. Galeev and R. Z. Sagdeev, "Transport phenomena in a collisionless plasma in a toroidal magnetic system", *Soviet Physics JETP-USSR* **26** (1968), no. 1, p. 233.
[17] — , "Theory of neoclassical diffusion", Reviews of Plasma Physics (M. Leontovich, éd.), vol. 7, Consultants Bureau, New York, 1979, p. 257–343.
[18] Z. Gao, K. Itoh, H. Sanuki and J. Q. Dong, "Eigenmode analysis of geodesic acoustic modes", *Physics of Plasmas* **15** (2008), no. 7, p. 072511.
[19] X. Garbet, J. Abiteboul, E. Trier, O. Gurcan, Y. Sarazin, A. Smolyakov, S. Allfrey, C. Bourdelle, C. Fenzi, V. Grandgirard, P. Ghendrih and P. Hennequin, "Entropy production rate in tokamaks with nonaxisymmetric magnetic fields", *Physics of Plasmas* **17** (2010), no. 7, p. 072505.
[20] A. L. Garcia-Perciante, J. D. Callen, K. C. Shaing and C. C. Hegna, "Time-dependent neoclassical viscosity", *Physics of Plasmas* **12** (2005), no. 5, p. 052516.

[21] H. Grad, "On the kinetic theory of rarefied gases", *Communications on Pure and Applied Mathematics* **2** (1949), no. 4, p. 331–407.

[22] K. Hallatschek, "Nonlinear three-dimensional flows in magnetized plasmas", *Plasma Physics and Controlled Fusion* **49** (2007), no. 12B, p. B137.

[23] R. Hazeltine, "Rotation of a toroidally confined, collisional plasma", *Physics of Fluids* **17** (1974), no. 5, p. 961–968.

[24] P. Helander and D. J. Sigmar, *Collisional transport in magnetized plasmas*, North-Holland, Amsterdam, 1988.

[25] F. L. Hinton and R. D. Hazeltine, "Theory of plasma transport in toroidal confinement systems", *Reviews of Modern Physics* **48** (1976), no. 2, p. 239–308.

[26] F. L. Hinton and C. W. Horton, "Amplitude limitation of collisional drift wave instability", *Physics of Fluids* **14** (1971), no. 1, p. 116.

[27] F. L. Hinton and J. A. Robertson, "Neoclassical dielectric property of a tokamak plasma", *Physics of Fluids* **27** (1984), no. 5, p. 1243–1247.

[28] F. L. Hinton and M. N. Rosenbluth, "Dynamics of axisymmetric (E $\times$ B) and poloidal flows in tokamaks", *Plasma Physics and Controlled Fusion* **41** (1999), p. A653–A662.

[29] F. L. Hinton and S. K. Wong, "Neoclassical ion-transport in rotating axisymmetric plasmas", *Physics of Fluids* **28** (1985), no. 10, p. 3082–3098.

[30] S. P. Hirshman, "Ambipolarity paradox in toroidal diffusion, revisited", *Nuclear Fusion* **18** (1978), no. 7, p. 917–927.

[31] — , "Moment equation approach to neoclassical transport-theory", *Physics of Fluids* **21** (1978), no. 2, p. 224–229.

[32] S. P. Hirshman and D. J. Sigmar, "Neoclassical transport of impurities in tokamak plasmas", *Nuclear Fusion* **21** (1981), no. 9, p. 1079–1201.

[33] C. T. Hsu, R. D. Hazeltine and P. J. Morrison, "A generalized reduced fluid model with finite ion-gyroradius effects", *Physics of Fluids* **29** (1986), no. 5, p. 1480–1487.

[34] C. T. Hsu, K. C. Shaing and R. Gormley, "Time-dependent parallel viscosity and relaxation rate of poloidal rotation in the banana regime", *Physics of Plasmas* **1** (1994), no. 1, p. 132–138.

[35] B. Kadomtsev and I. O. Pogutse, "Reviews of plasma physics", Reviews of Plasma Physics (M. Leontovich, éd.), vol. 5, Consultants Bureau, New York, 1970, p. 379.

[36] L. M. Kovrizhnykh, "Neoclassical theory of transport processes in toroidal magnetic confinement systems, with emphasis on non-

axisymmetric configurations", *Nuclear Fusion* **24** (1984), no. 7, p. 851–936.

[37] — , "Evolution of the neoclassical theory (historical review, 1951–1989)", *Physica Scripta* **43** (1991), p. 194–202.

[38] — , "Relaxation of plasma rotation in toroidal magnetic confinement systems", *Plasma Physics Reports* **29** (2003), no. 4, p. 279–289.

[39] L. Kovrizhnykh, "Transport phenomena in toroidal magnetic systems", *Soviet Physics JETP-USSR* **29** (1969), no. 3, p. 475.

[40] V. B. Lebedev, P. N. Yushmanov, P. H. Diamond, S. V. Novakovskii and A. I. Smolyakov, "Plateau regime dynamics of the relaxation of poloidal rotation in tokamak plasmas", *Physics of Plasmas* **3** (1996), no. 8, p. 3023–3031.

[41] G. R. McKee, D. K. Gupta, R. J. Fonck, D. J. Schlossberg, M. W. Shafer and P. Gohil, "Structure and scaling properties of the geodesic acoustic mode", *Plasma Physics and Controlled Fusion* **48** (2006), no. 4, p. S123–S136.

[42] A. V. Melnikov, V. A. Vershkov, L. G. Eliseev, S. A. Grashin, A. V. Gudozhnik, L. I. Krupnik, S. E. Lyscnko, V. A. Mavrin, S. V. Perfilov, D. A. Shelukhin, S. V. Soldatov, M. V. Ufimtsev, A. O. Urazbaev, G. Van Oost and L. G. Zimeleva, "Investigation of geodesic acoustic mode oscillations in the T-10 tokamak", *Plasma Physics and Controlled Fusion* **48** (2006), no. 4, p. S87–S110.

[43] A. B. Mikhailovskii, A. I. Smolyakov, E. A. Kovalishen, M. S. Shirokov, V. S. Tsypin and R. M. O. Galvao, "Generation of zonal flows by ion-temperature-gradient and related modes in the presence of neoclassical viscosity", *Physics of Plasmas* **13** (2006), no. 5, p. 052516.

[44] A. B. Mikhailovskii and V. S. Tsypin, "Transport-equations of plasma in a curvilinear magnetic-field", *Beitrage Aus Der Plasmaphysik-Contributions to Plasma Physics* **24** (1984), no. 4, p. 335–354.

[45] R. C. Morris, M. G. Haines and R. J. Hastie, "The neoclassical theory of poloidal flow damping in a tokamak", *Physics of Plasmas* **3** (1996), no. 12, p. 4513–4520.

[46] V. Naulin, A. Kendl, O. E. Garcia, A. H. Nielsen and J. J. Rasmussen, "Shear flow generation and energetics in electromagnetic turbulence", *Physics of Plasmas* **12** (2005), no. 5, p. 052515.

[47] S. V. Novakovskii, C. S. Liu, R. Z. Sagdeev and M. N. Rosenbluth, "The radial electric field dynamics in the neoclassical plasmas", *Physics of Plasmas* **4** (1997), no. 12, p. 4272–4282.

[48] F. I. Parra and P. J. Catto, "Transport of momentum in full f gyrokinetics", *Physics of Plasmas* **17** (2010), no. 5, p. 056106.

[49] A. G. Peeters, C. Angioni, A. Bortolon, Y. Camenen, F. J. Casson, B. Duval, L. Fiederspiel, W. A. Hornsby, Y. Idomura, T. Hein, N. Kluy, P. Mantica, F. I. Parra, A. P. Snodin, G. Szepesi, D. Strintzi, T. Tala, G. Tardini, P. de Vries and J. Weiland, "Overview of toroidal momentum transport", *Nuclear Fusion* **51** (2011), no. 9, p. 094027.

[50] I. O. Pogutse, A. I. Smolyakov and A. Hirose, "Magnetohydrodynamic equations for plasmas with finite-larmor-radius effects", *Journal of Plasma Physics* **60** (1998), p. 133–149.

[51] A. Rogister, "Revisited neoclassical transport-theory for steep, collisional plasma edge profiles", *Physics of Plasmas* **1** (1994), no. 3, p. 619–635.

[52] M. N. Rosenbluth, P. Rutherford, J. Taylor, E. Frieman and L. Kovrizhnykh, "Neoclassical effects on plasma equlibria and rotation", *Plasma Physics and Controlled Nuclear Fusion Research* (June 17–23, 1971, Madison, Wisconsin, USA), vol. 1, International Atomic Energy Agency, Vienna, 1971, p. 495–507.

[53] M. Rosenbluth and A. Simon, "Finite larmor radius equations with nonuniform electric fields and velocities", *Physics of Fluids* **8** (1965), no. 7, p. 1300.

[54] V. Rozhansky and M. Tendler, "Plasma rotation in tokamaks", Reviews of Plasma Physics (B. Kadomtsev, éd.), vol. 18, Consultants Bureau, N.Y. & London, 1996, p. 147.

[55] P. Rutherford, "Collisional diffusion in an axisymmetric torus", *Physics of Fluids* **13** (1970), no. 2, p. 482.

[56] B. D. Scott, "Energetics of the interaction between electromagnetic ExB turbulence and zonal flows", *New Journal of Physics* **7** (2005), p. 92.

[57] V. Shafranov, "Reviews of plasma physics", Reviews of Plasma Physics (M. Leontovich, éd.), vol. 2, Consultants Bureau, New York, 1966, p. 103.

[58] K. C. Shaing, "Magnetohydrodynamic-activity-induced toroidal momentum dissipation in collisionless regimes in tokamaks", *Physics of Plasmas* **10** (2003), no. 5, p. 1443–1448.

[59] K. C. Shaing and S. P. Hirshman, "Relaxation rate of poloidal rotation in the banana regime in tokamaks", *Physics of Fluids B-Plasma Physics* **1** (1989), no. 3, p. 705–707.

[60] A. N. Simakov and P. J. Catto, "Momentum transport in arbitrary mean-free path plasma with a maxwellian lowest order distribution function", *Plasma Physics and Controlled Fusion* **49** (2007), no. 6, p. 729–752.

[61] A. I. Smolyakov, "Gyroviscous forces in a collisionless plasma with temperature gradients", *Canadian Journal of Physics* **76** (1998), no. 4, p. 321–331.

[62] A. I. Smolyakov, X. Garbet and C. Bourdelle, "On the parallel momentum balance in low pressure plasmas with an inhomogeneous magnetic field", *Nuclear Fusion* **49** (2009), p. 125001.

[63] A. I. Smolyakov, X. Garbet, G. Falchetto and M. Ottaviani, "Multiple polarization of geodesic curvature induced modes", *Physics Letters A* **372** (2008), no. 45, p. 6750–6756.

[64] A. I. Smolyakov and E. Lazzaro, "On neoclassical effects in the theory of magnetic islands", *Physics of Plasmas* **11** (2004), no. 9, p. 4353–4360.

[65] W. M. Stacey and D. J. Sigmar, "Viscous effects in a collisional tokamak plasma with strong rotation", *Physics of Fluids* **28** (1985), no. 9, p. 2800–2807.

[66] T. E. Stringer, "Equilibrium diffusion rate in a toroidal plasma at intermediate collision frequencies", *Physics of Fluids* **13** (1970), no. 3, p. 810.

[67] H. Sugama and T. H. Watanabe, "Collisionless damping of geodesic acoustic modes", *Journal of Plasma Physics* **72** (2006), p. 825–828.

[68] H. Sugama, T. H. Watanabe, M. Nunami and S. Nishimura, "Momentum balance and radial electric fields in axisymmetric and nonaxisymmetric toroidal plasmas", *Plasma Physics and Controlled Fusion* **53** (2011), no. 2, p. 024004.

[69] M. Taguchi, "Relaxation rate of poloidal rotation in the banana regime", *Plasma Physics and Controlled Fusion* **33** (1991), no. 7, p. 859–868.

[70] I. Tamm and A. Sakharov, "Theory of magnetic thermonuclear reactor", Plasma physics and the problem of controlled thermonuclear reactions (M. Leontovich, éd.), vol. 1, Pergamon Press, 1959, London & New York, 1959, p. 3.

[71] T. Watari, Y. Hamada, T. Notake, N. Takeuchi and K. Itoh, "Geodesic acoustic mode oscillation in the low frequency range", *Physics of Plasmas* **13** (2006), no. 6, p. 062504.

[72] N. Winsor, J. L. Johnson and J. M. Dawson, "Geodesic acoustic waves in hydromagnetic systems", *Physics of Fluids* **11** (1968), no. 11, p. 2448.

[73] S. K. Wong and V. S. Chan, "The neoclassical angular momentum flux in the large aspect ratio limit", *Physics of Plasmas* **12** (2005), p. 092513.

[74] — , "Fluid theory of radial angular momentum flux of plasmas in an axisymmetric magnetic field", *Physics of Plasmas* **14** (2007), p. 112505.

[75] — , "Kinetic theory of radial angular momentum flux of collisional plasmas in an axisymmetric magnetic field", *Physics of Plasmas* **14** (2007), p. 122501.

[76] — , "Self-consistent poloidal electric field and neoclassical angular momentum flux", *Physics of Plasmas* **16** (2009), p. 122507.

[77] P. N. Yushmanov, "Reviews of plasma physics", Reviews of Plasma Physics (M. Leontovich, éd.), vol. 16, Consultants Bureau, New York, 1990, p. 117.

[78] F. Zonca, L. Chen and R. Santoro, "Kinetic theory of low-frequency Alfven modes in tokamaks", *Plasma Physics and Controlled Fusion* **38** (1996), no. 11, p. 2011–2028.

Chapter 8

The General Fishbone Like Dispersion Relation

Fulvio Zonca

Associazione Euratom-ENEA sulla Fusione, C.R. Frascati, C.P. 65 - 00044 - Frascati, Italy
Institute for Fusion Theory and Simulation, Zhejiang University, Hangzhou 310027, People's Republic of China

8.1 Introduction

In a burning plasma, high energy alpha particles can interact with the main thermal plasma. The following lecture presents a general framework where such an interaction (and more generally, the interaction of energetic particles with the thermal plasma) can be described within the context of a linear description of plasma dynamics, yielding the so-called *general fishbone-like dispersion relation* [9, 11, 14, 39, 43, 45, 47, 49].

Although a purely linear analysis is presented below, it should be noted that the fishbone-like dispersion relation can also be used as a starting point for nonlinear developments [48, 49], which will not be discussed in the present lecture.

This chapter is an offspring of lecture notes. In this spirit *suggested exercises* are indicated conventionally with $\leftrightarrow$. Readers are kindly suggested to consider them as important complement to the notes themselves.

8.2 Motivation and outline

There exist several motivations for the study of energetic particle physics. The main one is to identify fluctuations of the Alfvén branch [1, 2, 5] that are most important for causing wave-induced fast ion transports in burning toroidal plasmas of fusion interest. More generally, fast ion transport is intimately related with rotation and momentum transport via generation of radial electric fields (linear and nonlinear).

Two fundamental ideas will be developed in the following presentation, which apply within the optimal wavelength and frequency ordering for studying collective Alfvén mode excitation in burning plasmas [47]:

- Both resonant and non-resonant wave-particle interaction with fast ions are determined by the magnetic drift curvature coupling.
- A wide class of shear Alfvén waves of fusion interest can be described within the unified framework of one single fishbone-like dispersion relation.

8.3 Fundamental equations

The detailed and accurate description of the Alfvén spectrum requires the use of a kinetic modeling for both the thermal plasma and the energetic population [51]. In order to derive the equations describing the fluctuations belonging to the Shear Alfvén spectrum in Section 8.5, one must couple to the electromagnetic fields using Maxwell equations an appropriate gyrokinetic response of the particles at frequency ω, in the limit of low frequency plasma oscillations.[1]

$$|\omega| \ll |\omega_{ci}| \,, \tag{8.1}$$

where the notation $\omega_{cs} = (eB/mc)_s$ stands for the cyclotron frequency of a species s, in the $C.G.S.$ unit framework.

More precisely, the Coulomb gauge is used where three scalar fields ($\delta\phi$, $\delta B_\parallel$, $\delta\psi$) completely describe the electromagnetic fields, and therefore the plasma response, with $\delta\psi$ defined by

$$\delta A_\parallel \equiv -\mathrm{i}\left(\frac{c}{\omega}\right)\mathbf{b}\cdot\nabla\delta\psi \,. \tag{8.2}$$

[1]Here I closely follow [10] and [21].

The determination of the three scalar fields is done using three field equations: quasi-neutrality, vorticity (curl of momentum equation), and perpendicular Ampère's law.

8.3.1 *The collisionless gyrokinetic equation*

In the gyrokinetic description, the perturbed distribution function δf_s of a species s, $f_s = F_{0s} + \delta f_s$ where F_{0s} is the reference equilibrium distribution function at ($\delta\phi = 0$, $\delta B_\| = 0$, $\delta\psi = 0$), can be decomposed into a part trivially reducing to the fluid response in the appropriate limits, and the kinetic compression response δK_s [10],

$$\delta f_s = \left(\frac{\mathrm{e}}{m}\right)_s \left[\frac{\partial F_0}{\partial \mathcal{E}}\delta\phi - \frac{QF_0}{\omega}\mathrm{e}^{\mathrm{i}L_k} J_0(k_\perp \rho_s)\delta\psi\right]_s$$
$$+ \left(\frac{\mathrm{e}}{m}\right)_s \frac{\partial F_0}{B\partial\mu}\Bigg[\left(1 - J_0(k_\perp\rho_s)\mathrm{e}^{\mathrm{i}L_k}\right)\left(\delta\phi + \mathrm{i}\left(\frac{v_\|}{\omega}\right)\mathbf{b}\cdot\nabla\delta\psi\right)$$
$$- \frac{v_\perp}{k_\perp c} J_1(k_\perp\rho_s)\mathrm{e}^{\mathrm{i}L_k}\delta B_\|\Bigg]_s + \mathrm{e}^{\mathrm{i}L_{ks}}\delta K_s\,, \tag{8.3}$$

where δK_s is obtained via the gyrokinetic equation, which can be written

$$\left[\omega_{tr}\partial_\theta - \mathrm{i}(\omega - \omega_d)\right]_s \delta K_s = \mathrm{i}\left(\frac{\mathrm{e}}{m}\right)_s QF_{0s}\Bigg[J_0(k_\perp\rho_s)\left(\delta\phi - \delta\psi\right)$$
$$+ \left(\frac{\omega_d}{\omega}\right)_s J_0(k_\perp\rho_s)\delta\psi + \frac{v_\perp}{k_\perp c}J_1(k_\perp\rho_s)\delta B_\|\Bigg] \tag{8.4}$$

in the space of the extended poloidal angle variable ($qR\nabla_\| \to \partial_\theta$, see [16]). The above equation combines particularities of the toroidal geometry and of the associated particle motion, as well as the effect of a perturbation on the particle distribution function, included in terms $\propto QF_{0s}$. Note that the gyrokinetic distribution function is that of guiding centers while that required in the Maxwell equations for the distribution of charges and currents is the actual particle distribution function, their relationship being governed by finite Larmor radius effects embedded in the Bessel functions.

The signification of the various terms is given below.

$\boldsymbol{\kappa} = \boldsymbol{b}\cdot\boldsymbol{\nabla}\boldsymbol{b}$ — the magnetic field curvature.

$L_{ks} = \omega_{cs}^{-1}(\mathbf{k}\times\mathbf{b})\cdot\mathbf{v}$ — generator of coordinate transform from guiding-center to particle position.

J_0, J_1 — the gyroaverage Bessel operators.

$\omega_{tr} = v_\parallel/qR$ — the transit frequency.

$\omega_{ds} = \omega_{cs}^{-1}(\boldsymbol{k}\times\boldsymbol{b})\cdot(\mu\boldsymbol{\nabla}B + v_\parallel^2\boldsymbol{\kappa}) \equiv (m_s c/\mathrm{e}_s)(\mu\Omega_B + v_\parallel^2\Omega_\kappa/B)$ — the magnetic drift frequency.

$QF_{0s} = (\omega\partial_{\mathcal{E}} + \hat{\omega}_*)_s F_{0s}$ — contains free energy gradients of the particle equilibrium distribution function.

$$\text{with}: \quad \mathcal{E} = v^2/2, \qquad \hat{\omega}_{*s}F_{0s} = \omega_{cs}^{-1}(\mathbf{k}\times\mathbf{b})\cdot\nabla F_{0s}\,. \tag{8.5}$$

Note that terms $\propto \boldsymbol{b}\cdot\boldsymbol{\nabla}J_0(k_\perp\rho_s)$ of the gyrokinetic equation have been neglected: see [10] for justification.

8.3.2 *Vorticity equation*

One readily derives the vorticity equation in the form originally given in Ref. [47]. Schematic derivation, definition of symbols and interpretation of terms are given below (see also [4] for a recent detailed discussion of these issues).

$$\begin{aligned}
&\boldsymbol{B}\cdot\boldsymbol{\nabla}\left(\frac{k_\perp^2}{k_\vartheta^2 B^2}\sigma_k \boldsymbol{B}\cdot\boldsymbol{\nabla}\delta\psi\right) + \frac{4\pi\omega^2}{k_\vartheta^2 c^2}\sum_s\left\{\langle \mathrm{e}_s\delta f_s\rangle + \mathrm{i}\frac{\boldsymbol{v}_E\cdot\boldsymbol{\nabla}}{\omega}\langle \mathrm{e}_s F_{0s}\rangle\right.\\
&\left.-\left\langle\frac{\mathrm{e}_s^2}{m_s}\frac{QF_{0s}}{\omega}\left(1-J_0^2(k_\perp\rho_s)\right)\right\rangle\delta\phi\right\} + \left\langle\sum_s\frac{4\pi}{k_\vartheta^2c^2}J_0^2(k_\perp\rho_s)\omega\omega_{ds}\right.\\
&\left.\times\frac{\mathrm{e}_s^2}{m_s}\frac{QF_{0s}}{\omega}\right\rangle\delta\psi - \left\langle\sum_s\frac{4\pi\mathrm{e}_s}{k_\vartheta^2c^2}J_0(k_\perp\rho_s)\omega\omega_{ds}\delta K_s\right\rangle\\
&+\frac{4\pi\omega^2}{k_\vartheta^2c^2}\sum_s\left\{\left\langle\frac{v_\perp}{k_\perp c}J_0(k_\perp\rho_s)J_1(k_\perp\rho_s)\frac{\mathrm{e}_s^2}{m_s}\frac{QF_{0s}}{\omega}\right\rangle\delta B_\parallel\right.\\
&\left.-\left\langle\frac{\mathrm{e}_s^2}{m_s}\frac{\partial F_{0s}}{B\partial\mu}\left[\left(1-J_0^2(k_\perp\rho_s)\right)\delta\phi - \frac{v_\perp}{k_\perp c}J_0(k_\perp\rho_s)J_1(k_\perp\rho_s)\delta B_\parallel\right]\right\rangle\right\} = 0\,.
\end{aligned} \tag{8.6}$$

8.3.2.1 *Schematic derivation of vorticity equation*

The vorticity equation is derived from the quasineutrality equation, $\boldsymbol{\nabla}\cdot\boldsymbol{J}=0$, in the asymptotic limit $k\lambda_D \ll 1$ where k is the characteristic wavelength of interest and λ_D the Debye length. The following formal steps in obtaining 8.6 are then made: i) both sides of the gyrokinetic equation 8.4 are multiplied by $(4\pi\mathrm{i}\omega)/(k_\vartheta^2c^2)\mathrm{e}_sJ_0(k_\perp\rho_s)$ where the Bessel function $J_0(k_\perp\rho_s)$ must be introduced to recover the perturbation to the actual particle distribution function, ii) sum on *all* particle species and integrate over velocity space to compute the local current density.

Hereafter, angular brackets $\langle ...\rangle$ indicate velocity space integration, hence over $v_\parallel$ and $v_\perp$. Taking into account $\mu=v_\perp^2/(2B)$, $\mathcal{E}=v^2/2$ one then obtains:

$$\langle ...\rangle = 2\pi\sum_{v_\parallel/|v_\parallel|=\pm}\int\frac{B}{|v_\parallel|}d\mathcal{E}d\mu\,(...)$$

The vorticity equation can be compared with the usual terms which appear in the MHD energy principle, or reduced MHD vorticity equation. The total equilibrium current $J_\parallel$, and terms $\propto \boldsymbol{b}\cdot\boldsymbol{\nabla}J_0(k_\perp\rho_s)$ are dropped since they are usually negligible for the physical parameter ordering considered here [10].

8.3.2.2 *The field line bending term (FLB)*

The field line bending term (FLB) accounts for the restoring force of bended field lines $\propto \omega_{tr}\partial_\theta\delta K_s$, namely:

$$\begin{aligned}\mathrm{FLB} = &\frac{4\pi}{ck_\vartheta^2}\boldsymbol{B}\cdot\boldsymbol{\nabla}\left[\frac{\boldsymbol{b}}{B}\cdot\boldsymbol{\nabla}\left(\frac{J_\parallel}{B}\right)\times\boldsymbol{\nabla}\delta\psi\right]+\frac{4\pi\mathrm{i}\omega}{k_\vartheta^2c^2}\boldsymbol{B}\cdot\boldsymbol{\nabla}\left\langle\sum_s\mathrm{e}_s\frac{v_\parallel}{B}\left(\delta f\right.\right.\\ &\left.\left.-\mathrm{i}\frac{\mathrm{e}}{m}\frac{\partial F_0}{B\partial\mu}\frac{v_\parallel}{\omega}\left(1-J_0^2(k_\perp\rho)\right)\boldsymbol{b}\cdot\boldsymbol{\nabla}\delta\psi\right)_s\right\rangle, \qquad (8.7)\\ \simeq &\,\boldsymbol{B}\cdot\boldsymbol{\nabla}\left(\frac{k_\perp^2}{k_\vartheta^2B^2}\sigma_k\boldsymbol{B}\cdot\boldsymbol{\nabla}\delta\psi\right),\end{aligned}$$

$$\text{with}\quad \sigma_k = 1+\frac{4\pi}{k_\perp^2c^2}\left\langle\sum_s\frac{e_s^2}{m_s}\frac{\partial F_{0s}}{B\partial\mu}v_\parallel^2\left(1-J_0^2(k_\perp\rho_s)\right)\right\rangle. \qquad (8.8)$$

8.3.2.3 *The inertia term*

The inertia – $\propto \omega\delta K_s$ – is naturally combined with the parallel electric field – $\propto (\delta\phi-\delta\psi)$ – on the right hand side (RHS). We call this term the inertia charge uncovering term (ICU). Given the electric drift, $\boldsymbol{v}_E = (\mathrm{i}c/B)(\boldsymbol{b} \times \boldsymbol{k})\delta\phi$, The ICU term is:

$$
\begin{aligned}
\mathrm{ICU} &= \frac{4\pi\omega^2}{k_\vartheta^2 c^2} \sum_s \mathrm{e}_s \left\langle J_0(k_\perp \rho_s)\delta K_s + \frac{\mathrm{e}_s}{m_s}\frac{QF_{0s}}{\omega} J_0^2(k_\perp \rho_s)(\delta\phi - \delta\psi) \right\rangle \\
&= \frac{4\pi\omega^2}{k_\vartheta^2 c^2} \sum_s \left\{ \langle \mathrm{e}_s \delta f_s \rangle + \mathrm{i}\frac{\boldsymbol{v}_E \cdot \boldsymbol{\nabla}}{\omega} \langle \mathrm{e}_s F_{0s} \rangle - \left\langle \frac{\mathrm{e}_s^2}{m_s}\frac{QF_{0s}}{\omega} \right.\right. \\
&\quad \times \left(1 - J_0^2(k_\perp \rho_s)\right) \Big\rangle \delta\phi - \left\langle \frac{\mathrm{e}_s^2}{m_s}\frac{\partial F_{0s}}{B\partial\mu} \left[\left(1 - J_0^2(k_\perp \rho_s)\right)\delta\phi \right.\right. \\
&\quad \left.\left.\left. - \frac{v_\perp}{k_\perp c} J_0(k_\perp \rho_s) J_1(k_\perp \rho_s) \delta B_\parallel \right] \right\rangle \right\} . \qquad (8.9)
\end{aligned}
$$

Assuming quasi-neutrality for both the equilibrium and fluctuation contributions, the first two terms on the RHS are vanishingly small. The third term on the RHS describes the well known response due to plasma inertia or, equivalently, the divergence of the polarization current.

8.3.2.4 *The MHD non-adiabatic particle compression*

The MHD non-adiabatic particle compression (MPC), $\propto \omega_{ds}\delta\psi$, is readily derived in the form

$$
\mathrm{MPC} = \left\langle \sum_s \frac{4\pi}{k_\vartheta^2 c^2} J_0^2(k_\perp \rho_s)\omega\omega_{ds} \frac{\mathrm{e}_s^2}{m_s}\frac{QF_{0s}}{\omega} \right\rangle \delta\psi . \qquad (8.10)
$$

8.3.2.5 *The kinetic non-adiabatic particle compression*

In the present approach, a kinetic contribution is added to the compression term. The kinetic – *non MHD* – non-adiabatic particle compression (KPC), $\propto \omega_{ds}\delta K_s$, reads

$$
\mathrm{KPC} = -\left\langle \sum_s \frac{4\pi \mathrm{e}_s}{k_\vartheta^2 c^2} J_0(k_\perp \rho_s)\omega\omega_{ds}\delta K_s \right\rangle . \qquad (8.11)
$$

Note that this term contains the most important dynamic effects for Energetic Particle Modes (EPM) [11]. These modes arise from the existence of energetic particles, see below, including both *resonant* as well as *non-resonant* particle responses.

8.3.2.6 *The magnetic field compression*

The magnetic field compression (MFC), $\propto \delta B_{\parallel}$, can be written as

$$\mathrm{MFC} = \left\langle \sum_s \frac{4\pi\omega^2}{k_\vartheta^2 c^2} \frac{v_\perp}{k_\perp c} J_0(k_\perp \rho_s) J_1(k_\perp \rho_s) \frac{\mathrm{e}_s^2}{m_s} \frac{QF_{0s}}{\omega} \right\rangle \delta B_{\parallel} \,. \tag{8.12}$$

Generally, it is convenient to combine MPC and MFC terms to demonstrate well known simplifications due to perpendicular pressure balance [10]. This point amounts to replacing expressions containing $\Omega_B \delta\psi + (\omega/c)\delta B_{\parallel}$ by $\Omega_\kappa \delta\psi$, owing to the MHD equilibrium conditions and perpendicular pressure balance (where Ω_κ and Ω_B have been defined in Eq. (8.5), and represent the grad-B and curvature drifts). Failing to do so, e.g. dropping $\delta B_{\parallel}$ without modifying the grad-B drift appropriately, is not consistent and may generate errors.

↔ *Show that simplifications due to perpendicular pressure balance reduce to the simple rule: drop $\delta B_{\parallel}$ and substitute $\boldsymbol{\nabla} B$ with curvature drift* [10, 27].

8.3.3 *Quasi-neutrality condition*

Without enforcing the charge quasi-neutrality condition specifically for the equilibrium or the fluctuations, the quasi-neutrality equation reads:

$$\begin{aligned}
&\sum_s \langle \mathrm{e}_s \delta f_s \rangle + \mathrm{i} \frac{\boldsymbol{v}_E \cdot \boldsymbol{\nabla}}{\omega} \sum_s \langle \mathrm{e}_s F_{0s} \rangle \\
&\quad = \sum_s \mathrm{e}_s \left\langle J_0(k_\perp \rho_s) \delta K_s \frac{\mathrm{e}_s}{m_s} \frac{QF_{0s}}{\omega} J_0^2(k_\perp \rho_s)(\delta\phi - \delta\psi) \right\rangle \\
&\quad + \sum_s \left[\left\langle \frac{\mathrm{e}_s^2}{m_s} \left(\frac{QF_{0s}}{\omega} + \frac{\partial F_{0s}}{B\partial\mu} \right) \left(1 - J_0^2(k_\perp \rho_s)\right) \right\rangle \delta\phi \right. \\
&\quad \left. - \left\langle \frac{\mathrm{e}_s^2}{m_s} \frac{\partial F_{0s}}{B\partial\mu} \frac{v_\perp}{k_\perp c} J_0(k_\perp \rho_s) J_1(k_\perp \rho_s) \right\rangle \delta B_{\parallel} \right] .
\end{aligned} \tag{8.13}$$

↔ *Derive this equation term by term* (see [10]).

8.3.4 *Perpendicular Ampère's law*

From the low-frequency Ampère's law, one readily derives [10]

$$k_\perp^2 \delta B_{\parallel} = -\frac{4\pi}{c} k_\perp \sum_s \left\langle \mathrm{e}_s v_\perp J_1(k_\perp \rho_s) \left[\delta K_s - \frac{\mathrm{e}_s}{m_s} \frac{QF_{0s}}{\omega} J_0(k_\perp \rho_s) \delta\psi \right] \right\rangle$$

$$+\frac{4\pi}{c}k_\perp \sum_s \left\langle \frac{e_s^2}{m_s}\frac{\partial F_{0s}}{B\partial\mu}[5pt]v_\perp J_1(k_\perp\rho_s)\left[J_0(k_\perp\rho_s)\delta\phi \right.\right.$$

$$\left.\left. +\frac{v_\perp}{k_\perp c}J_1(k_\perp\rho_s)\delta B_\parallel\right]\right\rangle .$$

$\leftrightarrow$ *Derive this equation term by term* (see [10]).

The equations derived in the previous paragraphs are extremely general and complicated to solve. For this reason, one can develop reduced models on the basis of relative ordering of the various quantities in the problem of interest [4, 10, 47]. Incidentally, it is worthwhile noting that – in the appropriate limit – they reduce to the form used in [38] for theoretical investigations of Kinetic Ballooning Modes (KBMs).

8.4 Studying collective modes in burning plasmas

We follow here the theoretical model used in [10]. Thus, we assume an axisymmetric toroidal plasma with two components in the distribution function: the bulk or core (referred to with the letter c), made of electrons (e) and thermal ions (i), which can be generally assumed Maxwellian and isotropic; and an energetic ion component (E), which is typically anisotropic and/or non-Maxwellian.

In the following Sections, 8.4.1 to 8.4.4, the relevant approximations for the study of collective modes in burning plasmas are introduced. They are based on an ordering with respect to

$$\delta = L_p/R_0 \tag{8.14}$$

as a reference small parameter, where L_p is a typical radial pressure scale length of the plasma equilibrium and R_0 is the plasma major radius on the magnetic axis.

8.4.1 *Ideal plasma equilibrium in the low-β limit*

Given $P_\perp$ and $P_\parallel$, the *total* perpendicular and parallel plasma pressures, they satisfy the ideal MHD equilibrium conditions [27]

$$\frac{\partial P_\parallel}{\partial B} = \frac{P_\parallel - P_\perp}{B}, \tag{8.15}$$

$$\tau \boldsymbol{\nabla}_\perp \ln B + \frac{4\pi}{B^2}\tilde{\boldsymbol{\nabla}} P_\perp = \sigma \boldsymbol{\kappa}, \tag{8.16}$$

where we define:

$$\tau \equiv 1+(4\pi/B)(\partial P_\perp/\partial B)\,,\tilde{\nabla} \equiv \nabla\psi\partial_\psi\,,\sigma \equiv 1+(4\pi/B^2)(P_\perp - P_\parallel)\,. \quad (8.17)$$

↔ *Derive these equilibrium conditions assuming anisotropic diagonal pressure tensor,* $\mathbf{P} = P_\perp \mathbf{I} + (P_\parallel - P_\perp)\mathbf{bb}$*, with* $\mathbf{I}$ *the unit tensor, and no equilibrium flow.*

In a somewhat different manner from that of Ref. [10], where space plasmas are considered, we address tokamak conditions such that $\beta \equiv 2P/(B^2/(4\pi)) \ll 1$. For convenience we then set:

$$\beta_c \approx \delta\,. \quad (8.18)$$

Given this low-β ordering, $\sigma_k \simeq \sigma \equiv 1 + (4\pi/B^2)(P_\perp - P_\parallel) \simeq 1$ and $\tau \equiv 1 + (4\pi/B)(\partial P_\perp/\partial B) \simeq 1$ in the ideal MHD equilibrium conditions [27].

8.4.2 *Approximations for the energetic population*

The core component is assumed cold compared to the energetic ions, *i.e.* $T_c/T_E \approx \delta$. The energetic component is tenuous, with $n_E/n_c \approx \delta^{3/2}$. It implies [47]

$$\beta_E \approx \delta^{1/2}\beta_c \quad (8.19)$$

consistently with $P_E \approx (\tau_{sd}/\tau_E)P_c$, with τ_{sd} is the energetic particle slowing down time and τ_E the energy confinement time, holding in thermonuclear plasmas where the power density source is dominated by charged fusion products. Equations derived below with this ordering are easily generalized to the case $\beta_E \approx \beta_c$ [10, 41]. Generally, energetic ions are anisotropic and characterized by $P_{E\parallel} \neq P_{E\perp}$.

8.4.3 *Characteristic frequencies of particle motions*

From the orderings of physical parameters, introduced above, it is possible to determine the characteristic frequencies of particle motions for thermal and energetic particles, *i.e.*, the relevant inverse time scales for resonant wave-particle energy transfers: the transit frequency of circulating particles, the bounce and toroidal precession frequencies of trapped particles.

From the previous orderings, it comes that $v_E \approx v_A$, $v_A \equiv (4\pi n_c m_i)^{-1/2}B$. Thus, transit frequencies scale as $\omega_{tE}/\omega_A \approx 1$, $\omega_{ti}/\omega_A \approx \delta^{1/2}$ and $\omega_{te}/\omega_A \approx \delta^{-1}$, where $\omega_A \equiv v_A/qR_0$. The bounce frequencies scale as: $\omega_{bE}/\omega_A \approx \delta^{1/2}$, $\omega_{bi}/\omega_A \approx \delta$ and $\omega_{be}/\omega_A \approx \delta^{-1/2}$, while the toroidal precession frequencies scale as: $\bar{\omega}_{dE}/\omega_A \approx k_\theta \rho_E$ and $\bar{\omega}_{di}/\omega_A \approx \bar{\omega}_{de}/\omega_A \approx \delta^{1/2} k_\theta \rho_i \approx \delta k_\theta \rho_E$.

8.4.4 *Alfvén wave frequency and wavelength orderings*

We now want to determine the types of modes which can be driven unstable by energetic particles. The energetic particle drive depends on several conditions, such as i) the resonance between the mode frequency and the characteristic frequencies of the energetic particle motion, ii) the existence of available free energy of the energetic population distribution function (at equilibrium for linear analysis), *i.e.* the "gradients" of the energetic particle distribution function, displayed in Eq. (8.5), iii) the radial mode structure. Such conditions reduce the extent of the mode frequencies and wavelengths, which can interact with energetic particles, as will clearly appear in the combination of Eq. (8.22) to Eq. (8.24).

An important category of modes, which fit these requirements is described below and verifies the following *frequency* properties:

- $\omega \lesssim \omega_A$.
  Reducing to this frequency range, typical of the Shear Alfvén spectrum, provides simplified kinetic equations, but clearly, much higher frequency Compressional Alfvén Eigenmodes (called CAEs – see, e.g., [23] – and corresponding to the usual MHD fast magneto-acoustic waves) are not taken into account.
- $\omega_{te} \gg \omega_{be} \gg \omega \gg \bar{\omega}_{de}$.
  With such approximations, electrons are massless and we can ignore trapped electron dynamics, *i.e.* $\delta K_e = 0$. Trapped ion dynamics can be neglected as well[2].

It follows that $|\omega_{*E}/\omega| \gg 1$, with ω_{*E} generically indicating the energetic particle diamagnetic frequency that can be easily recognized in Eq. (8.5). Hence, the drive is essentially due to gradients in the configuration space, more specifically ω_{*pE}, *i.e.* the energetic particle diamagnetic frequency associated with their pressure gradient, from which the modes tap the fast ion expansion free energy, $|\omega| \ll |\omega_{*pE}| \ll |\omega_{cE}|$.

[2]The importance of trapped particle dynamics at very low frequency has been recently emphasized and discussed in [7, 30, 51].

The *wavelength* of the fastest growing mode depends on this driving mechanism. It is determined according to the following rules. Because $\omega_{*pE} \propto n$, the drive is expected to be stronger for high-n modes, which are preferentially excited. Nevertheless, finite magnetic drift orbit width averaging (contained, e.g., in the phase-space integration of Eq. (8.6) and, more specifically, in the response weighted by the Bessel functions) imposes $k_\perp \rho_E < 1$, where ρ_E is the relevant fast ion magnetic drift orbit width (typically larger than the respective Larmor radius). Otherwise, orbit averaging would greatly reduce the interaction strength: thus, the shortest wavelength is set by drift/banana orbit width. Note that linear stability of Alfvén Modes with finite orbit widths is extensively studied and confirms $k_\perp \rho_E < 1$ for most unstable modes[3].

Non-linear dynamics and particle transport also require $k_\perp \rho_E < 1$ [13]. At shorter wavelengths, particle response is nearly adiabatic and small wave-induced particle transports are expected [13, 21].

More quantitatively, for the study of the Shear Alfvén spectrum, it is relevant to use the wavelength ordering [47]:

$$k_\theta \rho_E \approx \delta^{1/2} . \tag{8.20}$$

This ordering naturally encompasses and is completely consistent with those originally introduced before [10, 11, 39]:

$$\bar{\omega}_{di} \ll \omega_{bi} \ll \omega_{*i} \approx \omega_{ti} \approx \bar{\omega}_{dE} \approx \omega_{bE} \lesssim \omega \lesssim \omega_A \approx \omega_{tE} \ll \omega_{*E} .$$

This ordering implies $k_\theta \rho_i \approx \delta$, $\omega_{*E}/\omega_A \approx \delta^{-1/2}$, $\omega_{*c}/\omega_A \approx \delta^{1/2}$, $\bar{\omega}_{dE}/\omega_A \approx \delta^{1/2}$ and $\bar{\omega}_{di}/\omega_A \approx \delta^{3/2}$. It consistently describes the excitation of the Shear Alfvén frequency spectrum by energetic ions precession, precession-bounce and transit resonances in the range $\omega_{*i} \approx \omega_{ti} \lesssim \omega \lesssim \omega_A$, covering the whole frequency range from Kinetic Ballooning Mode (KBM) / Beta Alfvén Eigenmode (BAE) to higher frequency Toroidal Aflvén Eigenmode (TAE). The general and comprehensive theoretical picture anticipated previously is demonstrated[4].

8.4.5 *Applications of the general theoretical framework*

Within the present general theoretical framework and orderings of physical parameters, tractable expressions of quasi-neutrality and vorticity can be

[3]See Ref. [14] for a recent review.

[4]The notion of Beta Alfvén Acoustic Eigenmode (BAAE), introduced recently [24, 25], is yet another particular case of the general theoretical framework described in this lecture: see Ref. [51].

obtained. Solving for $\delta B_\parallel$ via $\perp$-pressure balance and letting

$$
\begin{aligned}
\delta A_\parallel &\equiv -\mathrm{i}(c/\omega)\boldsymbol{b}\cdot\boldsymbol{\nabla}\delta\psi\,, \\
\hat{\omega}_{ds} &= (m_s c/\mathrm{e}_s)(\mu + v_\parallel^2/B)\Omega_\kappa\,, \\
\Omega_B &= (\boldsymbol{k}\times\boldsymbol{b})\cdot\boldsymbol{\nabla}\ln B\,, \\
\Omega_\kappa &= (\boldsymbol{k}\times\boldsymbol{b})\cdot\boldsymbol{\kappa}\,,
\end{aligned}
\tag{8.21}
$$

vorticity and quasineutrality become [47]

$$
\begin{aligned}
&\mathbf{B}\cdot\nabla\left(\frac{k_\perp^2}{k_\vartheta^2 B^2}\mathbf{B}\cdot\nabla\delta\psi\right) + \frac{\omega(\omega-\omega_{*pi})}{v_A^2}\frac{k_\perp^2}{k_\vartheta^2}\delta\phi \\
&\quad - \left\langle \sum_{s\neq e}\frac{4\pi\mathrm{e}_s}{k_\vartheta^2 c^2}J_0(k_\perp\rho_s)\omega\hat{\omega}_{ds}\delta K_s \right\rangle \\
&\quad + \frac{4\pi}{k_\vartheta^2 B^2}\mathbf{k}\times\mathbf{b}\cdot\tilde{\nabla}\left(P_\perp + P_\parallel\right)\Omega_\kappa\delta\psi = 0\,,
\end{aligned}
\tag{8.22}
$$

$$
\left\langle \sum_{s\neq E}\frac{\mathrm{e}_s^2}{m_s}\frac{\partial F_{0s}}{\partial\mathcal{E}}\right\rangle(\delta\phi-\delta\psi) + \sum_{s=i}\mathrm{e}_s\left\langle J_0(k_\perp\rho_s)\delta K_s\right\rangle = 0\,,
\tag{8.23}
$$

and are to be coupled to the gyrokinetic equation

$$
\begin{aligned}
&\left[\omega_{tr}\partial_\theta - \mathrm{i}\left(\omega-\omega_d\right)\right]_s\delta K_s = \mathrm{i}\left(\frac{\mathrm{e}}{m}\right)_s QF_{0s}\left[J_0(k_\perp\rho_s)\left(\delta\phi-\delta\psi\right)\right. \\
&\quad \left. + \left(\frac{\hat{\omega}_d}{\omega}\right)_s J_0(k_\perp\rho_s)\delta\psi - 4\pi m_E\frac{m_s}{\mathrm{e}_s}\frac{2J_1(k_\perp\rho_s)}{k_\perp\rho_s}\mu\left\langle\frac{2J_1(k_\perp\rho_E)}{k_\perp\rho_E}\mu\delta K_E\right\rangle\right].
\end{aligned}
\tag{8.24}
$$

One then finds that in the ideal MHD approximation $\delta E_\parallel \simeq 0$, or $\delta\phi = \delta\psi$. Without the kinetic compression terms $\propto \delta K_s$, quasineutrality yields the ideal Ohm's Law, $\delta\phi = \delta\psi$ while vorticity reduces to that obtainable from the MHD momentum equation.

$\leftrightarrow$ *Show that this is the case, taking into account that terms* $\propto \boldsymbol{\nabla}(J_\parallel/B)$ *are dropped.*

$\leftrightarrow$ *Show that the vorticity equation is the kinetic extension of* $\boldsymbol{\nabla}\cdot\delta\boldsymbol{J} = 0$.

Such an approximation makes sense in the high frequency limit, $\omega_A \gtrsim \omega$, $\omega \gtrsim \omega_{*pi} \gg \omega_{ti}$, where the core ion non-adiabatic response can be neglected due to the ordering $\delta K_i \approx (\omega_{di}/\omega)(\mathrm{e}_i/m_i)(QF_{0i}/\omega)\delta\psi$. Thus, the quasineutrality condition reduces to the ideal MHD approximation. With $\delta E_\parallel \simeq 0$ and the $\propto \omega_{*pi}$ term neglected, the model becomes totally equivalent to [35], *i.e.* to the so called pressure coupling equation in the hybrid MHD-gyrokinetic approach.

One also finds that both resonant as well as non-resonant fast ion responses are determined by magnetic drift curvature coupling [47]. This is consistent with fast ions carrying pressure but not inertia : $v_{dE} \propto (v_\perp^2/2 + v_\parallel^2)/R_0\omega_{cE}$. In particular, Alfvén wave excitations by fast ions in toroidal systems is an evidence of the crucial role of magnetic drifts. This comes from the fact that the plasma possesses large parallel conductivity and negligible finite ion Larmor radius effects. Hence, as already pointed out, $\delta E_\parallel = 0$ and the gyro-averaged forces depend on $\mathbf{v}_d$: with low-β, the effect of $\delta B_\parallel$ is small and

$$\langle \mathbf{v} \times \boldsymbol{\delta}\mathbf{B} \rangle_g \simeq -\mathrm{i}\mathbf{v}_d \cdot \mathbf{k} \left\langle \delta A_\parallel \right\rangle_g \; ; \tag{8.25}$$

with $\langle \ldots \rangle_g$ standing for the gyro-averaging. Furthermore, the impact of the fields on the energetic particle population resonant dynamics (the RHS of Eq. (8.4), which determines in particular wave-particle power exchanges, is proportional to the drift velocity $\mathbf{v}_d$:

$$\langle \mathbf{v} \cdot \boldsymbol{\delta}\mathbf{E} \rangle_g = \mathbf{v}_d \cdot \langle \boldsymbol{\delta}\mathbf{E} \rangle_g \; . \tag{8.26}$$

Further simplification is possible for optimal frequency and wavelength ordering of collective mode excitation, developed for the Shear Alfvén spectrum in Section 8.4.4. It leads to the Reduced kinetic Shear Alfvén Wave (SAW) model equations:

$$\begin{aligned}
&\mathbf{B} \cdot \boldsymbol{\nabla} \left(\frac{k_\perp^2}{k_\vartheta^2 B^2} \mathbf{B} \cdot \boldsymbol{\nabla} \delta\psi \right) + \frac{\omega(\omega - \omega_{*pi})}{v_A^2} \frac{k_\perp^2}{k_\vartheta^2} \delta\phi \\
&\quad - \left\langle \sum_{s=i} \frac{4\pi \mathrm{e}_s}{k_\vartheta^2 c^2} \omega \hat{\omega}_{ds} \delta K_s \right\rangle + \frac{4\pi}{k_\vartheta^2 B^2} \mathbf{k} \times \mathbf{b} \cdot \tilde{\boldsymbol{\nabla}} \left(P_\perp + P_\parallel \right) \Omega_\kappa \delta\psi \\
&= \left\langle \frac{4\pi \mathrm{e}_E}{k_\vartheta^2 c^2} J_0(k_\perp \rho_E) \omega \hat{\omega}_{dE} \delta K_E \right\rangle ,
\end{aligned} \tag{8.27}$$

$$\left\langle \sum_{s \neq E} \frac{\mathrm{e}_s^2}{m_s} \frac{\partial F_{0s}}{\partial \mathcal{E}} \right\rangle (\delta\phi - \delta\psi) + \sum_{s=i} \mathrm{e}_s \langle \delta K_s \rangle = 0 \, , \tag{8.28}$$

$$\begin{aligned}
&\left[\omega_{tr} \partial_\theta - \mathrm{i} \left(\omega - \omega_d \right) \right]_s \delta K_s \\
&= \mathrm{i} \left(\frac{\mathrm{e}}{m} \right)_s Q F_{0s} \left[J_0(k_\perp \rho_s) \left(\delta\phi - \delta\psi \right) + \left(\frac{\hat{\omega}_d}{\omega} \right)_s J_0(k_\perp \rho_s) \delta\psi \right] .
\end{aligned} \tag{8.29}$$

There now remains to apply the above equation to specific modes of the Shear Alfvén Wave spectrum. In general, such perturbations are generally characterized by a two-scale radial structure, with a singular structure/layer concentrating most kinetic, inertial effects, and a more regular (ideal MHD) structure. For this reason, via asymptotic analysis it is possible to derive a general fishbone-like dispersion relation.

8.5 The general fishbone like dispersion relation

8.5.1 *Properties of the fishbone like dispersion relation*

8.5.1.1 *General properties*

Provided that modes are characterized by two-scale radial structures, it is possible to demonstrate that the dispersion relation of Shear Alfvén modes can be written in the form of a fishbone-like dispersion relation [9, 11, 39, 45].

$$-\mathrm{i}\Lambda + \delta W_f + \delta W_k = 0\,, \tag{8.30}$$

where δW_f and δW_k play the role of fluid (core plasma) and kinetic (fast ion) contribution to the potential energy, while Λ represents a generalized inertia term.[5]

In general[6], Alfvén perturbations exhibit a two-scale radial structure. The general fishbone-like dispersion relation can be derived by asymptotic matching the regular (ideal MHD) mode structure with the general (known) form of the Shear Alfvén Wave field in the singular (inertial) region, as the spatial location of the Shear Alfvén resonance, $\Lambda^2 = k_\parallel^2 q^2 R_0^2$ (reducing to $\omega^2 = k_\parallel^2 v_A^2$ in the appropriate limit), is approached.

Examples for the generalized inertia are:

$\Lambda^2 = \omega(\omega - \omega_{*pi})/\omega_A^2$, for $|k_\parallel q R_0| \ll 1$
typically used for Kinetic Ballooning Modes

$\Lambda^2 = (\omega_l^2 - \omega^2)/(\omega_u^2 - \omega^2)$, for $|k_\parallel q R_0| \approx 1/2$
typically used for Toroidal Alfvén Eigenmodes

where ω_l and ω_u are called the lower and upper accumulation points of the toroidal gap in the Shear Alfvén continuous spectrum [11].

[5]This simple form of Eq. (8.30) assumes modes with kinetic/inertial layer at $k_\parallel q R_0 = 0$ and finite magnetic shear. Other forms of Eq. (8.30), valid under less restrictive assumptions, are discussed in Ref. [51] and references therein.

[6]Not always, as discussed in Ref. [4].

δW_f is generally real and independent of the mode frequency, whereas δW_k is characterized by complex values and is frequency dependent, the real part accounting for non-resonant and the imaginary part for resonant wave particle interactions, e.g. with energetic ions.

8.5.1.2 *Alfvén Eigenmodes (AE) and Energetic Particle Modes (EPM)*

The fishbone-like dispersion relation demonstrates the existence of two types of modes [47] (note: $\Lambda^2 = k_\|^2 q^2 R_0^2$ represents the Shear Alfvén Wave continuum), defined as

- Discrete gap modes, or Alfvén Eigenmodes (AE), for $\mathbb{R}\text{e}\Lambda^2 < 0$,
- Energetic Particle continuum Modes (EPM), for $\mathbb{R}\text{e}\Lambda^2 > 0$.

Λ^2 naturally appears when solving the Shear Alfvén mode equations, and it is real when resonances with the thermal plasma are not taken into account. In the fishbone dispersion relation, its root square needs to be taken. When Λ is real ($\Lambda^2 > 0$), the $i\Lambda$ term in Eq. (8.30) can be associated to the so-called continuum damping. Continuum damping can be understood as a collisionless dissipation phenomenon due to resonant mode absorption, where the mode frequency coincides with the local Shear Alfvén Wave continuous spectrum [8], and corresponding to mode conversion to small scales of a dense spectrum [26]. Continuum damping characterizes a continuous spectrum of modes, or continuum, with radial singular structures. Because of continuum damping, modes with $\mathbb{R}\text{e}\Lambda^2 > 0$ can only survive in the presence of a free energy source. For Energetic Particle continuum Modes, the source is the energetic particle drive, accounted for by δW_k. On the other hand, Alfvén Eigenmodes with $\Lambda^2 < 0$ are discrete and, at the lowest order, they are not subject to continuum damping. Thus, they are characterized by a lower excitation threshold in the energetic particle drive. Thus, if they exist, Alfvén Eigenmodes are more easily driven unstable by energetic particles than Energetic Particle continuum Modes. However, there are conditions in which Alfvén Eigenmodes either do not exist or are unfavored (e.g., by spatial distributions of the energetic particles), where Energetic Particle continuum Modes are expected to play a dominant role and have detrimental effects on the energetic particle transport, due to their intrinsic resonant nature.

A particular role is played by the solutions of $\Lambda^2 = 0$, corresponding to the so-called accumulation points. For both Alfvén Eigenmodes and

Energetic Particle continuum Modes, the Shear Alfvén Wave accumulation point is the natural gateway through which modes are born at marginal stability [14]. Note however, that unstable points of the continuum may exist [43] and describe a convective instability that generates shorter and shorter scale-lengths while growing quasi-exponentially in time, till it eventually dissipates by mode conversion to damped shorter wavelengths.

For Energetic Particle Modes, the oscillation frequency is set by the relevant energetic ion characteristic frequency, for which the wave-particle resonant interaction is maximized [13][7]. Mode excitation requires the drive exceeding a threshold, due to continuum damping. However, the non-resonant fast ion response is crucially important as well, since it provides the compression effect that is necessary for balancing the positive MHD potential energy of the wave. Near marginal stability [9, 11], $\omega = \omega_r + i\gamma$ with $|\gamma/\omega_r| \ll 1$ and

$$\begin{aligned} &\mathbb{R}\mathrm{e}\delta W_k(\omega_r) + \delta W_f = 0\,, \\ &\gamma/\omega_r = (-\omega_r \partial_{\omega_r} \mathbb{R}\mathrm{e}\delta W_k)^{-1} (\mathbb{I}\mathrm{m}\delta W_k - \Lambda)\,. \end{aligned} \tag{8.31}$$

For Energetic Particle Modes, the iΛ term represents continuum damping.

For Alfvén Eigenmodes, the non-resonant fast ion response provides a real frequency shift, *i.e.* it removes the degeneracy with the continuum accumulation point, while the resonant wave-particle interaction gives the mode drive. Note that a choice needs to be done for the square root of Λ to be used in the dispersion relation Eq. (8.30). Physical constraints need to be used and the causality condition (see below) can be invoked [47], which enforces

- $\delta W_f + \mathbb{R}\mathrm{e}\delta W_k > 0$ when the Alfvén Eigenmode frequency is above the continuum accumulation point: inertia in excess with respect to field line bending
$$\Lambda^2 = \lambda_0^2(\omega_\ell - \omega)\,; \qquad \omega > \omega_\ell \Rightarrow \Lambda \to -\mathrm{i}\sqrt{-\Lambda^2}\,.$$
- $\delta W_f + \mathbb{R}\mathrm{e}\delta W_k < 0$ when the Alfvén Eigenmode frequency is below the continuum accumulation point: inertia is lower than field line bending
$$\Lambda^2 = \lambda_0^2(\omega - \omega_u)\,; \qquad \omega < \omega_u \Rightarrow \Lambda \to \mathrm{i}\sqrt{-\Lambda^2}\,.$$

For Alfvén Eigenmodes, iΛ represents the shift of mode frequency from the accumulation point.

[7] By arguments based on breaking of invariants of particle motion, this condition corresponds also to maximizing the particle and energy transport.

8.5.2 *Derivation of the fishbone like dispersion relation*

The main steps of the derivation are depicted here, using the ballooning representation[8]. As specified above, when introducing Eq. (8.30), the simple form of the general fishbone-like dispersion relation, discussed here, assumes modes with kinetic/inertial layer at $k_{\|}qR_0 = 0$ and finite magnetic shear. Other forms of Eq. (8.30), valid under less restrictive assumptions, are discussed in Ref. [51] and references therein. In particular, here we derive the explicit form of Λ^2 for the simple case in which $|\omega_{di}| \ll |\omega| \ll \omega_A$.

The vorticity equation is

$$\mathbf{B}\cdot\boldsymbol{\nabla}\left(\frac{k_\perp^2}{k_\vartheta^2 B^2}\mathbf{B}\cdot\boldsymbol{\nabla}\delta\psi\right)+\frac{\omega(\omega-\omega_{*pi})}{v_A^2}\frac{k_\perp^2}{k_\vartheta^2}\delta\phi$$
$$-\left\langle\sum_{s=i}\frac{4\pi e_s}{k_\vartheta^2 c^2}\omega\hat{\omega}_{ds}\delta K_s\right\rangle+\frac{4\pi}{k_\vartheta^2 B^2}\mathbf{k}\times\mathbf{b}\cdot\tilde{\nabla}\left(P_\perp+P_\|\right)\Omega_\kappa\delta\psi$$
$$=\left\langle\frac{4\pi e_E}{k_\vartheta^2 c^2}J_0(k_\perp\rho_E)\omega\hat{\omega}_{dE}\delta K_E\right\rangle. \tag{8.32}$$

The equation is seen to have two very different behaviors for small and large θ. It is convenient to solve for the two regions separately and to match the two regions in a second step.

8.5.2.1 *The inertial layer solution, or large θ solution $\theta \sim \omega_A/|\omega|$*

For "large" values of $\theta \sim \omega_A/|\omega|$, the ballooning interchange pressure term in Eq. (8.32) becomes negligible. Moreover, since the ballooning transform (at the lowest order) can be interpreted as the Fourier transform of the radial mode structure, large-θ values correspond to fine radial structure perturbations with a large radial wave number k_r. For fast ions (with a large Larmor radius ρ_E), it is reasonable to consider $\rho_E k_r \gg 1$. As a consequence, the response of the energetic particles is nearly adiabatic [47] and can be effectively neglected in the inertial layer region[9].

It is possible to solve the remaining equation in a perturbative framework, noticing the existence of a two-scale structure (two relevant scales for ∂_θ). Recalling that in the ballooning representation, $qR_0(\mathbf{B}/B)\cdot\boldsymbol{\nabla}\to\partial_\theta$, the comparison of the order of magnitude of the field line bending term (the first term in Eq. (8.32)) with the other terms shows of Eq. (8.32), allows assessing the two scale structures of $\delta\psi$ and $\delta\phi$: i.e., $\theta_0 \sim 1$,

[8] See H. R. Wilson Lecture.

[9] For further information, see Ref. [41].

reflecting the periodic poloidal dependences of equilibrium quantities, and $\theta_1 \sim \omega_A/|\omega| \gg 1$.

With this in mind, it is possible to systematically make a two-scale analysis of Eq. (8.32), i.e., to derive its solution as an asymptotic expansions in the smallness parameter $\theta_0/\theta_1 \sim |\omega|/\omega_A$; e.g., $\delta\psi = \delta\psi^{(0)} + \delta\psi^{(1)} + \ldots$, $\delta\phi = \delta\phi^{(0)} + \delta\phi^{(1)} + \ldots$

At the lowest order, the expansion shows that the solution is a function of the long scale only:

$$\delta\phi^{(0)}(\theta_1) = \delta\psi^{(0)}(\theta_1) \text{ and } \delta K_i^{(0)} \propto \left(\delta\phi^{(0)}(\theta_1) - \delta\psi^{(0)}(\theta_1)\right).$$

With the notation $\overline{(\ldots)} = (2\pi)^{-1} \oint (\ldots) d\theta_0$ and the normalized function $\delta\Phi^{(0)} = \overline{(k_\perp/k_\vartheta)\delta\phi^{(0)}}$, the vorticity equation can be recast in the form

$$\partial^2_{\theta_1} \delta\Phi^{(0)} + \Lambda^2 \delta\Phi^{(0)} = 0\,, \tag{8.33}$$

$$\Lambda^2 \delta\Phi^{(0)} = \frac{\omega(\omega - \omega_{*pi})}{\omega_A^2} \delta\Phi^{(0)} - \overline{\left\langle \sum_{s=i} \frac{4\pi q^2 R_0^2 e_s}{k_\vartheta k_\perp c^2} \omega \hat{\omega}_{ds} \delta K_s^{(1)} \right\rangle}. \tag{8.34}$$

↔ *Derive these equations step by step.*

↔ *How general are they? Could you derive the results for arbitrary geometry or only for simple equilibria, e.g. the $(s-\alpha)$ model equilibrium often used in the ballooning formalism [16]?* Here, $s = rq'/q$ is the magnetic shear and $\alpha = -R_0 q^2 \beta'$ is the normalized pressure gradient.

Equation (8.33) is simply a wave-like equation and has the obvious solutions $\delta\Phi^{(0)} = \delta\Phi_0 \exp(i\Lambda|\theta_1|)$, with Λ the square root of Λ^2 and $\delta\Phi_0$ a constant. In linear theory, the amplitude of the mode is not constrained and, in the following, it is assumed normalized such that $\delta\Phi_0 = 1$.

In order to determine the correct root of Λ^2, one needs to make use of a physical argument, or causality constraint, as mentioned above. With the wave structure of the equation, a physical argument is the choice of an outgoing wave boundary condition, that is a positive group velocity $\partial_{\omega_r}\mathrm{Re}\Lambda > 0$. In θ space (i.e., in the radial Fourier space), this is equivalent to propagation to small scales.

8.5.2.2 *Asymptotic matching with the small θ-solution*

At small θ, the inertia term (the second term in the LHS of the vorticity equation) becomes negligible and the solution simply depends on the short scale only. It is possible to match the large and small θ regions, by matching the logarithmic derivatives (recall that $\delta\Phi_0 = 1$).

Introducing arbitrary values $\theta = \pm\Theta_1$, such that $|\Theta_1| \gg 1$, $|\Lambda\Theta_1| \ll 1$, multiplication of the vorticity equation by $\delta\phi^{(0)*}$ and integration in θ-space leads to:

$$\frac{1}{2}\left[\delta\Phi^{(0)*}\partial_\theta\delta\Phi^{(0)}\right]_{-\Theta_1}^{\Theta_1} = \delta W_f + \delta W_k$$

$$\equiv \frac{q^2 R_0^2}{2}\int_{-\Theta_1}^{\Theta_1} d\theta\delta\phi^{(0)*}\left\langle\sum_{s=i}\frac{4\pi e_s}{k_\vartheta^2 c^2}\omega\hat{\omega}_{ds}\delta K_s\right\rangle$$

$$-\frac{q^2 R_0^2}{2}\int_{-\Theta_1}^{\Theta_1} d\theta\delta\phi^{(0)*}\frac{4\pi}{k_\vartheta^2 B^2}\mathbf{k}\times\mathbf{b}\cdot\tilde{\nabla}\left(P_\perp + P_\parallel\right)\Omega_\kappa\delta\phi^{(0)}$$

$$+\frac{q^2 R_0^2}{2}\int_{-\Theta_1}^{\Theta_1} d\theta\delta\phi^{(0)*}\left\langle\frac{4\pi e_E}{k_\vartheta^2 c^2}J_0(k_\perp\rho_E)\omega\hat{\omega}_{dE}\delta K_E\right\rangle$$

$$+\frac{1}{2}\int_{-\Theta_1}^{\Theta_1} d\theta\left(|\partial_\theta\frac{k_\perp}{k_\vartheta}\phi^{(0)}|^2 - \frac{k_\perp'' k_\perp}{k_\vartheta^2}|\phi^{(0)}|^2\right), \qquad (8.35)$$

where all contributions related to the resonant response of the energetic are gathered in δW_k, whereas the remaining ones are kept in δW_f, the ideal "plasma fluid response".

In the inertial region, the long wave-length solution is $\delta\Phi^{(0)} = \exp(i\Lambda|\theta_1|)$, and

$$\frac{1}{2}\left[\delta\Phi^{(0)*}\partial_\theta\delta\Phi^{(0)}\right]_{-\Theta_1}^{\Theta_1} = i\Lambda\,. \qquad (8.36)$$

Asymptotic matching finally returns the general Fishbone Like Dispersion Relation (FLDR): $i\Lambda = \delta W_f + \delta W_k$. This form, derived on the basis of Refs. [9, 11, 14, 39, 45, 47], is essentially the same derived by Nguyen et al. [33] and used for the interpretation of observation of Beta Alfvén Eigenmode modes excited by ion cyclotron resonance heating in Tore Supra [34].

↔ *Repeat the derivation of the general Fishbone Like Dispersion Relation. What terms are missing? Can you include kink drive ($\propto \boldsymbol{\nabla} J_\parallel$)?*

↔ *Derive the general Fishbone Like Dispersion Relation for arbitrary geometry (drop the large aspect ratio assumption). Then generalize the derivation to waves with finite but small $|k_\parallel q R_0| \ll 1$. Can you solve the problem generally also for finite $|k_\parallel q R_0|$ (e.g. Toroidal Alfvén Eingenmodes)?*

↔ *Establish a link between δW_f and the well known expression of MHD δW [33, 34]). Then, assuming that the mode structure is known, give a close expression for δW_k. At this point, use a (local) gyrokinetic code for* *determining numerically the expression of* Λ*. Which dispersion relation can you investigate in this way? Can you recover known analytical results?*

With the expression of Λ given in Eq. (8.34), it is possible to easily derive the following expression, valid for a simple low-pressure magnetic equilibrium with circular flux surfaces and neglecting the contribution from trapped particles:[10]

$$\begin{aligned}\Lambda^2 = & \frac{\omega^2}{\omega_A^2}\left(1-\frac{\omega_{*pi}}{\omega}\right)+q^2\frac{\omega_{ti}^2}{2\omega_A^2}\left[\left(1-\frac{\omega_{*ni}}{\omega}\right)\left((\omega/\omega_{ti}^{(+)})F(\omega/\omega_{ti}^{(+)})\right.\right.\\ & \left.+(\omega/\omega_{ti}^{(-)})F(\omega/\omega_{ti}^{(-)})\right)-\frac{\omega_{*Ti}}{\omega}\left((\omega/\omega_{ti}^{(+)})G(\omega/\omega_{ti}^{(+)})\right.\\ & \left.+(\omega/\omega_{ti}^{(-)})G(\omega/\omega_{ti}^{(-)})\right)-\left((\omega/\omega_{ti}^{(+)})N_m(\omega/\omega_{ti}^{(+)})\frac{N_{m-1}(\omega/\omega_{ti}^{(+)})}{D_{m-1}(\omega/\omega_{ti}^{(+)})}\right.\\ & \left.\left.+(\omega/\omega_{ti}^{(-)})N_m(\omega/\omega_{ti}^{(-)})\frac{N_{m+1}(\omega/\omega_{ti}^{(-)})}{D_{m+1}(\omega/\omega_{ti}^{(-)})}\right)\right]. \end{aligned} \tag{8.37}$$

Here, the subscripts $m, m\pm 1$ denote the poloidal mode number to be used in the evaluation of diamagnetic terms of the complex functions F, G, N_m, D_m defined below and the following notations are adopted: $\omega_{*ni} = (T_i c/eB)(\boldsymbol{k}\times\boldsymbol{b})\cdot\boldsymbol{\nabla}(n_i)/n_i$ and $\omega_{*Ti} = (T_i c/eB)(\boldsymbol{k}\times\boldsymbol{b})\cdot\boldsymbol{\nabla}(T_i)/T_i$, $\omega_{*pi} = \omega_{*ni}+\omega_{*Ti}$, $\omega_{ti} = (2T_i/m_i)^{1/2}/qR_0$ and $\omega_{ti}^{(\pm)} = (2T_i/m_i)^{1/2}\left[1\pm(nq-m)\right]/qR_0$. Meanwhile,

$$\begin{aligned} F(x) &= x\left(x^2+3/2\right)+\left(x^4+x^2+1/2\right)Z(x)\,, \\ G(x) &= x\left(x^4+x^2+2\right)+\left(x^6+x^4/2+x^2+3/4\right)Z(x) \\ N_m(x) &= \left(1-\frac{\omega_{*ni}}{\omega}\right)\left[x+\left(1/2+x^2\right)Z(x)\right] \\ &\quad -\frac{\omega_{*Ti}}{\omega}\left[x\left(1/2+x^2\right)+\left(1/4+x^4\right)Z(x)\right]\,, \\ D_m(x) &= \left(\frac{1}{x}\right)\left(1+\frac{1}{\tau}\right)+\left(1-\frac{\omega_{*ni}}{\omega}\right)Z(x) \\ &\quad -\frac{\omega_{*Ti}}{\omega}\left[x+\left(x^2-1/2\right)Z(x)\right]\,. \end{aligned} \tag{8.38}$$

and $Z(x) = \pi^{-1/2}\int_{-\infty}^{\infty} e^{-y^2}/(y-x)dy$ is the plasma dispersion function. Equation (8.37) accounts for finite diamagnetic frequency splitting, due to

[10]Please, see [51] for details. As stated previously, the effects of trapped particles have been recently discussed by [7, 30, 51].

finite mode number, as well as for finite $k_{\|}qR_0$, as discussed in [30, 50, 51]. Furthermore, it accounts for Beta Alfvén Eigenmodes (BAE), Kinetic Ballooning Modes (KBM), Beta Alfvén Acoustic Eigenmodes (BAAE) [24, 25] and other oscillations of the low frequency Alfvén and Alfvén-acoustic fluctuation spectrum in a single unified theoretical framework that reduces to the case originally discussed in [43, 45] for high mode number modes with $k_{\|}qR_0 \to 0$ at the kinetic singular layer [51].

8.6 Special cases of the fishbone like dispersion relation

8.6.1 *Toroidal Alfvén Eigenmodes (TAE)*

Toroidal Alfvén Eigenmodes (TAE) are the canonical paradigm of Alfvén Eigenmodes (AE) in toroidal plasmas.

For Toroidal Alfvén Eigenmodes (TAE), the generalized inertia term reads

$$\Lambda^2 = (\omega_l^2 - \omega^2)/(\omega_u^2 - \omega^2) \tag{8.39}$$

with $\omega_{l,u} = (v_A/2qR_0)(1 \mp \epsilon_0/2)$ the lower and upper accumulation points of the Shear Alfvén continuous spectrum toroidal gap, $\epsilon_0 = 2(r/R_0 + \Delta')$, Δ' accounting for the Shafranov shift and having assumed, for simplicity, a large aspect ratio tokamak equilibrium with circular shifted magnetic flux surfaces. Moreover, considering a local $(s - \alpha)$ model equilibrium [16] $(s = rq'/q, \alpha = -R_0q^2\beta')$, a reduced expression for δW_f can be derived for $|s|, |\alpha| < 1$ in the form

$$\delta W_f = (|s|\pi/4)(1 - \alpha/\alpha_c)\,, \tag{8.40}$$

where the "critical" value $\alpha = \alpha_c$, above which δW_f changes sign, is a known function of the magnetic shear s [12].

From the previous comments on Alfvén Eigenmodes (AE) (in particular concerning the consequences of the causality constraint), we can draw the following conclusions.

If $\alpha < \alpha_c$, $\delta W_f > 0$. For small δW_k, a discrete mode can exist with $\omega \gtrsim \omega_l$, consistently with the previous existence conditions (causality constraints) for a discrete Alfvén Eigenmode (AE) gap mode to exist just above the lower accumulation point. Such a mode is shifted toward ω_l and eventually stabilized for $\alpha > \alpha_c$, due to the increased coupling with the Shear Alfvén continuous spectrum.

As α increases above α_c, another Toroidal Alfvén Eigenmode (TAE) can exist for $\omega \lesssim \omega_u$ provided the appropriate existence condition for discrete Alfvén Eigenmode (AE) gap mode is satisfied, *i.e.* $\delta W_f + \mathbb{R}\mathrm{e}\delta W_k < 0$. (ballooning unstable). Such condition is less interesting for practical applications, for this type of Toroidal Alfvén Eigenmode (TAE) would be coexisting with ideal MHD instabilities. At even higher values of α (second stability region) other discrete Alfvén Eigenmodes (AE) may exist, as recently discussed in Ref. [4].

8.6.2 *Alfvén Cascades*

Alfvén Cascades (AC) are modes which have been observed in the presence of a flat or reversed q-profile, and are characterized by a sweeping oscillation frequency (see Fig. 8.1).

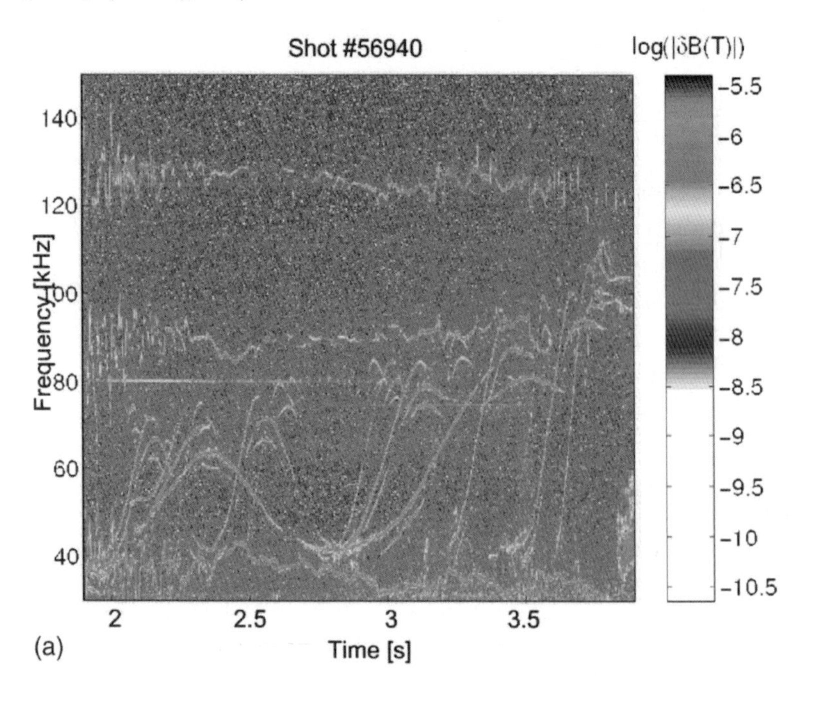

Fig. 8.1 Alfvén Cascades, for a color version see Fig. 6.6.

The original theoretical interpretation of Alfvén Cascade (AC) excitation by large orbit trapped energetic ion tails generated by Ion Cyclotron Resonant Heating (ICRH) on JET can be found in Ref. [6].

Alfvén Cascades can be interpreted as Alfvén Eigenmode (AE) gap modes in the natural frequency gap that arises in the Alfvén continuous spectrum at the radial location, r_0, where $q(r_0) = q_0$ has a minimum, *i.e.* $S^2 \equiv r_0^2 q''(r_0)/q_0^2 > 0$. Consistently with theoretical predictions, Alfvén Cascades (AC) can be described by a fishbone-like dispersion relation with:

$$\Lambda^2 = (nq_0 - m)^{-1}(\omega^2 - k_\parallel^2 v_A^2)/\omega_A^2\,, \tag{8.41}$$

$$\delta W_f = -(\pi/4)(S/n^{1/2})\,. \tag{8.42}$$

Applying the above causality constraints, or existence conditions for a discrete Alfvén Eigenmode (AE) gap mode, $\mathbb{R}\text{e}\Lambda^2 < 0$ with $\omega^2 > k_\parallel^2 v_A^2$, it is necessary to have $nq_0 - m < 0$ and

$$\delta W_f + \mathbb{R}\text{e}\delta W_k > 0\,. \tag{8.43}$$

This condition can be provided both by fast ions as well as by core plasma equilibrium effects [22, 37, 46].

8.7 Summary and discussions

One of the key experimental evidences [32] is the observation of low frequency Shear Alfvén Wave (SAW) in a wide range of mode numbers, driven by both energetic (long wavelength) as well as thermal (short wavelength) ions, confirming theoretical predictions [43, 45].

For interpretation of these observations, as shown above, it is necessary to use kinetic theories for the proper treatment of wave-particle interactions with circulating as well as trapped thermal plasma particles [7, 30]. Meanwhile, for sufficiently strong energetic particle drive (ions and electrons), experimental observations show evidence of a continuous transition between various Shear Alfvén Wave (SAW) and MHD fluctuation branches [17, 19, 20, 31, 36, 50], consistent with theoretical predictions [9, 11, 12, 15, 39, 44] based on the general Fishbone Like Dispersion Relation.

The present theoretical framework suggests that low-frequency Shear Alfvén Wave (SAW) and acoustic oscillations should be seen as either EPM or Alfvén Eigenmodes, (AE) localized within the Kinetic Thermal Ion (KTI) frequency gap [14]. The intrinsic limitations of MHD/fluid analyses at low frequency in collisionless tokamak plasmas of fusion interest suggest developing new numerical investigation tools, with aim at bridging simulations of meso- and micro-scale phenomena [14, 29].

The linear gyrokinetic code LIGKA is capable to address these issues [29, 30]. Meanwhile, an eXtended version of the HMGC code (XHMGC [41], nonlinear hybrid MHD-gyrokinetic code), has been developed to handle kinetic thermal ion compressibility effects. Successful comparisons of XHMGC numerical studies of kinetic Beta Alfvén Eigenmode excited by energetic particles vs. analytic predictions have been already carried out. Global non-perturbative (full-f) electromagnetic gyrokinetic simulations are also becoming available, such as for the GTC [18, 28, 42] and GYRO [3, 40] codes. Work is ongoing to complete the present theoretical-analytical framework; e.g., extending the Λ^2 expression including trapped particle dynamics with finite $k_{\|}qR_0$ and diamagnetic sideband generation, as well as finite Larmor radius and finite magnetic-orbit width effects in the whole frequency range. With these results, the further generalization of the general Fishbone Like Dispersion Relation at short scales [45] will complete the present theoretical framework and allow verification of local and global numerical simulations vs. local analytic theories, consolidating the basis for a complete toolbox for numerical computation and application of the general Fishbone Like Dispersion Relation in situations of practical interest.

Acknowledgments

Precious discussions with Liu Chen and his contribution to collecting the material reported in this lecture are kindly acknowledged. Thanks to Christine Nguyen and Emanuele Tassi for their help in editing the lecture notes.

References

[1] H. Alfvén, (1942). *Nature* **150**, 405.
[2] H. Alfvén, *Cosmical Electrodynamics*, Clarendon 1950, Oxford, UK.
[3] E. Bass and R. E. Waltz (2010). *Phys. Plasmas* **17**, 112319.
[4] A. Bierwage, L. Chen and F. Zonca (2010). *Plasma Phys. Control. Fusion* **52**, 015004 and 015005.
[5] D. Biskamp, *Nonlinear Magnetohydrodynamics*, Cambridge University Press 1993, Campbridge United Kingdom, page 51.
[6] B. N. Breizman, M. S. Pekker and S. E. Sharapov (2005). *Phys. Plasmas* **12**, 112506.
[7] I. Chavdarovski and F. Zonca (2009). *Plasma Phys. Control. Fusion* **51**, 115001.

[8] L. Chen and A. Hasegawa (1974). *Phys. Fluids* **17**, 1399.
[9] L. Chen, R. B. White and M. N. Rosenbluth (1984). *Phys. Rev. Lett.* **52**, 1122.
[10] L. Chen and A. Hasegawa (1991). *J. Geophys. Phys. Res.* **96**, A2, 1503.
[11] L. Chen (1994). *Phys. Plasmas* **1**, 1519.
[12] L. Chen and F. Zonca (1995). *Phys. Scr.* **T60**, 81.
[13] L. Chen (1999). *J. Geophys. Res.* **104**, 2421.
[14] L. Chen and F. Zonca (2007). *Nucl. Fusion* **47**, S727.
[15] C. Z. Cheng, N. N. Gorelenkov and C. T. Hsu (1995). *Nucl. Fusion* **35**, 1639.
[16] J. W. Connor, R. J. Hastie and J. B. Taylor (1978). *Phys. Rev. Lett.* **40**, 396.
[17] D. S. Darrow et al. (2008). *Nucl. Fusion* **48**, 084004.
[18] W. Deng, Z. Lin, I. Holod, X. Wang, Y. Xiao and W. L. Zhang (2010). *Phys. Plasmas* **17**, 112504.
[19] E. D. Fredrickson et al. (2006). *Phys. Plasmas* **13**, 056109.
[20] E. D. Fredrickson et al. (2009). *Phys. Plasmas* **16**, 122505.
[21] E. A. Frieman and L. Chen (1982). *Phys. Fluids.* **25**, 502.
[22] G. Y. Fu and H. L. Berk (2006). *Phys. Plasmas* **13**, 052502.
[23] N. N. Gorelenkov, C. Z. Cheng, E. Fredrickson, E. Belova, D. Gates, S. Kaye, G. J. Kramer, R. Nazikian and R. White (2002). *Nucl. Fusion* **42**, 977.
[24] N. N. Gorelenkov, H. L. Berk, E. Fredrickson and S. E. Sharapov (2007). *Phys. Lett. A* **370**, 70.
[25] N. N. Gorelenkov, H. L. Berk, N. A. Crocker, E. D. Fredrickson, S. Kaye, S. Kubota, H. Park, W. Peebles, S. A. Sabbagh, S. E. Sharapov, D. Stutmat, K. Tritz, F. M. Levinton, H. Yuh, the NSTX Team and JET EFDA Contributors (2007). *Plasma Phys. Control. Fusion* **49**, B371.
[26] A. Hasegawa and L. Chen (1976). *Phys. Fluids* **19**, 1924.
[27] A. Hasegawa and T. Sato (1989). *Space Plasma Physics (Stationary Processes)*, vol. 1 (Springer, New York).
[28] Z. Lin, T. S. Hahm, W. W. Lee, W. M. Tang and R. B. White (1998). *Science* **281**, 1835.
[29] Ph. Lauber and S. Günter (2008). *Nucl. Fusion* **48**, 084002.
[30] Ph. Lauber, M. Brüdgam, D. Curran, V. Igochine, K. Sassenberg, S. Günter, M. Maraschek, M. García-Muñoz, N. Hicks and the ASDEX Upgrade Team (2009). *Plasma Phys. Control. Fusion* **51**, 124009.

[31] F. Nabais et al. (2005). *Phys. Plasmas* **12**, 102509.

[32] R. Nazikian et al. (2006). *Phys. Rev. Lett.* **96**, 105006.

[33] C. Nguyen, X. Garbet and A. I. Smolyakov (2008). *Phys. Plasmas* **15**, 112502.

[34] C. Nguyen, X. Garbet, R. Sabot, L.-G. Eriksson, M. Goniche, P. Maget, V. Basiuk, J. Decker, D. Elbèze, G. T. A. Huysmans, A. Macor, J.-L. Ségui and M. Schneider (2009). *Plasma Phys. Control. Fusion* **51**, 095002.

[35] W. Park, S. Parker, H. Biglari, M. Chance, L. Chen, C. Z. Cheng, T. S. Hahm, W. W. Lee, R. Kulsrud, D. Monticello, L. Sugiyama and R. B. White (1992). *Phys. Fluids B* **4**, 2033.

[36] M. Podestà et al. (2009). *Phys. Plasmas* **16**, 056104.

[37] S. E. Sharapov, B. Alper, H. L. Berk, D. N. Borba, B. N. Breizman, C. D. Challis, A. Fasoli, N. C. Hawkes, T. C. Hender, J. Mailloux, S. D. Pinches and D. Testa (2002). *Phys. Plasmas* **9**, 2027.

[38] W. M. Tang, J. W. Connor and R. J. Hastie (1980). *Nucl. Fusion* **20**, 1439.

[39] T. S. Tsai and L. Chen (1993). *Phys. Fluids B* **5**, 3284.

[40] R. E. Waltz, J. M. Candy and M. N. Rosenbluth (2002). *Phys. Plasmas* **9**, 1938.

[41] X. Wang, S. Briguglio, L. Chen, C. Di Troia, G. Fogaccia, G. Vlad and F. Zonca (2011). An extended hybrid magnetohydrodynamics gyrokinetic model for numerical simulation of shear Alfvén waves in burning plasmas, *to be published in Phys. Plasmas*; *arXiv:1012.5388* [physics.plasma-ph].

[42] H. S. Zhang, Z. Lin, I. Holod, X. Wang, Y. Xiao and W. L. Zhang (2010). *Phys. Plasmas* **17**, 112505.

[43] F. Zonca, L. Chen and R. A. Santoro (1996). *Plasma Phys. Control. Fusion* **38**, 2011.

[44] F. Zonca and L. Chen (1996). *Phys. Plasmas* **3**, 323.

[45] F. Zonca, L. Chen, J. Q. Dong and R. A. Santoro (1999). *Phys. Plasmas* **6**, 1917.

[46] F. Zonca, S. Briguglio, L. Chen, S. Dettrick, G. Fogaccia, D. Testa and G. Vlad (2002). *Phys. Plasmas* **9**, 4939.

[47] F. Zonca and L. Chen (2006). *Plasma Phys. Control. Fusion* **48**, 537.

[48] F. Zonca, S. Briguglio, L. Chen, G. Fogaccia, T. S. Hahm, A. V. Milovanov and G. Vlad (2006). *Plasma Phys. Control. Fusion* **48**, B15.

[49] F. Zonca and L. Chen (2007). The general fishbone-like dispersion relation: a unified description for shear Alfvén Mode excitations, in emp-Proceedings of the 34th EPS Conference on Plasma Physics (Warsaw, Poland), *ECA* **31F**, CD-ROM file P4.071, `http://epsppd.epfl.ch/Warsaw/html/contents.htm`.

[50] F. Zonca, L. Chen, A. Botrugno, P. Buratti, A. Cardinali, R. Cesario, V. Pericoli Ridolfini and JET-EFDA contributors (2009). *Nucl. Fusion* **49**, 085009.

[51] F. Zonca, A. Biancalani, I. Chavdarovski, L. Chen, C. Di Troia and X. Wang (2010). *J. Phys.: Conf. Ser.* **260**, 012022.

[49] J. Zaanen and [illegible] (1988) [illegible]
[illegible]
[illegible]
[illegible]
[illegible]

[50] [illegible]
[illegible]
[illegible]

[51] [illegible]
[illegible]

Chapter 9

What is a Reversed Field Pinch?

D. F. Escande

Aix-Marseille Université, CNRS, PIIM,
UMR 7345, 13013 Marseille, France

9.1 Introduction

The reversed field pinch (RFP) is a magnetic configuration germane to the tokamak, that produces most of its magnetic field by the currents flowing inside the plasma; external coils provide only a small edge toroidal field whose sign is reversed with respect to the central one, whence the name of the configuration. Because of the presence of magnetic turbulence and chaos, the RFP had been considered for a long period as a terrible confinement configuration. Then strong enhancements were triggered in a transient way in the MST machine [1]. However, recently a change of paradigm occurred for this device. This was symbolically exhibited by the cover story of the August 2009 issue of Nature Physics: "Reversed-field pinch gets self-organized" [2]. Indeed, when the toroidal current is increased in the RFX-mod RFP in Padua (Italy), a self-organized helical state with an internal transport barrier (ITB) develops, and a broad zone of the plasma becomes hot (above 1 keV for a central magnetic field above 0.8 T).

The possibility of this helical state and of the corresponding improvement in confinement had been theoretically predicted. The present theoretical picture of the RFP mainly comes from three-dimensional nonlinear visco-resistive magnetohydrodynamic (MHD) simulations. They have exhibited dynamics having strong similarities with the experimental one, which triggered the experimental search for RFP states with improved confinement. The RFP ohmic state involves, as the nonlinear tearing mode, a helical electrostatic potential generating, as an electric drift, the so-called dynamo velocity field. The magnetic topology can bifurcate from a magnetic island to kink-like magnetic surfaces with higher resilience to magnetic chaos. This theoretical scenario was found to be relevant when internal transport barriers (ITB) enclosing a broad hot domain were discovered [3]. The internal transport barriers occur in the vicinity of a maximum of the safety factor.

The new paradigm for the RFP supports its reappraisal as a low-external field, non-disruptive, ohmically heated approach to magnetic fusion, exploiting both self-organization and technological simplicity. The RFP comes with interesting transversal physics involving the other magnetic configurations. This chapter provides an introduction to the present knowledge about the RFP, and tries to bridge the large gap in its description that developed since the book by Ortolani and Schnack [4].

9.2 Short description

As a device, the reversed field pinch (RFP) looks very much like a tokamak: a toroidally symmetric vacuum vessel is surrounded by a toroidally symmetric set of identical toroidal field coils; a central solenoid provides the loop voltage necessary to trigger a discharge, and to drive a current in the toroidal plasma inside the vessel. Furthermore a RFP may be operated as a tokamak, but a low field one. Indeed toroidal field coils provide only a small toroidal field (an order of magnitude smaller than the central one). The two magnetic configurations differ in several other respects:

– Most of the RFP magnetic field is produced by currents flowing inside the plasma, and for the same central toroidal field, the plasma current is an order of magnitude larger in the RFP than in the tokamak. As a result, the RFP equilibrium magnetic field has toroidal and poloidal components of comparable amplitudes. Their typical radial profiles are shown in Fig. 9.1.

– Since for the same central toroidal field, the plasma current is an order of magnitude larger in the RFP than in the tokamak, for the same resistivity in both configurations ohmic heating is two orders of magnitude larger in the RFP. While very high magnetic fields are required to reach thermonuclear temperatures in an ohmically heated tokamak (see Sec. 5.2 of [5]), much smaller fields should be required for a RFP at parity of confinement. Therefore a RFP should be able to reach thermonuclear temperatures without any additional heating, a dramatic simplification. As yet no temperature saturation has been observed in the largest present RFP when current was increased up to 1.8 MA in optimized discharges to reach temperatures above 1.5 keV (see Fig. 3 of [6]).

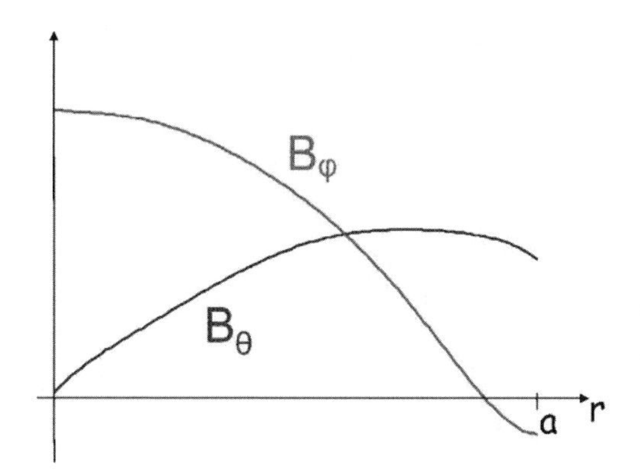

Fig. 9.1 Radial profile of the toroidal and poloidal magnetic fields in a RFP.

– The safety factor profile is typically decreasing. It starts from about $q_0 \simeq a/2R$ in the center (a and R are the plasma minor and major radii) to reach a small value at the plasma edge related to the weak reversal of the toroidal field (see Fig. 9.2). Therefore in contrast to the tokamak, magnetic field lines in a RFP are almost poloidal. While the tokamak is close to a curved theta-pinch, the RFP is close to a curved Z-pinch.

– At variance with the tokamak, the toroidal current is pushed to values above the Kruskal-Shafranov threshold. This triggers a helical MHD instability (resistive kink) of the plasma annulus that nonlinearly saturates at finite amplitude with the help of a strong radial decrease of the toroidal field that reverses at the edge (see Sec. 9.12.3). This self-organization mechanism makes the RFP configuration disruption-free because it corresponds to a full MHD relaxation of the magnetic field. This is at variance with the tokamak where the central toroidal field is maintained close to its edge value, but may relax disruptively.

– In the RFP the edge radial magnetic field needs to be controlled either by a thick shell or preferably by feedback controlled saddle coils [7]. This is important to avoid resistive wall modes which are unstable even at zero β, and to improve confinement. However this brings an additional complexity

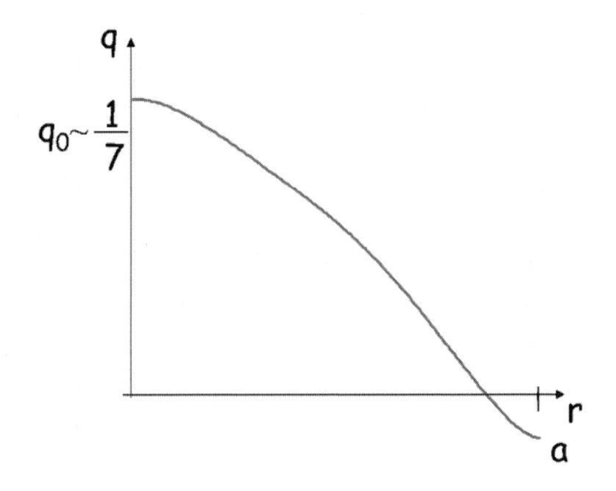

Fig. 9.2 Typical radial profile of the safety factor in RFX-mod.

to the front-end with respect to a tokamak without magnetic ELM control, as explained in Sec. 9.5.

– The helical deformation due to the saturated kink makes RFP plasmas similar to stellarator ones. However the two configurations are lying symmetrically with respect to the tokamak, as far as magnetic self-organization is concerned: in a stellarator the magnetic field is almost completely defined by imposed outer currents, while in a RFP it is mostly produced by inner currents.

– Since the toroidal field coils only produce a weak outer negative toroidal field, the much stronger positive one in the plasma core is produced by poloidal currents inside the plasma. Since such currents cannot be driven by the toroidal loop voltage alone, a plasma flow is necessary for providing a $\mathbf{v} \times \mathbf{B}$ electromotive force in the poloidal direction. In other words a dynamo is acting in the plasma. As explained in Sec. 9.8, the plasma flow is now understood as a mere electrostatic drift due to the helical distortion of the plasma, a feature similar to that of a nonlinear tearing mode. From an astrophysicist viewpoint, the RFP dynamo is only half a dynamo since the toroidal current is directly driven by an externally imposed electromotive force.

There are two ways of producing strong magnetic fields with little heat dissipation: using currents in superconducting magnets, or in hot plasmas: the resistivity of a 20 keV plasma is about one tenth of that of copper at room temperature. The RFP has the unique feature among confinement devices to choose this second path. The remnant ohmic dissipation turns out to be useful to reach thermonuclear temperatures. This is in contrast with the power used to operate the cryogenic system required by superconducting magnets. In the latter case one must extract ohmic and nuclear heat out of the "refrigerator"!

Finally since the outer magnetic field applied to a RFP plasma is weak, this configuration comes close to a minimum requirement for the virial theorem, which states that self-confinement of a magnetized plasma by inner currents is impossible, and that some currents flowing in outer conductors are necessary. This is in contrast with the tokamak and the stellarator where the outer and inner magnetic fields have a comparable amplitude.

Presently the largest RFP is the RFX-mod device (R = 2 m, a = 0.459 m, maximum current 2 MA) in Padua (Italy); this corresponds to a 10 m^3 plasma with a maximum central magnetic field above 1 T [8]. Other present machines are MST (R= 1.5 m, a = 0.5 m, nominal current 0.6 MA) in Madison (USA), EXTRAP T2R (R = 1.24 m, a = 0.18 m, nominal current 300 kA) in Stockholm (Sweden), and RELAX (R = 0.51 m, a = 0.25 m, nominal current 80 kA) in Kyoto (Japan). The KTX machine is under construction in Hefei (China) with a size close to MST and RFX and should operate in 2015.

9.3 Usefulness of the RFP configuration for fusion science and dynamo physics

The progress in tokamak performance was an incentive to design and to build ITER, in order to address the issue of burning thermonuclear plasmas. This project should be followed by a demonstration fusion plant called DEMO. However, important physics issues need to be solved both for ITER and for the definition of DEMO: what is the origin of the Greenwald density limit [9], how do transport barriers form and stay, what is the origin of plasma rotation, what is the effect of additional heating, how to scale reactor parameters out of smaller experiments, how dangerous may be fast particle driven MHD modes, what is the role of ambipolar electric fields, what is the benefit of radiating layers, of a helical deformation of the magnetic field, etc.? RFP experiments may bring transversal views about all these issues and help sorting out their many interpretations in the tokamak frame. They may bring a similar help to stellarator physics too.

In particular, understanding the density limit in magnetic confinement might enable to come closer to this limit or to overcome it, would increase considerably the reactivity of thermonuclear plasmas, which would dramatically increase the prospects of magnetic fusion. Important progress in this direction might be done by taking advantage that this limit is the same Greenwald limit in the tokamak and the RFP. With respect to the tokamak and the stellarator, the RFP has a low imposed external field. It has a helical magnetic field like the stellarator, but it is more self-organized than a tokamak, and much more than a stellarator.

Apart from physics issues, there are also operational issues in common with the tokamak like the control of MHD modes [7]. Some coordinated activity already brought fruitful results like the possibility of controlling the (2,1) mode in tokamaks and of lowering q(a) below 3 [10]. Another possibility yet to explore would be to turn a disruptive discharge in a tokamak into a RFP one.

Since (half) a dynamo is acting in the RFP, there is a natural resonance with the astrophysical dynamos. The corresponding communities have been interacting for several years, in particular in the frame of the Center for Magnetic Self-Organization (CMSO) in the United States. The von Karman Sodium (VKS) experiment in Cadarache came with a striking result: an incompressible fluid dynamo can drive a RFP magnetic state all by itself! This experiment studies dynamo action in the flow generated inside a cylinder filled with liquid sodium by the rotation of coaxial impellers (von Karman geometry). It evidenced the self-generation of a stationary dynamo when the impellers do not rotate with the same angular velocity, when at least one is made of a ferromagnetic material, and when the velocity fluctuations are rather small. Furthermore the dynamo with two rotating impellers is not the superposition of that with single rotating impellers. The magnetic field averaged over a long enough time corresponds to a RFP magnetic state with a large $m = 0$ mode (see Fig. 7 of [11]). It is striking that the incompressible turbulent flow produced by the impellers leads to the same magnetic equilibrium as in a current driven pinch whose plasma is compressible. Understanding the universality of the configuration might lead to a large leap forward of dynamo theory.

9.4 Attractivity of the RFP configuration for a reactor

We just explained that the RFP configuration is disruption-free and should be able to reach thermonuclear temperatures without any additional heating. Here are further attractive aspects of a possible RFP reactor:

– Since most of the magnetic field is produced by currents flowing inside the plasma, the maximum magnetic field is not bounded by the maximum value imposed by superconductor technology like for tokamaks and stellarator, but by the ability to drive a high plasma current. A priori this should enable to reach higher magnetic fields than provided by Nb-Sn cables. Therefore improving performances may be sought not only by increasing the size of the device, but also by increasing its current.

– Since most of the magnetic field is produced by currents flowing inside the plasma, copper toroidal field coils might be used. This also lowers the requirements on the thickness of their neutron shielding, and on the growth of the radial build up. The coils are also subjected to lower forces than in a tokamak or a stellarator, which decreases the mechanical constraints. All this would bring a dramatic technological simplification coming with large money savings. However a superconducting central solenoid might still be advisable.

– β's as high as 0.25 are already possible, which reduces the amplitude of the confining magnetic field for a given thermonuclear reactivity. With the low external toroidal field this means a high engineering beta (it is already up to 26% in present experiments).

– The RFP configuration is relatively compact, which leads to a high mass power density, and possibly to a single-piece maintenance thanks to the low field toroidal field coils.

– This comes with efficient assembly and disassembly, and possibly a free choice of aspect ratio.

– Since the RFP does not suffer from disruptions, fast ramp-up and ramp-down of the current are possible. This opens the prospect of a non-stationary reactor where rapidly succeeding (on the order of 1 s) current flat-tops with opposite signs of the current would be produced by a double swing poloidal circuit. Indeed the rapid succession of flat-tops would minimize the mechanical fatigue of materials due to thermal cycles which is a concern for tokamaks, since ramp down and ramp up must be slow to avoid disruptions.

– Though non axisymmetric, the magnetic field should not require optimization like in the stellarator. Indeed, because of ohmic heating, alpha particles do not need to transfer their energy efficiently to the plasma center. Furthermore the loss of axisymmetry decreases when going towards the plasma edge, which decreases the trapped fraction of fast particles and the related transport. Finally with the amplitude of the helical field seen in experiments, super-banana orbits are present in negligible fraction, which helps fast particles to be better confined [12]. All this should prevent alpha particles to be lost rapidly to the wall.

– Because of ohmic heating, the enhancement of alpha particle transport due to unstable Alfvenic or geodesic acoustic modes is less an issue than in the tokamak or the stellarator.

These features would lead to a simpler, more robust, and less expensive reactor than the tokamak or stellarator configurations. It, is worth recording that, for ITER, disruptions are the main physical challenge, the neutral beam heating system is still under development, and the large superconducting coils are a very expensive high tech part of the machine whose failure would require a long shut down, especially whenever the machine would be nuclearly activated.

9.5 Challenges ahead

As just explained, the RFP configuration has many potentially interesting features for a thermonuclear reactor. However it must face several challenges to become a credible candidate. Fortunately there is no first-principles-based showstopper. The main challenge for the RFP is the quality of confinement, since its helical magnetic self-organization comes with a magnetic chaos which is as yet higher than in a stellarator. Improvements are now sought by a better control of the edge magnetic field [13, 14], and of plasma-wall interaction [15], and by a further increase of the plasma current.

Another challenge is core fueling, since wall-recycling is not efficient enough to provide high densities at high plasma current [15]. Pellet injection needs a fine tuning, since big pellets decrease too much the bulk temperature, while small ones do not feed the plasma center. The need for this fine tuning is a consequence of the self-organized feature of the RFP plasma.

A path to escape the constraints of full self-organization of the RFP plasma is to drive the plasma either impulsively or by sustained current drive [17, 18]. This line of research is being very active on the MST machine [19].

The need of feedback controlled saddle coils to control the edge radial magnetic field brings an additional complexity to the front-end with respect to a tokamak. RFX-mod has 192 saddle coils to avoid resistive wall modes and to provide adequate boundary conditions to the helical magnetic field inside the plasma. The number of such coils that would be necessary for a

RFP reactor is still an open issue, but is likely to be higher than the one considered for ELM control in ITER. The peak electric power used by the RFX feedback control system is on the order of 25% of the ohmic power, but the mean power is about 10%.

9.6 Lawson criterion

Reaching thermonuclear temperatures without any additional heating means that a RFP cannot work in the ignition regime. Indeed decreasing ohmic heating would mean decreasing the magnetic field and confinement too. This makes the usual "Lawson criterion" of tokamaks and stellarators irrelevant for the RFP.

In order to yield a net production of electricity a RFP power station should verify condition $e(P_{\text{ohm}}+P_{\text{fusion}}(1+m_{\text{blanket}})) > P_{\text{ohm}}+P_{\text{coils}}$ where P_{ohm}, P_{fusion} and P_{coils} are respectively the ohmic power, the power produced by fusion reactions, and the power dissipated in the coils and other electrical devices of the reactor; e is the power station efficiency and m_{blanket} the multiplication factor of the fusion power due to exothermic nuclear reactions in the tritium breeding blankets. This implies

$$P_{\text{fusion}} > [(1/e-1)P_{\text{ohm}} + P_{\text{coils}}/e]/(1+m_{\text{blanket}})\,. \qquad (9.1)$$

The total energy in the plasma is $W = 3\langle nT\rangle V$ where n is density and T the temperature, V the plasma volume, and the brackets mean "volume averaging over V". Let τ_{E} be the energy confinement time. Then the power loss W/τ_{E} should be balanced by ohmic heating and alpha heating in a stationary regime: $W/\tau_{\text{E}} = P_{\text{ohm}} + P_{\text{fusion}}/4$. Using Eq. (9.1) this yields $P_{\text{fusion}}(3/4 + m_{\text{blanket}} + 1/4e) > (1/e-1)W/\tau_{\text{E}} + P_{\text{coils}}/e$. Taking into account the fact that P_{fusion} scales like n^2, this condition yields again a condition on $n\tau_{\text{E}}$ which is reminiscent of that originally derived by Lawson which did not consider ignition [20]. The minimum value of $n\tau_{\text{E}}$ depends on the choice of the parameters. Neglecting P_{coils}/e, for $m_{\text{blanket}} = 0.2$, and $e = 1/3$, one finds a value about 1.2 times the classical value for ignition in a tokamak 1.5 $10^{20}m^{-3}s$ (Sec. 1.5 of [5]). This value decreases if the efficiency increases.

9.7 Intuitive model of magnetic self-reversal

Since magnetic field reversal may sound mysterious it is useful to first introduce a simple a toy model which makes it intuitive [21, 22]. Consider a current-carrying resistive wire initially placed on the axis of a cylindrical flux conserver (see Fig. 9.3). A finite, but small, axial magnetic field is present inside the cylinder. This means an azimuthal current is flowing in the cylinder (Fig. 9.3(a)). The wire is in unstable equilibrium, and a small perturbation triggers a kink. Imagine there is a "demon" forcing the absolute value of the pitch, but not its sign. Then the kink develops with a pitch whose sign is such that the azimuthal part of the current flowing in the wire has the same orientation as the one in the cylinder, which brings a mutual attractive force (Fig. 9.3(b)). Such a pitch brings also a solenoidal effect which increases the magnetic field and flux inside the kinked wire. The flux conserver imposes accordingly a decrease of the magnetic field and flux outside. As long as the current in the cylinder keeps its sign, the instability cannot quench. The continuing growth of the magnetic field and flux inside the kinked wire forces the outer magnetic field and the current in the cylinder to reverse. Eventually the wire finds an equilibrium where it is trapped in a sheared magnetic field (Fig. 9.3(c)). This model exhibits a self-organized magnetic system with field reversal where the loss of cylindrical symmetry is essential. It is a variant of a model for the saturation of the ideal kink proposed by Kadomtsev where the wire is superconducting, and where the current vanishes at saturation [23].

One of the scenarios for a RFP discharge, "self-reversal scenario", is quite similar to that of the above wire model. We show later that this picture is confirmed by two dimensional resistive MHD numerical simulations of the RFP, which are usually run with the axial flux conservation rule. Since the RFP is a self-organized magnetic system resulting from the nonlinear saturation of a (resistive) kink mode involving field reversal, it turns out to be resilient to disruptions. In RFP experiments, aided reversal is generally preferred to self-reversal in order to save volt-seconds and to help the formation of the poloidal part of the current [24]. However most of the current corresponds to the paramagnetic pinch component of the configuration as discussed in Sec. 9.12.3.

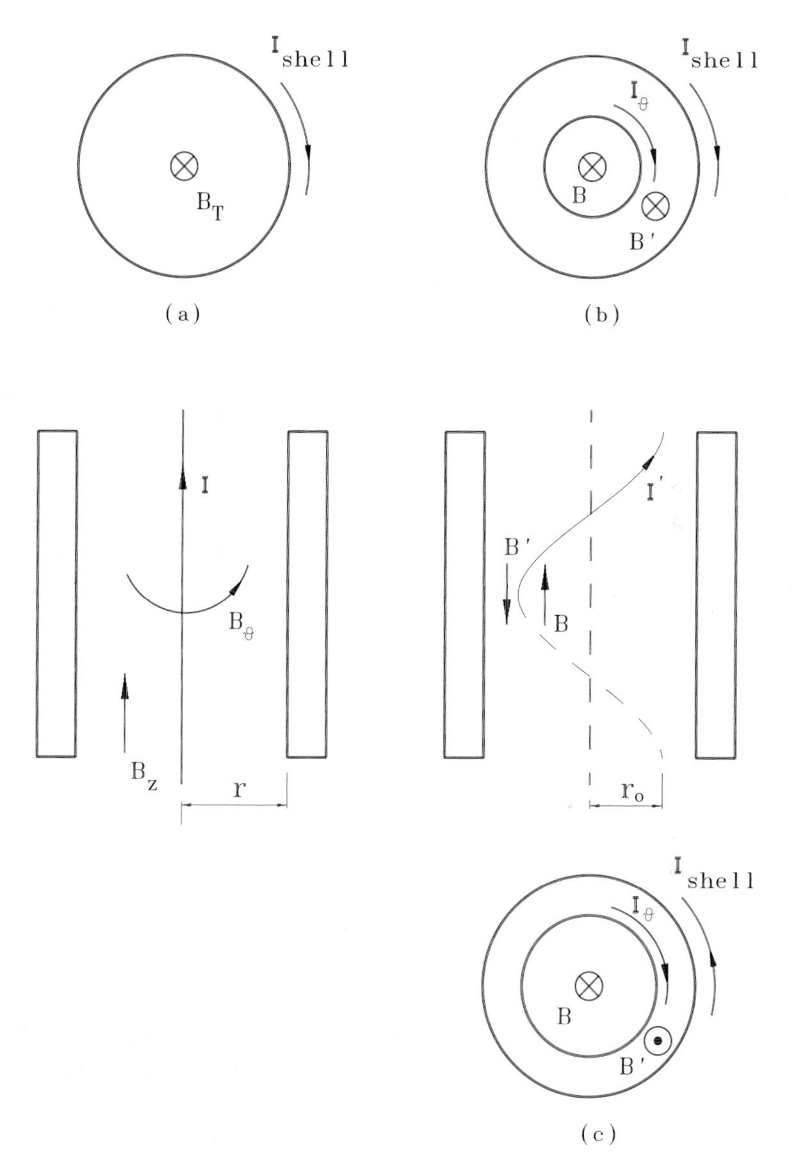

Fig. 9.3 Wire model (reproduced by permission from [22] doi:10.1088/0741-3335/42/12B/319).

9.8 Intuitive description of the dynamo

Another element of RFP physics which may sound mysterious is the dynamo process underlying the configuration. This process turns out to be a mere consequence of magnetic relaxation, as is revealed by nonlinear visco-resistive MHD simulations of the RFP where stationary pure single helicity (SH) states are found [25]: (i) Since the toroidal and poloidal magnetic fields have the same order of magnitude in a RFP, the current flowing in the plasma is much higher than in a tokamak for the same field; indeed $q \ll 1$ over the whole cross-section of the plasma. This means that the Kruskal-Shafranov threshold is crossed during the start-up phase of the discharge. In analogy with the wire model, a resistive kink develops, bringing a helical distortion of the magnetic field (see Fig. 9.4) (ii) This distortion

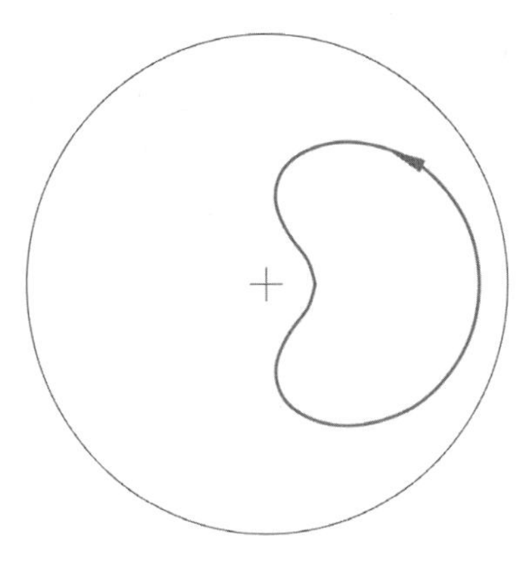

Fig. 9.4 Helical magnetic surface.

requires a modulation of the current density along current lines. (iii) This modulation requires the existence a modulated electric field, which must be electrostatic since the single helicity state is stationary. (iv) The corresponding electrostatic potential brings a component of the electric field perpendicular to the magnetic field. (v) This component drives an $\mathbf{E} \times \mathbf{B}$ motion whose non axisymmetric part is exactly the dynamo velocity field. This picture is backed up by the analytical calculations of Secs. 9.12.2 and 9.12.5.

In reality such a description holds for any plasma column with a single helicity perturbation of the magnetic field. This holds in particular for a saturated tearing mode. Therefore the RFP dynamo belongs in the same family as the tearing mode dynamo.

9.9 Necessity of a helical deformation

Up to this point the RFP configuration was linked with a helical deformation of the plasma. We now show that there cannot be an axisymmetric RFP.

Imagine there is an axisymmetric RFP with a vanishing pressure gradient next to the reversal radius r_r. Then the pitch of current lines reverses together with that of the magnetic field at r_r. For a given loop voltage the poloidal current must reverse at the reversal radius, since the sign of the toroidal current is fixed. But this implies the toroidal field is minimum at r_r (see Fig. 9.5), which contradicts our assumption that it reverses. A $\mathbf{v} \times \mathbf{B}$ electromotive force cannot help since $\mathbf{B}$ is poloidal at reversal. If there is a finite pressure gradient next to r_r, this modifies only the toroidal current at reversal, and leaves unaffected the reasoning on the poloidal current.

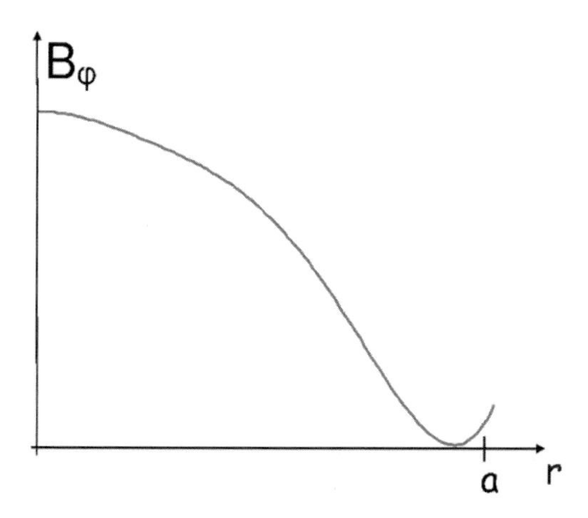

Fig. 9.5 Impossibility of reversal in a cylindrical RFP.

Because of the importance of this result, we prove it again by an alternative method. If an axisymmetric RFP would exist, a finite azimuthal current would be necessary at r_r to produce the reversal of the toroidal field. However it could not be driven neither by the inductive electric field due to the loop voltage, nor by a $\mathbf{v} \times \mathbf{B}$ electromotive force since $\mathbf{B}$ is poloidal at reversal.

Therefore no dynamo can sustain an axisymmetric RFP. This statement is germane to Cowling's theorem in astrophysics [26] which states that no axisymmetric dynamo can maintain a symmetric magnetic field. By extension one often speaks of a "Cowling theorem" for the RFP. The wire model shows that the RFP fulfills this theorem in a quite natural way: field reversal is a consequence of the plasma kink, and thus of the loss of axisymmetry. From both its magnetic and velocity fields, the RFP may be viewed as a helically distorted paramagnetic pinch. As shown in [27], the helical deformation of the plasma is already present for ultra-low safety factor configurations.

9.10 MHD simulations

9.10.1 *From single to multiple helicity*

The analytical description of the RFP configuration is still at its infancy, but a lot of information about RFP self-organization has been provided by numerical simulations using the visco-resistive compressible nonlinear MHD model in the constant-pressure, constant-density approximation [4, 28–30]. Though its limited physics contents, this model displays a dynamics with strong similarities with the experimental one.

The first theoretical proof of existence of single helicity states was their discovery in 1990 in 3D numerical simulations where *no helicity was a priori forced upon the dynamics* [28, 31–33]. Later on, the existence of single helicity states was understood as corresponding to the high dissipation limit of the system [29, 34], as defined by the product d of the central viscosity and resistivity (the corresponding dimensionless number is the Hartmann number scaling as $1/\sqrt{d}$). When d decreases, secondary modes with other helicities develop in the system. Together with this, the magnetic configuration becomes time-dependent, and switches intermittently between two fluctuating states: a quasi single helicity (QSH) state where secondary modes are smaller than the original one, and a multiple helicity (MH) state where at least two modes have similar amplitudes. When d decreases, the duration of the quasi single helicity phases decreases, as well as the percentage of time where quasi single helicity dominates. As will be shown in Sec. 9.11 the numerical scenario is quite similar to the experimental one in RFX-mod: a decrease of d in the former is equivalent to a decrease of the current in the latter, and switches the transition from quasi single helicity to multiple helicity states. Historically the numerical scenario was an incentive to revisit the RFX data base in 1999. Shots with long lasting quasi single helicity states were found to be present for quite a time in the data base [35], but they had been neglected because they were out of the traditional multiple helicity paradigm of the machine!

When performed in toroidal geometry, 3D nonlinear visco-resistive MHD simulations show that toroidal coupling prevents the system to reach a pure single helicity state when d increases, but that magnetic chaos due to toroidal coupling stays limited in single helicity states for the aspect ratios of the largest present RFP's [36].

On top of the bifurcation leading from multiple helicity to single helicity, there is another one which modifies the magnetic topology. When the amplitude of the single helicity mode is low enough, there is a magnetic island. Therefore the magnetic field displays two magnetic axes: the unperturbed axisymmetric one and the one related to the island O-point. When the amplitude of the single helicity mode increases there is a bifurcation that changes the topology of the magnetic field through the coalescence of the axisymmetric O-point with the island X-point [35]. Then the former island O-point becomes the only magnetic axis, and the magnetic topology is like that of a non resonant kink. This coalescence may be intuitively understood by noting that an increase of the single helicity mode means that the inner loop of the separatrix becomes tighter and tighter about the former axisymmetric O-point and vanishes when its inner area vanishes. Then the safety factor q profile goes through a maximum located in the vicinity of the former separatrix [37]. Furthermore the resilience to chaotic perturbations of a one parameter one degree of freedom Hamiltonian dynamics increases when its corresponding separatrix vanishes due to a saddle-node bifurcation. As a result, single helicity states without separatrix are more resilient to chaos than those with a magnetic island [35, 38–40].

The next subsections provide a more precise picture of the results provided by the visco-resistive compressible nonlinear MHD model in the constant-pressure, constant-density approximation [28–30]. Its equations are

$$\frac{\partial \mathbf{B}}{\partial t} = \nabla \times (\mathbf{v} \times \mathbf{B}) - \nabla \times (\eta \mathbf{J}) \,, \tag{9.2}$$

$$\frac{\partial \mathbf{v}}{\partial t} + (\mathbf{v} \cdot \nabla)\mathbf{v} = \mathbf{J} \times \mathbf{B} + \nabla^2(\nu \mathbf{v}) \,, \tag{9.3}$$

where η and ν are the radial distributions of respectively the resistivity and of the kinematic viscosity, and where $\mathbf{J} = \nabla \times \mathbf{B}$ and $\nabla \cdot \mathbf{B} = 0$. Here $\mathbf{B}$ is normalized to the value B_0 of the axial magnetic field on axis, time and velocity are normalized to the Alfvén time τ_A and velocity v_A

respectively computed with B_0, and the position to the plasma radius a. In these units η is the inverse Lundquist number, $\eta = \tau_A/\tau_R \equiv S^{-1}$, and $\nu = \tau_A/\tau_V \equiv R_m^{-1}$. The RFP is simulated as a straight periodic cylinder with axial periodicity $2\pi R$.

The plasma current and the axial magnetic flux are taken as constant, which implies the constancy of the pinch parameter $\Theta \equiv B_\theta(a)/\langle B_z\rangle$ (in this paragraph the averages are done over the toroidal and poloidal angles). This type of model has been widely used for RFP simulations (see [4] and [34], and references therein). Usually numerical simulations start from a paramagnetic pinch state with $B_z(a)$ close to 0. For $\Theta \gtrsim 1.55$ the system which is (resistive) kink unstable, relaxes toward a RFP state where the reversal parameter $F \equiv B_z(a)/\langle B_z\rangle$ is in the range $0 < F < -0.5$.

9.10.2 *Single helicity*

The possibility of having a RFP plasma in a pure single helicity state was put forward since 1983 through two-dimensional numerical simulations [41–45] with $\beta = 0$ where a stationary RFP state was found by retaining only one ratio m/n for the Fourier harmonics beyond $(0,0)$, and thus forcing the final RFP state to be in the single helicity state. Some indication of this possibility was already present in the first numerical simulation of the RFP [46]. In the simulations, the magnetic field reversal is much smaller than in the wire model, as only a small part of the axial current is involved in the drive of the helical deformation.

Let $\eta_0 = \eta(0)$, $\nu_0 = \nu(0)$, $\bar\eta = \eta/\eta_0$, and $\bar\nu = \nu/\nu_0$. Applying to Eqs. (9.2) and (9.3) the rescaling $t \to \bar t = \sqrt{\frac{\eta_0}{\nu_0}}\, t$ with the corresponding rescaling of velocity $\mathbf{v} \to \bar{\mathbf{v}} = \sqrt{\frac{\nu_0}{\eta_0}}\, \mathbf{v}$ yields

$$\frac{\partial \mathbf{B}}{\partial \bar t} = \nabla \times (\bar{\mathbf{v}} \times \mathbf{B}) - \nabla \times (H^{-1}\, \bar\eta \mathbf{J}), \tag{9.4}$$

$$P^{-1}\left[\frac{\partial \bar{\mathbf{v}}}{\partial \bar t} + (\bar{\mathbf{v}} \cdot \nabla)\bar{\mathbf{v}}\right] = \mathbf{J} \times \mathbf{B} + \nabla^2 (H^{-1}\, \bar\nu \bar{\mathbf{v}}), \tag{9.5}$$

where $P = \nu_0/\eta_0$ is the magnetic Prandtl number, and $H = (\eta_0\nu_0)^{-1/2}$ is the Hartmann number. The rescaled dynamics depends only on the value of the Hartmann number when the inertia term becomes negligible. This happens to be the case for a large range of simulation parameters. Indeed simulations reveal that H is the right control parameter independently of P and Θ [29, 34] (see also Fig. 9.6); there the aspect ratio was $R/a = 4$.

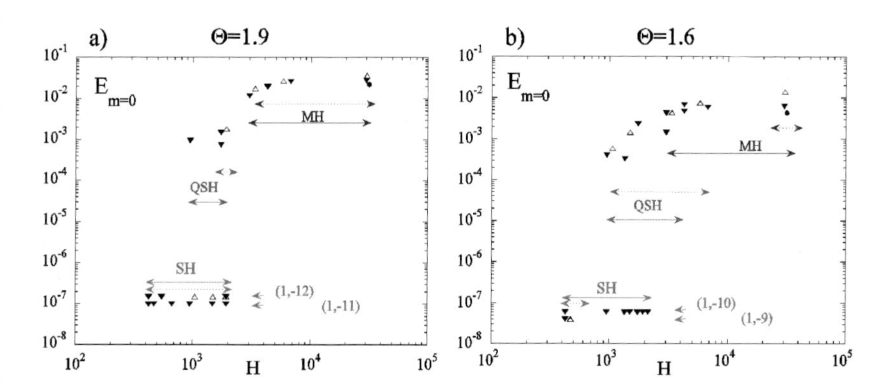

Fig. 9.6 Transition diagrams at two values of the pinch parameter: (a) $\Theta = 1.9$, (b) $\Theta = 1.6$. $E_{m=0}$ is the time averaged magnetic energy of the $m = 0$ modes and H is the Hartmann number. The symbol $\triangle$ corresponds to $S = 3.3 \times 10^3$ with $P = [0.012, 50]$, $\blacktriangledown$ to $S = 3.0 \times 10^4$ with $P = [1, 5000]$, and $\bullet$ to $S = 10^5$ with $P = 10$. The vanishing single helicity $m = 0$ modes energy is represented as a finite conventional value with different offsets associated to the different preferred helicities (reproduced by permission from [34] doi:10.1088/0741-3335/46/12B/027).

The Hartmann number was introduced in the fusion context in [47, 48], and is present implicitly through the product d of dissipations introduced in Sec. 9.10.1 in [49] and in [50] for a bifurcation approach of tearing modes.

Therefore the previous "high viscosity" simulations turn out to be in fact "high dissipation" or "small H" simulations. The upper bound in H to reach single helicity is about 1000. As usual in nonlinear dynamics an increase of dissipation (decrease of H) is favorable to a laminar behavior of the system. Depending on the initial conditions, two nearby different helicities were found to be selected by the plasma when relaxing to single helicity (SH) [29, 34]. The corresponding magnetic states are stable fixed points of the dynamics defined by Eqs. (9.2) and (9.3).

These numerical simulations enabled the description of the dynamo provided in Sec. 9.8 [25, 51, 52]. The electric field $\mathbf{E}$ in the single helicity state can be computed from the simulation variables by using Ohm's law (9.14). It is curl-free, since the final state is stationary. Its divergence provides an electric charge distribution which takes on a dipolar helical structure, and whose amplitude is small, in agreement with the assumptions leading to MHD equations. Then the perpendicular part of the velocity is provided by

$$v_{\perp es} = -\boldsymbol{\nabla}\phi \times \frac{\mathbf{B}}{\mathrm{B}^2}\,. \tag{9.6}$$

Since $\nabla \times \mathbf{E} = 0$, then $\partial E_z/\partial r = \partial E_r/\partial z$ and the $2\pi R$ periodicity in z imply that the average over z of E_z is a constant E_0 over space.

Numerical simulations bring the new information that the velocity field in the single helicity state is topologically equivalent to the radial pinch velocity field of the paramagnetic pinch, as given by Eq. (9.41) for an axisymmetric magnetic field. The single helicity field corresponds to a helically distorted version of the paramagnetic one, where the pinch is now toward the helical magnetic axis. Therefore the dynamo corresponds to a velocity field which is rather confining. In MHD codes where density is free to evolve, such a pinch velocity leads to a build up of the plasma density on the helical axis [53]. In a genuine plasma this triggers some transport mechanism opposing the growing density gradient. This effect is modified in turn by a pressure gradient, and by heat transport.

9.10.3 *Multiple helicity*

When H increases (dissipation decreases), another helicity becomes unstable in the previous single helicity equilibrium, and after a transition region for $10^3 \lesssim H \lesssim 10^4$ to be described later, the system enters a non stationary multiple helicity regime [29, 31–33]. This is shown in figure 9.6 where the energy of the $m = 0, n \neq 0$ modes is used as an order parameter for the single helicity multiple helicity transition; indeed these modes vanish in the single helicity state, since they result of the beating of at least two different helicities. From the point of view of nonlinear dynamics this regime corresponds to a new basin of self-organization. A lot of numerical simulations have been devoted to the multiple helicity regime (see [4, 28] for a review). Most of the features reported below about multiple helicity can be found in [28]. Spectral analysis of the multiple helicity state for $H > 10^4$ reveals that several $m = 1$ modes with different axial mode numbers $n > 0$ and comparable amplitudes are simultaneously present in the plasma, and that the $m = 1$ mode spectrum extends up to the maximum available n number of the simulation, but with a decreasing amplitude. This yields a broad range of values to the safety factor m/n of the modes. As in single helicity, the harmonics of the $m = 1$ modes are present too. Since Eqs. (9.2)–(9.3) are nonlinear, the beating of these modes produces a wide spectrum of modes with other m's, and in particular $m = 0, n \neq 0$ modes. All Fourier modes display temporal fluctuations. Indeed it is impossible to solve the parallel Ohm's law (Eq. (9.16)) for ϕ if multiple helicity is assumed for the spatial modulation of a static magnetic field [33]: a static multiple helicity ohmic state is impossible.

A *large helical deformation* of magnetic field lines is present in a finite axial (toroidal) domain. As a result the spatial modulation of the magnetic field is far from being random, even if the temporal fluctuations of each mode look turbulent: the nonlinear interaction drives a *strong mode locking* among the $m = 1$ modes. Order is also found temporally when monitoring the reversal parameter F. Indeed $F(t)$ displays a series of oscillations whose pseudo-period is about $\tau_R/10$. The phases where $F(t)$ decreases (increase of magnetic field reversal) are more rapid than those where it increases [28, 54]. The amplitude of the oscillations grows with Θ as in spheromak simulations [36]. They are called *fast relaxation events*. The fast relaxation events come together with a higher amplitude of the magnetic fluctuations, which shows that a solenoidal effect of the helical deformation is present in multiple helicity too. They correspond to the formation of thin current sheets with a rapid energy dissipation. The amplitude δB of magnetic fluctuations is found to scale like [55]

$$\delta B \sim H^{-\frac{1}{4}}(1 + P^{-1})^{\frac{1}{8}} , \tag{9.7}$$

consistently with the Sweet Parker expectation [28, 56]. For multiple helicity regimes confinement scaling is rather poor [57].

An important consequence of the multiple helicity character of the magnetic field spectrum is the fact that the dynamics of field lines is no longer integrable, and that *magnetic chaos* sets in [58, 59]. The magnetic mode spectrum corresponds mainly to $q \geq 0$ modes. Therefore the corresponding magnetic chaos is present in this domain when the field lines are reconstructed from the code outputs. This domain does not correspond to a well-defined domain in r because of the above-mentioned large helical deformation which brings a bulging of chaos at its axial position. In the $q > 0$ domain no magnetic island is visible. However for $q = 0$ a chain of magnetic islands is present. The $q > 0$ magnetic chaos bulges through one of the X-points of this island chain. The $m = 0$ mode increases chaotic transport in the $q \gtrsim 0$ domain, but the $m = 0$ chain of islands decreases the size of the chaotic domain, and plays the role of a transport barrier in the chaotic region [34, 60]. Equation (9.7) shows that the amplitude of magnetic fluctuations increases when H decreases. This does not mean that chaos is steadily increasing. Indeed when H decreases, at some moment the system reaches the single helicity state where the single helicity nature of the magnetic field makes the magnetic field lines non chaotic.

The large scale magnetic chaos of multiple helicity states has an interesting consequence, as was shown by Rusbridge [61] on the basis of an idea by Kadomtsev and Moffatt: (i) In the radial domain where the magnetic field is chaotic, transport is fast, and the equilibrium is almost force-free; therefore $\mathbf{J} = \mu\mathbf{B}$ where μ may be space-dependent. (ii) Setting this in $\nabla \cdot \mathbf{J} = 0$, implies $\nabla\mu \cdot \mathbf{B} = 0$, which shows that μ must be constant along field lines; thus μ is constant in the chaotic radial domain. This straightforward derivation yields a result in full agreement with the fact that in MHD simulations μ is almost constant in most of the $q > 0$ domain, but not outside [62]. In the past this domain of almost constant value of μ was rather considered as a hint of the validity of Taylor relaxation theory [63].

9.10.4 *Quasi single helicity*

When H increases, the *transition from single helicity to multiple helicity turns out to be continuous*, in analogy with a second order phase transition [29]. In the vicinity of $H = 2000$ the system displays a temporal intermittency whose laminar phases are of quasi single helicity type. Then the $m = 1$ magnetic field fluctuations spatial spectrum is made up almost entirely by one individual MHD mode. Single helicity or multiple helicity transients may last over long times [64]. The temporal intermittency in the transition from single helicity to multiple helicity is like critical opalescence in a second order phase transition, and is the signature of a bifurcation of the dynamical system of interest.

As a result of the continuity between single helicity and multiple helicity states, the large axially localized helical deformation in the latter is the result of a continuous deformation of the axially uniform helical shape of the former. The continuity bears on the dynamo as well. Indeed, though a fluctuating inductive electric field be present in the multiple helicity state, the velocity field is still dominated by the electrostatic contribution [25, 51, 52]. Magnetic chaos sets in progressively during the transition too.

9.11 Experimental results

Historically, the RFP configuration was discovered in an empirical way [65]. During three decades the plasmas were found in a state of magnetic turbulence. In such a state weak confinement resulted from the chaos of magnetic field lines and strong plasma-wall interaction due to a toroidally localized large helical deformation of the plasma. Since 1993 plasmas with transient states where magnetic chaos decreases were found in the data base of all large RFP's. In 1999 these states were found to last most of the flat top of discharges with a current about 1 MA in the RFX machine, but in a non stationary way. After an upgrade of RFX which provided in particular a good control of the edge radial magnetic field, higher currents were reached and a topological bifurcation occurred in the magnetic field providing a broad hot central domain bounded by an internal transport barrier [2, 68]. The following subsections address these successive steps.

9.11.1 *Multiple helicity*

At low current, measurements display a wide spectrum of magnetic fluctuations. The various spatial Fourier modes are temporally fluctuating too. They correspond to resistive kink and tearing modes. Many $m = 1$ and $m = 0$ modes with different toroidal mode numbers n and comparable amplitudes are simultaneously present in the plasma, which motivates to qualify this regime *Multiple Helicity* (MH). A locking is observed between these modes, leading to a toroidally localized *large helical deformation* (LHD) of the plasma and of the magnetic field lines. As a result the plasma bulges and has a localized strong interaction with the first wall.

Images of the plasma core obtained through soft X-ray (SXR) tomography display a poloidally symmetric emissivity. This indicates that a wide region of strong heat transport is present in the plasma core, in agreement with the existence of a chaotic magnetic field. The temperature gradient is localized in the domain where the toroidal field reverses.

9.11.2 *Quasi single helicity*

Transient states where the magnetic field fluctuations spatial spectrum is dominated by one individual ($m = 1, n \sim 2R/a$) MHD mode, were first detected in all large RFP's [69–74], and found to last for several energy confinement times. Later on, the single helicity (SH) states found in MHD simulations [29], triggered the discovery in the data base of RFX of similar states lasting over the whole flat top at about 1 MA [76]. The dominant magnetic mode has an amplitude 1 to few percents of the central magnetic field, and several times that of the other modes [75–77]. This motivates to qualify such regimes *Quasi Single Helicity* (QSH).

In contrast with multiple helicity plasmas, a *"bean"-like hot structure* is evident in SXR tomographic images in the quasi single helicity case in three large RFP's [75–79]. The "bean" structure is recovered when magnetic field lines are reconstructed from magnetic measurements, and turns out to correspond to a magnetic island. It comes with a decrease of magnetic chaos with respect to the multiple helicity case.

The local improvement of confinement due to magnetic flux surfaces in the plasma core shows up also in direct electron temperature profiles measurements performed by multipoint Thomson scattering [76, 78]. On top of the one found in the reversal region, a strong temperature gradient is found at the edges of the magnetic island: an *internal transport barrier* is found to exist.

In three large RFP's there was a trend for an increased probability of the occurrence of quasi single helicity states when the current is increased [78–80].

9.11.3 *Upgrade of the RFX device*

Till the end of the 90's, the RFX device had a thick and distant conducting shell: vertical magnetic field penetration time $\tau_{\text{wall}} = 450$ ms, shell to wall proximity $b/a \approx 1.24$. The discharge duration was about 100 ms, with plasma currents up to 1 MA. In the early 2000s the machine was upgraded and renamed RFX-mod: a thin closer shell was installed and completely covered with 192 independently fed saddle coils (4 in the poloidal and 48 in the toroidal directions) for feedback control [81]: vertical magnetic field penetration time: $\tau_{\text{wall}} = 50$ ms, shell to wall proximity: $b/a \approx 1.12$. The discharge duration is now up to 0.5 s, with plasma currents up to the 2 MA nominal value. Resistive wall modes can be completely avoided [82].

In this device, the RFP operation at currents $I \geq 0.6$ MA is characterized preponderantly by quasi single helicity (QSH) regimes. In fact, a quasi single helicity persistency up to 85% of the discharge flat top is obtained [83]. High current operation and high quasi single helicity persistency were made possible by the use and progressive optimization of the feedback control of the saddle coils aiming at creating a clean boundary analogous to an ideal virtual shell close to the plasma [7, 84, 85]. More recently, this clean boundary has included the possibility to excite a finite reference helicity [86, 87]. The standard high current operation shows the existence of time intervals with strong helical character, with duration and amplitudes growing with the plasma current [85, 88]. The normalized amplitude of the dominant mode tends to become a constant [83, 88, 89].

9.11.4 *From double to single magnetic axis*

During most of the duration of the flat top of RFX-mod high current discharges ($I \geq 1.5$ MA), the plasma self-organizes into a helical state characterized by nested magnetic surfaces winding around a single helical axis, but enclosed in an almost axisymmetric boundary [68]. This state is the result of two successive bifurcations occurring when the current is progressively increased. The first one is of MHD type, and makes the plasma to leave the Multiple Helicity (MH) state, characterized by the presence of several resonant modes with similar amplitudes and to reach the Quasi Single Helicity state which displays a single dominant mode and secondary ones with smaller amplitudes. At the lowest current intensities providing quasi single helicity states, the magnetic topology includes a magnetic island. Therefore the magnetic field displays two magnetic axes: the unperturbed axisymmetric one and the one related to the island O-point. For such states, termed Double Axis (DAx) states, a thermal helical structure winding around the unperturbed magnetic axis is observed [91], characterized by an electron internal transport barrier [92, 93], and the maximum electron temperature (up to 1.2 keV in 2011) is tightly correlated with the location of the magnetic island as reconstructed by external measurements. This internal transport barrier is the strongest in the rising phase of the quasi single helicity state [94].

The second bifurcation changes the topology of the magnetic field through the coalescence of the axisymmetric O-point with the island X-point. Then the former island O-point becomes the only magnetic axis, which motivates to term Single Helical Axis (SHAx) this kind of quasi single helicity state [2, 3]. As explained in Sec. 9.10.1 this topological bifurcation was first described theoretically. When broad electron temperature profiles were found in quasi single helicity states of RFX-mod, experimentalists confirmed immediately it was present [3]. The same phenomenology was found in MST [95, 96], and the Double Axis (DAx) state was also found in EXTRAP T2R [97]. Quasi single helicity states were also found in TPE-RX [98] and in RELAX [99, 100]. In RFX-mod there are also plasmas displaying external transport barriers close to the reversal surface [92].

In Single Helical Axis (SHAx) states, the region inside the above mentioned internal transport barrier has a flat temperature profile. It spans a significantly bigger volume than in Double Axis states [89], and the maximum electron temperature gradient at the internal transport barrier is a decreasing function of the amplitude of secondary modes [90]. Since this amplitude is reduced when plasma current is increased, it is expected that higher current plasmas will display even steeper thermal gradients and hotter helical cores. Plasma properties such as electron temperature, SXR emissivity and electron density have been found to be constant on helical magnetic surfaces [2] reconstructed with independent measurements, indicating that Single Helical Axis states are described by a MHD equilibrium characterized by almost invariant magnetic surfaces, in contrast with the low current multiple helicity states. The internal transport barriers are located in the region where the magnetic shear profile of the helical quasi single helicity equilibria reverses [37, 68]. The improvement of the magnetic reconstructions with the V3FIT-VMEC code makes it even clearer now (see Fig. 9.7). In agreement with numerical simulations, providing proper helical boundary conditions to RFX-mod, strongly increases the persistence of the helical equilibrium [86]. As indicated by numerical simulations [40], it is even possible to excite a non resonant quasi single helicity $n = 6$ state in RFX-mod by providing a corresponding helical boundary condition [87].

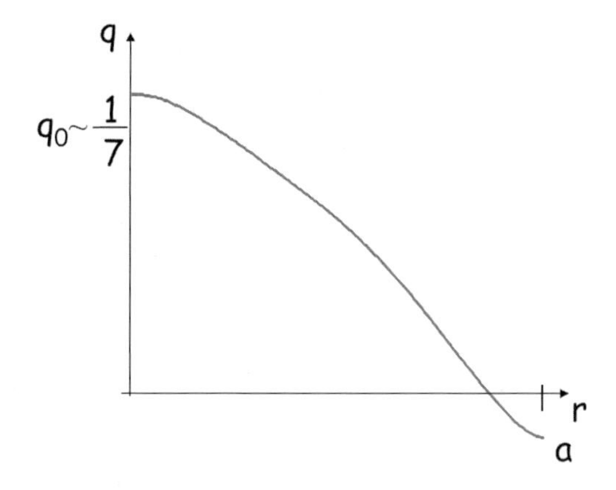

Fig. 9.7 q profile in a Single Helical Axis (SHAx) state of RFX-mod reconstructed by V3FIT-VMEC; courtesy by L. Marrelli and D. Terranova.

When a pellet is launched inside the hot domain, the density is found to peak there [101]. Nickel Laser Blow Off experiments show that impurities do not accumulate inside the helical structure [102]. The diffusion coefficient inside the island is of the same order as in multiple helicity, while a radial domain characterized by an increased outward pinch is found at the internal transport barrier. High edge transport barrier have also been observed in the $q = 0$ domain [68].

As yet the Single Helical Axis regimes are usually obtained at low densities ($n/n_G < 0.35$, with n_G Greenwald density) [15]. The possibility of obtaining hot Single Helical Axis structures at higher densities relies on a more efficient fueling of the core by pellet injection accompanied by a better control of the recycling from the wall [10, 15]. A particle confinement time in Single Helical Axis states of 12 ms was obtained with pellet injection [101].

9.12 Analytical description of the single helicity RFP

In 1958 RFP states were found by chance in the ZETA machine at Culham. In 1974, while computers were still unable to run a reasonable MHD simulation of the configuration, J. B. Taylor came with a relaxation theory [63] whose predictions had a fairly good agreement with experimental results till the late 90's. His work was the backbone of RFP theory till the emergence of the single helicity (SH) picture described in this chapter [4]. Now a whole series of experimental and theoretical results disagree with Taylor relaxation theory (see Sec. 7 of [103] for more details).

We now describe the present status of the analytical calculation of the ohmic single helicity equilibrium magnetic field of RFP's, by using resistive MHD in cylindrical geometry and in the force-free limit. A large part of this section is a rephrasing of the analytical parts of Refs. [105] and [39].

9.12.1 *Helical Grad-Shafranov equation*

Let (r, θ, z) be the usual cylindrical coordinates, and $\hat{\mathbf{e}}_x = \boldsymbol{\nabla} x$, with $x = r, \theta, z$ be the corresponding contravariant basis. The single helicity state of interest is defined by "toroidal" and "poloidal" mode numbers n and m, such that all quantities depend on θ and z only through $u = m\theta + kz$, where $k = -n/R$, and $2\pi R$ is the axial periodicity. Let $\boldsymbol{\sigma} = \boldsymbol{\nabla} r \times \boldsymbol{\nabla} u = m\hat{\mathbf{e}}_z/r - k\hat{\mathbf{e}}_\theta$, and $f(r) = r/(m^2 + k^2r^2)$. Calculations are simplified by introducing the auxiliary force-free magnetic field $\mathbf{h} = f(r)\boldsymbol{\sigma}$ (indeed it verifies $\boldsymbol{\nabla} \cdot \mathbf{h} = 0$ and $\boldsymbol{\nabla} \times \mathbf{h} = \beta\mathbf{h}$ with $\beta = -2mk/(m^2 + k^2r^2)$). In a gauge where its radial component vanishes, the vector potential can be written as $\mathbf{A} = \psi\boldsymbol{\nabla} z + \Phi\boldsymbol{\nabla}\theta$, where ψ and Φ are called poloidal and toroidal flux functions. This yields

$$\mathbf{B} = \boldsymbol{\nabla} \times \mathbf{A} = \boldsymbol{\nabla}\psi \times \boldsymbol{\nabla} z + \boldsymbol{\nabla}\Phi \times \boldsymbol{\nabla}\theta\,. \tag{9.8}$$

$\mathbf{A}$ can be equivalently written as the sum of a component along $\boldsymbol{\nabla} u$ and of $\chi\mathbf{h}$, where

$$\chi = m\psi - k\Phi\,, \tag{9.9}$$

is called helical flux. This yields

$$\mathbf{B} = \boldsymbol{\nabla}\chi \times \mathbf{h} + g\mathbf{h}\,, \tag{9.10}$$

where

$$g = r\boldsymbol{\sigma}\cdot\mathbf{B} = mB_z - krB_\theta\,, \tag{9.11}$$

is the helical magnetic field. Using Eq. (9.10) one finds that $\mathbf{B}\cdot\nabla\chi = 0$, i.e., χ is a constant along magnetic field lines. Therefore such lines lie on magnetic surfaces. We call S_χ the surface with helical flux χ, and we restrict to surfaces S_χ's enclosing $r = 0$, which is the case for the vicinity of the reversal surface we will focus on.

In Gaussian units Eq. (9.10) implies $\mathbf{J} = \boldsymbol{\nabla}\times\mathbf{B} = (\beta g - \triangle_h\chi)\mathbf{h} + \boldsymbol{\nabla} g\times\mathbf{h}$, when noticing that $\boldsymbol{\nabla}\times(\boldsymbol{\nabla}\chi\times\mathbf{h}) = -\triangle_h\,\chi\mathbf{h}$, where $\triangle_h\chi = f^{-1}[\partial_r(f\partial_r\chi) + r^{-1}\partial_u^2\chi]$; this is obtained by the successive estimates of the components of the vectors of interest. The force-free condition $\mathbf{J} = \lambda\mathbf{B}$ then requires $\boldsymbol{\nabla} g - \lambda\boldsymbol{\nabla}\chi$ to be colinear to $\mathbf{h}$. However, since g and χ depend on r and u only, their gradients are in the plane defined by $\boldsymbol{\nabla} r$ and $\boldsymbol{\nabla} u$ which is orthogonal to $\mathbf{h}$. Therefore $\boldsymbol{\nabla} g - \lambda\boldsymbol{\nabla}\chi = 0$, which implies equation

$$g = g(\chi) \text{ and } \mathrm{d}g/\mathrm{d}\chi = \lambda\,. \tag{9.12}$$

Then the component parallel to $\mathbf{h}$ of the force-free condition implies the *helical Grad-Shafranov equation* [33, 104]

$$\triangle_h\,\chi = g[\beta - \mathrm{d}g/\mathrm{d}\chi]\,, \tag{9.13}$$

where $\triangle_h\chi = f^{-1}[\partial_r(f\partial_r\chi) + r^{-1}\partial_u^2\chi]$, as shown in Appendix A of Ref. [105]. In the case $m = 1$ and $n = 0$ this is the classical equation for axisymmetric plasmas.

9.12.2 *Parallel Ohm's law*

We describe the plasma by a general Ohm's law

$$\mathbf{E} + \mathbf{v}\times\mathbf{B} = \eta\mathbf{J} + \frac{c_1\mathbf{J}\times\mathbf{B} - c_2\boldsymbol{\nabla} p_e}{ne} \tag{9.14}$$

where η is the resistivity, and the c_i's are constants which may take on independently the values 0 or 1 depending on the type of Ohm's law of interest. If $c_2 = 1$ we restrict the electronic pressure p_e to be constant along magnetic field lines and to be a fixed fraction of the total pressure. We consider the electric field to be stationary

$$\mathbf{E} = E_0 \hat{\mathbf{e}}_z - \boldsymbol{\nabla}\phi \,, \tag{9.15}$$

where E_0 is the component related to the toroidal loop voltage of a real RFP. By performing the scalar product of Ohm's law with $\mathbf{B}$, one obtains the *parallel Ohm's law*

$$\mathbf{B} \cdot \boldsymbol{\nabla}\phi = E_0 \mathbf{B} \cdot \hat{\mathbf{e}}_z - \eta \mathbf{B} \cdot \mathbf{J} \,. \tag{9.16}$$

We note that Eqs. (9.16) and (9.14) do not depend on the c_i's, and therefore on the type of Ohm's law of interest. We obtain the *average parallel Ohm's law*

$$\langle \mathbf{B} \cdot \boldsymbol{\nabla}\phi \rangle_\chi = E_0 \langle B_z \rangle_\chi - \langle \eta \mathbf{B} \cdot \mathbf{J} \rangle_\chi = 0 \,, \tag{9.17}$$

by performing the flux surface average of Eq. (9.16) over S_χ, the magnetic surface with flux χ. This average is defined for a quantity Q as

$$\langle Q \rangle_\chi = \frac{\mathrm{d}I(Q)}{\mathrm{d}I(1)} \,, \tag{9.18}$$

where

$$I(Q) = \int_{D(\chi)} Q \, \mathrm{d}^3\mathbf{r} \,, \tag{9.19}$$

and $D(\chi)$ is the volume of axial length $2\pi R$ within S_χ. The derivation of Eq. (9.17) uses the force-free condition and relation $\langle \mathbf{B} \cdot \boldsymbol{\nabla}\phi \rangle_\chi = 0$ resulting from the divergence theorem: $\int_{D(\chi)} \mathbf{B} \cdot \boldsymbol{\nabla}\phi \, dV = \int_{S_\chi} \phi \mathbf{B} \cdot \mathbf{d\Sigma} = 0$. The expression for $\langle Q \rangle_\chi$ is simpler when going from Cartesian coordinates to (χ, u, θ). The corresponding Jacobian J is given by

$$J^{-1} = \boldsymbol{\nabla}\chi \times \boldsymbol{\nabla}\theta \cdot \boldsymbol{\nabla}u = \frac{k}{r} \partial_r \chi \,. \tag{9.20}$$

Then

$$\langle Q \rangle = \langle JQ \rangle_0 / \langle J \rangle_0 \,, \tag{9.21}$$

where the average $\langle F \rangle_0$ of a quantity $F(\chi, u)$ is defined by

$$\langle F(\chi, u) \rangle_0 = \frac{1}{2\pi} \oint F(\chi, u) \mathrm{d}u \,. \tag{9.22}$$

If the force-free condition $\mathbf{J} = \lambda \mathbf{B}$ is satisfied, the average parallel Ohm's law becomes

$$E_0 \langle B_z \rangle_\chi = \lambda \langle \eta \mathrm{B}^2 \rangle_\chi \,. \tag{9.23}$$

9.12.3 *Pinch-stellarator equation*

Here single helicity ohmic RFP states are described analytically as small helical perturbations of axisymmetric ohmic pinches with small edge conductivity and axial magnetic field, called *ultimate pinches*. At this point one might think about applying directly perturbation theory to Eqs. (9.13) and (9.23), taking an ultimate as zeroth order equilibrium, with the perturbation amplitude of the asymmetric magnetic flux as a small parameter of the ordering. However, the calculation must be done to second order, and the analytical calculation of the solution of the corresponding differential equations looks formidable. Fortunately interesting new physics may be found without computing the full magnetic field, but only $\langle B_z \rangle_\chi$ to second order. Indeed the safety factor of flux surface S_χ is proportional to $\langle B_z \rangle_\chi$. This is readily shown by noting that the "poloidal" (azimuthal) flux $F_{\text{pol}}(\chi)$ enclosed between S_χ and the perturbed magnetic axis is the volume integral of B_θ / r divided by 2π. Similarly the "toroidal" (axial) flux $F_{\text{tor}}(\chi)$ inside S_χ is the volume integral of B_z/R divided by 2π. Using the definition (9.18) of the surface average, this implies the safety factor is

$$q(\bar{\psi}) \equiv \mathrm{d}F_{\text{tor}}/\mathrm{d}F_{\text{pol}} = \frac{\langle B_z \rangle_\chi}{R \langle B_\theta / r \rangle_{\bar{\psi}}} \,. \tag{9.24}$$

The reversal parameter of a RFP is classically defined as $F = B_z^{0,0}(a)/\langle B_z \rangle_v$ where $\langle \rangle_v$ means "volume average inside $r = a$". We further notice that, whenever the plasma is bounded by an ideal shell, at the shell radius $r = a$ a magnetic surface must be cylindrical. Therefore a negative value of F means a negative value of $\langle B_z \rangle$ and of q at the wall.

In order to compute $\langle B_z \rangle_\chi$ to second order, we follow first closely Ref. [33]. Using Eq. (9.12) for the helical magnetic field g, we average Eq. (9.11) defining g over S_χ, we take its derivative over χ, and we use Ampere's law and the force-free relation. In the following it will be useful to parameterize flux surfaces by $\bar{\psi}(\chi) = \langle \psi(\chi, u) \rangle_0$. Various steps given in Appendix C of Ref. [105], yield the *pinch-stellarator equation*

$$\frac{\mathrm{d}}{\mathrm{d}\bar{\psi}} \langle B_z \rangle_{\bar{\psi}} = \frac{E_0}{\langle \eta \mathrm{B}^2 \rangle_{\bar{\psi}}} \langle B_z \rangle_{\bar{\psi}} + S(\bar{\psi}) \,, \tag{9.25}$$

with

$$S(\bar{\psi}) = k \langle (\partial_{\bar{\psi}} r) \, \partial_u B_r \rangle_0 + (k/m) \mathrm{d}(\langle r B_\theta \rangle - \langle r B_\theta \rangle_0)/\mathrm{d}\bar{\psi} \,. \tag{9.26}$$

Equation (9.25) is a slightly different version of Eq. (34) of Ref. [33] which was a new formulation of an equation already derived in Ref. [106].

Both references make the following remarks: (i) $S(\bar{\psi})$ vanishes for a pinch with axial symmetry; (ii) Eq. (9.25) is a first order linear ordinary differential equation for $\langle B_z \rangle_{\bar{\psi}}$, which may be formally integrated; (iii) for an axially symmetric pinch it holds for all radii, and implies that $\langle B_z \rangle_{\bar{\psi}}$ cannot reverse at the plasma edge, confirming the statement of Sec. 9.9, but may be small there; (iv) in a stellarator since $E_0 = 0$, $S(\bar{\psi})$ is the only term providing a radial variation of $\langle B_z \rangle_{\bar{\psi}}$; (v) since according to Eq. (9.8) $\bar{\psi}$ decreases from the core to the edge, for a positive $\langle B_z \rangle_{\bar{\psi}}$ in the center, a positive $S(\bar{\psi})$ in a finite edge radial domain in $\bar{\psi}$ may a priori provide a reversal of $\langle B_z \rangle_{\bar{\psi}}$ in that domain. Using the Eq. (10) of Ref. [106], which is the analogue of the present Eq. (9.25), Ref. [107] proved numerically the existence of reversal for a pinch with twisted elliptical flux surfaces. The tokamak corresponds to $S(\bar{\psi}) = 0$, and to E_0 small enough to provide a radially almost constant axial magnetic field.

It is interesting to note that the experimental profiles of the reversed axisymmetric part of the toroidal field in RFP's are rather flat in the reversal domain after a steady decrease in the non reversed one (see Fig. 2 of [108]). This is in harmony with the leading term going from the pinch one to stellarator one in the reversal domain. In this domain these experimental profiles are rather shallow with respect to what Taylor relaxation theory [63] predicts with the Bessel function model: $B_z \sim \mathbf{J}_0(\mu r)$ with μ a constant. In such a model a finite edge poloidal current is necessary to provide reversal, and this was thought to be the origin of the high V_{loop} in RFP's. In contrast the pinch-stellarator equation shows that no current is necessary in the highly resistive edge of a RFP to provide the reversal. This indicates that the decrease of the experimental V_{loop} in RFP's might be obtained by a mere decrease of the resistivity inside the internal transport barrier, which might result from operation at higher currents [90].

9.12.4 *Single helicity ohmic RFP states*

If the resistivity is high enough in a finite radial domain $D = [r*, a]$ enclosing the plasma edge $r = a$, the pinch-stellarator equation (9.25) shows that for the unperturbed ultimate pinch, B_z stays small in D if it is small at $r = a$. Then parallel Ohm's law (9.23) shows that λ is also small in D. Therefore, due to Eq. (9.12) for the helical magnetic field g, up to a correction of order $B_z(a)$, $g(\chi)$ is a constant g_0 in D; g_0 turns out to be positive since the axial current is positive, and since $k < 0$. From now on, for any quantity Q, Q' means $\partial_r Q$, or $\mathrm{d}Q/\mathrm{d}r$ if Q depends on r only. Then the helical Grad-Shafranov equation (9.13) can be readily integrated in D to give

$$\chi_0' = mg_0/kr\,, \tag{9.27}$$

up to a correction of order $B_z(a)$, by taking into account that λ is small in D, that the axial current and rB_θ are almost constant in D, and that Eq. (9.10) defining $\mathbf{B}$ in terms of χ and g implies

$$\partial_r \chi = -r\boldsymbol{\nabla} u \cdot \mathbf{B}\,. \tag{9.28}$$

We now perturb this ultimate pinch with a small (m, n) perturbation that is non resonant in D. Let $\epsilon\chi_1(r)\cos(u)$, with ϵ small, be the corresponding perturbation of $\chi_0(r)$. We describe $\chi_1(r)$ by the Newcomb equation obtained by linearization of the helical Grad-Shafranov equation (9.13) without any inhomogeneous term; this is natural if we consider the helical state to result from a tearing-like instability of the ultimate pinch. Newcomb equation is

$$\frac{\mathrm{d}}{\mathrm{d}r}\left(f\frac{\mathrm{d}\chi_1}{\mathrm{d}r}\right) - \frac{1}{r}\chi_1 = \left[\beta(r)\lambda - \lambda^2 - g\frac{\mathrm{d}\lambda}{\mathrm{d}\chi}\right]\chi_1\,, \tag{9.29}$$

where all functions of χ are evaluated at $\chi_0(r)$. Since g is assumed to be constant in D to zeroth order, the right hand side of Eq. (9.29) vanishes in this interval, and we are left with

$$\frac{\mathrm{d}}{\mathrm{d}r}\left(f\frac{\mathrm{d}\chi_1}{\mathrm{d}r}\right) - \frac{1}{r}\chi_1 = 0\,. \tag{9.30}$$

inside D. There it is convenient to introduce the flux label ρ such that $\chi = \chi_0(\rho)$. Since the range of χ may be larger than that of χ_0, we complement $\chi_0(r)$ for $r > a$ by a straight line tangent to $\chi_0(r)$ at $r = a$. Then on the flux surface labeled by ρ, r is given by

$$r(\rho, u) = \rho + \epsilon r_1(\rho)\cos u + O(\epsilon^2)\,, \tag{9.31}$$

where $r_1(\rho) = -\chi_1(\rho)/\chi_0'(\rho)$; the $O(\epsilon^2)$ terms contribute only at order higher than 2 in the following results. As shown in Appendix D of Ref. [105], with the above definitions $S(\rho)$ can be readily computed

$$S(\rho) = \frac{\epsilon^2 k^2[m - nq(\rho)]}{m\, g_0\, \rho\, \chi_0'(\rho)} P(\rho)\,, \tag{9.32}$$

where

$$P(\rho) = f^2(m^2 + 2k^2\rho^2)(\chi_1')^2 + 2\rho\chi_1'\chi_1 + \chi_1^2 \tag{9.33}$$

and $q(\rho)$ is the safety factor defined as follows. Using Eq. (9.8) the "poloidal" (azimuthal) flux enclosed between $S_{\bar{\psi}}$ and the perturbed magnetic axis (labeled by $\bar{\psi}_0$) of the ultimate pinch is readily computed to be $F_{\rm pol} = -2\pi R[\bar{\psi} - \bar{\psi}_0]$. Similarly the "toroidal" (axial) flux inside $S_{\bar{\psi}}$ is $F_{\rm tor} = 2\pi\bar{\Phi}(\bar{\psi})$ where $\bar{\Phi}(\bar{\psi}) = \langle\Phi(\bar{\psi}, u)\rangle_0$. As a result the safety factor is

$$q(\bar{\psi}) \equiv \mathrm{d}F_{\rm tor}/\mathrm{d}F_{\rm pol} = -R^{-1}\mathrm{d}\bar{\Phi}/\mathrm{d}\bar{\psi}\,. \tag{9.34}$$

Define the normalized logarithmic derivative of χ_1, $L = a\chi_1'/\chi_1$. Then for $S(\rho)$ to be positive $P(\rho)/\chi_1^2$, a quadratic polynomial in L, must be negative. Therefore a *necessary criterion for reversal* ([105]) is that $L(\rho)$ must be in between the two roots of this polynomial which are both negative,

$$L_\pm = -\frac{a(1+k^2\rho^2)}{\rho[1+(1\pm 1)k^2\rho^2]}\,, \tag{9.35}$$

where we made $m = 1$ for simplicity, and having in mind this corresponds to the most unstable modes in RFP's.

The criterion is suggestive that a finite edge radial magnetic field might be favorable for field reversal. In accordance with this, visco-resistive MHD simulations show that helical equilibria with low perturbation amplitudes achieve reversal when a finite edge radial magnetic field is applied [105]. Numerical simulations show that the criterion works for large perturbations of the pinch too, in particular those leading to states with a single helical axis. The necessary criterion is found to be satisfied in the RFX-mod experiment for reversed states with non zero reference edge radial magnetic field [105]. These experimental and numerical results show that the validity of the criterion is more general than suggested by the perturbative approach used for its derivation: nonlinear corrections look weak, and the condition that the edge temperature is much smaller than the central one sounds as a leading factor. Helical boundary conditions for the radial magnetic field, mimicking a helical modulation of the plasma magnetic boundary similar to the experimental one, are found to play an essential role for the qualitative and quantitative agreement of MHD simulations with respect to experimental observations in both reversed-field pinch (RFP) and tokamak magnetic configurations [40, 66, 67].

We now show how Eqs. (9.25), (9.26), and (9.32) can be used to exhibit single helicity ohmic RFP states. If $L_-(a) < L(a) < L_+(a)$, there is a finite domain $[r_S, a]$ where $S > 0$. Consider an ultimate pinch with a high resistivity in this domain, and such that $B_z(r_S)/B_z(0) = \mathrm{O}(\epsilon^{2+s})$ with $s > 0$. It may be for instance a Bessel function model (BFM) [109] where r_S is close enough to the value of ρ where the zeroth order Bessel function vanishes (the corresponding resistivity profile is obtained by solving Eq. (9.23) for η). When computing directly $\langle B_z \rangle$ by using the second order expressions for χ, λ and g, an appropriate choice of the integration constants occurring at second order enables to set $\langle B_z \rangle_{\rho=r_S} = B_z(r_S)$. Then in $[r_S, a]$ the r.h.s. of Eq. (9.25) is $\mathrm{O}(\epsilon^2) < 0$ for ϵ small. Integrating this equation from r_S to a yields $\langle B_z \rangle_{\rho=a} = \mathrm{O}(\epsilon^2) < 0$ for ϵ small. Therefore there is reversal of $\langle B_z \rangle$ at the edge provided the system can settle to a finite and small ϵ. For a tearing mode resonating inside the plasma, this is controlled by the stability index Δ' which depends on $L(a)$. In a cylindrical full MHD model $\Delta' \sim \sqrt{\epsilon}\ln(\epsilon)$ [110]. For the simple case where the ultimate pinch is a BFM inside of r_S, $\Delta'(L(a))$ can be analytically computed by an extension of the calculation in Ref. [111]. This yields a formula for $\Delta'(\mu, L(a))$ where μ is the constant value of λ for the BFM. Since $S(\rho)$ has a weak dependence on μ through $\chi_0'(\rho)$, it is possible to adjust μ such that $\Delta'(\mu, L(a))$ be small and provide a finite and small ϵ at saturation of the tearing mode. Therefore for this class of ultimate pinches the existence of perturbed single helicity ohmic RFP states is proved. For more general ultimate pinches one should resort to numerical calculations, but by continuity with the BFM case, one should expect a broader class of ultimate pinches to provide RFP states with a finite and small ϵ.

9.12.5 *Calculation of the dynamo*

Equation (9.8) implies that

$$\mathbf{B} \cdot \boldsymbol{\nabla} u = -(kJ)^{-1} . \tag{9.36}$$

Writing

$$\boldsymbol{\nabla} \phi = \partial_u \phi \, \boldsymbol{\nabla} u + \partial_\chi \phi \, \boldsymbol{\nabla} \chi , \tag{9.37}$$

the last expression in Eq. (9.8) yields

$$\mathbf{B} \cdot \boldsymbol{\nabla} \phi = -\partial_u \phi (kJ)^{-1} . \tag{9.38}$$

Equations (9.16) and (9.38) finally yield

$$\partial_u \phi = -r \partial_\chi r \, E_0 \mathbf{B} \cdot \hat{\mathbf{e}}_z - \eta \mathbf{B} \cdot \mathbf{J} , \tag{9.39}$$

where all quantities are computed at the position defined by (χ, u). This equation is thoroughly discussed in Sec. 2.1.2 of [39].

Equations (9.14) and (9.15) imply that the perpendicular component of the velocity can be written as

$$v_\perp = v_{\perp es} + v_{\perp nes} , \tag{9.40}$$

where the electrostatic component $v_{\perp es}$ is given by Eq. (9.6) and

$$v_{\perp nes} = [E_0 \hat{\mathbf{e}}_z - \eta \mathbf{J} - \frac{c_1 \mathbf{J} \times \mathbf{B} - c_2 \boldsymbol{\nabla} p_e}{ne}] \times \frac{\mathbf{B}}{\mathrm{B}^2}, \tag{9.41}$$

is the non electrostatic component of $v_\perp$. These equations show that the dynamo velocity field is slaved to the magnetic equilibrium.

Reference [39] shows that $v_{\perp es}$ may be written as

$$v_{\perp es} = -\partial_u \phi \, \boldsymbol{\nabla} u \times \frac{\mathbf{B}}{\mathrm{B}^2} - \partial_\chi \phi \, \boldsymbol{\nabla} \chi \times \frac{\mathbf{B}}{\mathrm{B}^2}. \tag{9.42}$$

By decomposing the estimate of $v_{\perp es}$ over $\boldsymbol{\nabla} u \times \mathbf{B}/\mathrm{B}^2$ and $\boldsymbol{\nabla} \chi \times \mathbf{B}/\mathrm{B}^2$, one can estimate both $\partial_u \phi$ and $\partial_\chi \phi$. The first one may be compared to its estimate provided by Eq. (9.39). The combination of both expressions provides an estimate of $\phi(\chi, u)$. Then taking the Laplacian of ϕ yields the charge density due to the ambipolarity constraint.

The importance of the electrostatic field for the RFP dynamo shows up when considering the spatial mean, over θ and z, of the parallel Ohm's law (9.16)

$$\langle \eta \mathbf{B} \cdot \mathbf{J} \rangle = E_0 \langle B_z \rangle - \langle \mathbf{B} \cdot \boldsymbol{\nabla} \phi \rangle. \tag{9.43}$$

The radial profiles of the terms in this equation in a single helicity simulation of the RFP are displayed in Fig. 5(a) of [112]. It shows the contribution from the induction electromotive force proportional to E_0 turns out to be larger than the one from the mean parallel current density in the plasma core, and smaller in the edge. The difference is balanced by the electrostatic term, which provides what are often named the anti-dynamo contribution in the core and direct dynamo contribution in the edge. It is worth noticing that there is no contribution from the dynamo term $\mathbf{v} \times \mathbf{B}$ in this case, since we are considering the scalar product of Ohm's law with the total magnetic field.

The standard picture where this dynamo term is important is recovered when taking the scalar product of Ohm's law with the mean axisymmetric magnetic field

$$\langle \eta \mathbf{B}_0 \cdot \mathbf{J} \rangle = E_0 \langle \eta \mathbf{B}_0 \cdot \hat{\mathbf{e}}_z \rangle - \langle \mathbf{B}_0 \cdot \mathbf{v} \times \mathbf{B} \rangle, \tag{9.44}$$

as shown in Fig. 5(b) of [112]. In this case, instead of the contribution from the electrostatic field (which vanishes since $\boldsymbol{\nabla} \phi$ has no axisymmetric component) the dynamo/anti-dynamo action is provided by the usual $\langle \mathbf{B}_0 \cdot \mathbf{v} \times \mathbf{B} \rangle$ term. In reality the average in (9.44) is generally done over the toroidal and poloidal angles. Therefore both the standard and the "electrostatic" picture are equivalent. However the former has been unable to explain the origin of the velocity field, while the latter does. The "electrostatic" picture applies as well to the nonlinear tearing mode which has a dynamo too [113]. Both this saturated mode and the RFP single helicity ohmic state belong in the same family of stationary ohmic states.

Finally it is worth noting that the helical potential generating the dynamo velocity field might bring an experimental issue. Indeed if the plasma is surrounded by a toroidally symmetric conducting wall, while the electrostatic equipotential surfaces are helical, then radial electrical currents are driven at the wall, which perturb both the local magnetic field and the electrostatic potential.

9.13 Conclusion

This chapter defined the RFP, presented the historical evolution of magnetic self-organization of these machines, and provided the present theoretical picture about the configuration. The latter is now recalled. The current-carrying wire model showed how a kink unstable system inside a flux conserver may be nonlinearly stabilized by field reversal. The RFP turns out to use a quite similar way to self-organize, as shown by MHD simulations. The latter proved the ohmic constraint to be of paramount importance for the RFP. This constraint requires in particular the magnetic field of an ohmic RFP to be helically distorted. This helical deformation is provided naturally by the nonlinear saturation of a resistive kink instability. This nonlinear feature of the RFP configuration prevents the occurrence of disruptions. It comes with a helical electrostatic electric field driving a drift which provides the main contribution to the dynamo velocity field, as occurs for the nonlinear tearing mode.

From both its magnetic and velocity fields, the RFP may be viewed as a helically distorted paramagnetic pinch: it is a helical self-organized ohmic magnetic configuration. The RFP may be in two different states which are continuously connected by changing dissipation: a single helicity (SH) and a multiple helicity (MH) state. In the single helicity state the helical deformation of the plasma is uniformly distributed along the plasma, and magnetic field lines are regular. In the multiple helicity state the helical deformation of the plasma is localized and magnetic field lines are chaotic. The helical shape of the single helicity plasmas introduces analogies with the stellarator [16]. However the helical distortion decreases when going close to the shell, which alleviates the issue of fast particle confinement. The main theoretical challenge now is the introduction of self-consistency in the resistivity profile as a result of heat transport, which rules the electron temperature profile. This new element of self-organization will be addressed with the PIXIE3D nonlinear MHD initial value code [114], recently benchmarked in tokamak and RFP contexts with the SpeCyl code [115]. The feedback of heat transport on MHD states is not trivial at all, since it was recently shown to modify the stability of pressure-driven resistive g-modes [116].

Experimental results in all present or recent RFP's confirm many of the features of the present theoretical picture. In RFX-mod at high plasma currents (above 1 MA) there is a new regime, where the plasma spontaneously self-organizes in a single-axis helical state, with an internal transport barrier. This regime comes close to the theoretically predicted chaos-free helical ohmic equilibrium with a single magnetic axis. Experiments clearly indicate that the plasma is naturally choosing this improved regime at high current if a clean magnetic boundary condition is provided. The reason why high current is beneficial is still awaiting a theoretical explanation. The main experimental challenge in RFX-mod is presently to reach higher particle densities by pellet injection and by reduction of wall recycling. This is a prerequisite for a meaningful scaling of energy confinement time with plasma parameters.

Present experimental and theoretical results prove a change of paradigm occurred for the RFP: this configuration is no longer dominated by magnetic turbulence and chaos. From a theoretical viewpoint this new paradigm stems from the existence of two bifurcations: that from multiple to single helicity, and that from a double to a single helical axis. This new paradigm comes with other good news. First, no poloidal currents need to be driven in the resistive edge of the plasma. Second, the dynamo mechanism is no longer mysterious: it is a mere electrostatic drift due to the helical deformation of the current lines, as occurs for the nonlinear tearing mode. The RFP used to be a terrible confinement configuration: might it become a terrific one?

The RFP comes with interesting transversal physics involving the other magnetic configurations. Analogies have already been found between the tokamak and the RFP as long as edge physics [117], the Greenwald limit [15, 118], and internal transport barriers (ITB) [68] are concerned. Furthermore the RFP is an excellent test bed for the efficient control of multiple resistive wall modes [14, 82]. Issues like internal transport barriers or the impact of non axisymmetry on confinement might provide synergies with the stellarator studies. As a result the RFP is also useful to bring support to the present two main lines of magnetic confinement.

Acknowledgments

I thank D. Bonfiglio, S. Cappello, G. Marchiori, and F. Sattin for useful comments, and P. Ghendrih for his help in editing the document.

References

[1] B. E. Chapman et al., *Phys. Rev. Lett.* **87**, 205001 (2001).
[2] R. Lorenzini et al., *Nature Phys.* **5**, 570 (2009).
[3] R. Lorenzini et al., *Phys. Rev. Lett.* **101**, 025005 (2008).
[4] S. Ortolani and D. D. Schnack, *Magnetohydrodynamics of Plasma Relaxation* (Singapore: World Scientific, 1993).
[5] J. A. Wesson, *Tokamaks* (Oxford University press, Oxford, 2004).
[6] P. Martin et al., *Nucl. Fusion* **51**, 094023 (2011).
[7] P. Zanca et al., *Phys. Plasmas* **20**, 056112 (2013).
[8] G. Rostagni, *Fus. Eng. Design* **25**, 301 (1995).
[9] M. Greenwald, *Plasma Phys. Contr. Fus.* **44**, R27 (2002).
[10] P. Piovesan, et al., *Phys. Plasmas* **20**, 056112 (2013).
[11] Monchaux et al., *Phys. Fluids* **21**, 035108 (2009).
[12] M. Gobbin, G. Spizzo, L. Marrelli and R. B. White, *Phys. Rev. Lett.* **105**, 195006 (2010).
[13] P. Zanca et al., *Plasma Phys. Contr. Fus.* **51**, 015006 (2009).
[14] P. Zanca et al., *Plasma Phys. Contr. Fus.* **54**, 124018 (2012).
[15] M. E. Puiatti, to be published in Plasma Phys. Control. Fusion.
[16] M. Gobbin et al., *Phys. Plasmas* **18**, 062505 (2011).
[17] K. J. McCollam, J. K. Anderson, A. P. Blair et al., *Phys. Plasmas* **17**, 082506 (2010).
[18] J.-E. Dahlin and J. Scheffel, *Nucl. Fusion* **47**, 1184 (2007).
[19] B. E. Chapman, *Nucl. Fusion* **49**, 104020 (2009).
[20] J. D. Lawson, *Proc. Phys. Soc.* **B70**, 6 (1957).
[21] D. E. Escande and D. Bénisti, in *Proc. of the 7th European Fusion Theory Conference*, ed. A. Rogister (Jülich: Forschungzentrum KFA) **1**, 127 (1997).
[22] D. F. Escande et al., *Plasma Phys. Contr. Fus.* **42**, B243 (2000).

[23] B. B. Kadomtsev, *Tokamak Plasma: A Complex Physical System* (Bristol: IOP, 1992), p. 34.
[24] V. Antoni et al., in *Proc. 15th IAEA Int. Conf. on Plasma Phys. and Contr. Nucl. Fus. Res., Sevilla* **2**, 405 (1994).
[25] D. Bonfiglio, S. Cappello and D. F. Escande, *Phys. Rev. Lett.* **94**, 145001 (2005).
[26] T. G. Cowling, *Magnetohydrodynamics*, (New York: Interscience, 1957).
[27] D. Bonfiglio, S. Cappello, R. Piovan, L. Zanotto and M. Zuin, *Nucl. Fusion* **48**, 115010 (2008).
[28] S. Cappello and D. Biskamp, *Nucl. Fusion* **36**, 571 (1996).
[29] S. Cappello and D. F. Escande, *Phys. Rev. Lett.* **85**, 3838 (2000).
[30] C. R. Sovinec et al., *Phys. Plasmas* **8**, 475 (2001).
[31] S. Cappello and R. Paccagnella, in *Proc. Workshop on Theory of Fusion Plasmas*, ed. E. Sindoni (Bologna: Compositori, 1990), p. 595
[32] S. Cappello and R. Paccagnella, *Phys. Fluids* **B4**, 611 (1992).
[33] J. M. Finn, R. A. Nebel and C. C. Bathke, *Phys. Fluids* **B4**, 1262 (1992).
[34] S. Cappello, *Plasma Phys. Contr. Fus.* **46**, B313 (2004).
[35] D. F. Escande et al., *Phys. Rev. Lett.* **85**, 1662 (2000).
[36] T. A. Sovinec et al., *Phys. Plasmas* **10**, 1727 (2003).
[37] M. Gobbin et al., *Phys. Rev. Lett.* **106**, 025001.
[38] D. Bonfiglio, M. Veranda et al., in *Proc. of Joint Varenna-Lausanne Workshop on the Theory of Fusion Plasmas 2010* eds. X. Garbet and O. Sauter (Varenna, Italy) **260**, 012003 (2010).
[39] S. Cappello et al., *Nucl. Fusion* **51**, 103012 (2011).
[40] M. Veranda et al., *Plasma Phys. Contr. Fus.* **55**, 074015 (2013).
[41] E. J. Caramana, R. A. Nebel and D. D. Schnack, *Phys. Fluids* **26**, 1305 (1983).
[42] A. Y. Aydemir and D. C. Barnes, *Phys. Rev. Lett.* **52**, 930 (1984).
[43] J. A. Holmes et al., *Phys. Fluids* **28** 261 (1985).
[44] D. D. Schnack et al., *Phys. Fluids* **28** 321 (1985).
[45] K. Kusano and T. Sato, *Nucl. Fusion* **26** 1051 (1986).
[46] A. Sykes and J. Wesson, *Proceedings of the 8th EPS Conference on Controlled Fusion and Plasma Physics*, Prague, Vol. 1 (European Physical Society, 1977), p. 80.
[47] D. Montgomery, *Plasma Phys. Control. Fusion* **34**, 1157 (1992).
[48] D. Montgomery, *Plasma Phys. Control. Fusion* **35**, B105 (1993).
[49] C. Tebaldi and M. Ottaviani, *J. Plasma Phys.* **62**, 513 (1999).

[50] D. Grasso, R. J. Hastie, F. Porcelli and C. Tebaldi, *Phys. Plasmas* **15**, 072113 (2008).

[51] D. Bonfiglio, S. Cappello and D. F. Escande, *Charge Separation as the basis of the Reversed Field Pinch dynamo*, RFP 10th IEA/RFP-Workshop, Padova, Italy, 2004, http://www.igi.pd.cnr.it/wwwevent/RFPWS/RFPWS_presentations.html.

[52] D. Bonfiglio et al., *Theory of Fusion Plasmas, AIP conference proceedings*, **871**, 3 (2006).

[53] M. Onofri, F. Malara and P. Veltri, *Phys. Rev. Lett.* **101**, 255002 (2008).

[54] K. Kusano and T. Sato, *Nucl. Fusion* **30**, 2075 (1990).

[55] D. Terranova et al., *Plasma Phys. Control. Fusion* **42**, 843 (2000).

[56] D. D. Biskamp, *Nonlinear Magnetohydrodynamics* (Cambridge: Cambridge University Press, 1993).

[57] J. Scheffel and D. D. Schnack, *Nucl. Fusion* **40**, 1885 (2000).

[58] S. Cappello, F. D'Angelo, D. F. Escande, R. Paccagnella and D. Bénisti, *Proc. 26th EPS Conference on Controlled Fusion and Plasma Physics, Maastricht* (European Physical Society) **23J**, 981 (1999).

[59] D. F. Escande, R. Paccagnella et al., *Phys. Rev. Lett.* **85**, 3169 (2000).

[60] G. Spizzo, S. Cappello, A. Cravotta, D. Escande, I. Predebon, L. Marrelli, P. Martin and R. White, *Phys. Rev. Lett.* **96**, 025001 (2006).

[61] M. G. Rusbridge, *Plasma Phys. Contr. Fus.* **33**, 1381 (1991).

[62] S. Cappello and D. Biskamp, *Proc. International Conf. on Plasma Phys.*, Nagoya (1996).

[63] J. B. Taylor, *Phys. Rev. Lett.* **33**, 1139 (1974).

[64] R. A. Nebel, J. M. Finn and C. C. Bathke, *Single helicity and quasi-single helicity states in reversed field pinches*, RFP 10th IEA/RFP-Workshop, Padova, Italy, 2004, http://www.igi.pd.cnr.it/wwwevent/RFPWS/RFPWS_presentations.html.

[65] H. A. B. Bodin, *Nucl. Fusion* **30**, 1717 (1990).

[66] D. Bonfiglio, M. Veranda et al., *Phys. Rev. Lett.* **111**, 085002 (2013).

[67] D. Bonfiglio et al., *Proceeding of the 40th EPS Conference on Plasma Physics* (2013), P2.145.

[68] M. E. Puiatti, *Plasma Phys. Control. Fusion* **51**, 124031 (2009).

[69] P. R. Brunsell, *Phys. Fluids* **B5**, 885 (1993).

[70] P. Nordlund and S. Mazur, *Phys. Plasmas* **1**, 4032 (1994).

[71] Y. Hirano, Y. Yagi et al., *Plasma Phys. Contr. Fus.* **39**, A393 (1997).

[72] S. Martini et al., *Plasma Phys. Contr. Fus.* **41**, A315 (1999).

[73] J. S. Sarff et al., *Phys. Rev. Lett.* **78**, 62 (1997).

[74] P. Martin, *Plasma Phys. Contr. Fus.* **41**, A247 (1999).
[75] P. Piovesan, G. Spizzo, Y. Yagi et al., *Phys. Plasmas* **11**, 151 (2004).
[76] D. F. Escande, P. Martin, S. Ortolani et al., *Phys. Rev. Lett.* **85**, 1662 (2000).
[77] P. Martin et al., *Phys. Plasmas* **7**, 1984 (2000).
[78] P. Martin et al., *Nucl. Fusion* **43**, 1855 (2003).
[79] L. Marrelli et al., *Phys. Plasmas* **9**, 2868 (2002).
[80] T. Bolzonella and D. Terranova, *Plasma Phys. Control. Fusion* **44**, 2569 (2002).
[81] P. Sonato et al., *Fusion Eng. Des.* **66**, 161 (2003).
[82] T. Bolzonella, *Fusion Eng. Des.* **82**, 1064 (2007).
[83] P. Piovesan et al., in *Proc. 35th EPS Conf. on Plasma Physics, Herssonissos*, Crete O4.029 (2008).
[84] P. Zanca et al., *Nucl. Fusion* **47**, 1425 (2007).
[85] L. Marrelli et al., *Plasma Phys. Contr. Fus.* **49**, B359 (2007).
[86] P. Piovesan et al., *Plasma Phys. Control. Fusion* **53**, 084005 (2011).
[87] S. Cappello et al., in *Proceedings of the 24th IAEA Fusion Energy Conference*, San Diego, USA (2012), http://www-naweb.iaea.org/napc/physics/FEC/FEC2012/papers/199_THP216.pdf.
[88] P. Piovesan, M. Zuin, A. Alfier et al., *Nucl. Fus.* **49**, 085036 (2009).
[89] M. Valisa et al., *Plasma Phys. Contr. Fus.* **50**, 124031 (2008).
[90] R. Lorenzini et al., *Nucl. Fusion* **52**, 062004 (2012).
[91] P. Franz et al., *Phys. Plasmas* **13**, 012510 (2006).
[92] M. E. Puiatti, M. Valisa et al., *Nucl. Fusion* **51**, 073038 (2011).
[93] A. Alfier et al., *Plasma Phys. Contr. Fus.* **50**, 035013 (2008).
[94] P. Franz, M. Gobbin, L. Marrelli, A. Ruzzon et al., *Nucl. Fusion* **53**, 053011 (2013).
[95] W. F. Bergerson, et al., *Phys. Rev. Lett.* **107**, 255001 (2011).
[96] F. Auriemma, P. Zanca et al., *Plasma Phys. Contr. Fus.* **53**, 105006 (2011).
[97] L. Frassinetti et al., *Phys. Plasmas* **14**, 112510 (2007).
[98] Y. Hirano et al., *Phys. Plasmas* **12**, 112501 (2005).
[99] S. Masamune, A. Sanpei et al., *J. Fusion Energ.* **28**, 187 (2009).
[100] R. Ikezoe, S. Masamune, K. Oki et al., *J. Phys. Soc. Jpn.* **81**, 115001 (2012).
[101] D. Terranova et al., *Nucl. Fusion* **50**, 035006 (2010).
[102] S. Menmuir et al., *Proc. 36th EPS Conf. on Plasma Phys.*, Sofia (2009).

[103] S. Cappello, Theory of Fusion Plasmas, *Joint Varenna-Lausanne International Workshop, AIP Conf. Proc.* **1069**, 27–39 (2008).
[104] J. L. Johnson et al., *Phys. Fluids* **1**, 281 (1958).
[105] D. Bonfiglio, D. F. Escande, P. Zanca and S. Cappello, *Nucl. Fusion* **51**, 063016 (2011).
[106] V. D. Pustovitov, *Pis'ma Zh. Eksp. Teor. Fiz.* **35**, 3 (1982) [*JETP Lett.* **35**, 1 (1982)].
[107] P. N. Vabishchevich et al., *Fiz. Plazmy* **9**, 484; transl. *Soviet J. Plasma Phys. JETP* **9**, 280 (1983).
[108] H. A. B. Bodin, *IAEA Fusion Energy Conference*, Lausanne, Vol. 1 (1984), p. 417.
[109] S. Lundquist, *Phys. Rev.* **83**, 307 (1951).
[110] N. Arcis, D. F. Escande and M. Ottaviani, *Phys. Plasmas* **14**, 032308 (2007).
[111] R. D. Gibson and K. J. Whiteman, *Plasma Phys.* **10**, 1101 (1968).
[112] S. Cappello, D. Bonfiglio and D. F. Escande, *Phys. Plasmas* **13**, 056102 (2006).
[113] D. F. Escande and M. Ottaviani, *Phys. Lett. A* **323**, 278 (2004).
[114] L. Chacón et al., *J. Comput. Phys.* **178**, 15 (2002); *J. Comput. Phys.* **188**, 573 (2003).
[115] D. Bonfiglio, L. Chacon and S. Cappello, *Phys. Plasmas* **17**, 082501 (2010).
[116] A. A. Mirza, J. Scheffel and T. Johnson, *Nucl. Fusion* **52**, 123012 (2012).
[117] V. Antoni, *Recent Res. Devel. Plasmas* **2**, 19 (2002).
[118] M. E. Puiatti et al., *Phys. Plasmas* **16**, 012505 (2009); *Nucl. Fusion* **49**, 045012 (2009).

Index

Bold font entries indicate Chapter and Section titles, *see* the Foreword

ABC flow, 128, 133
  forcing, 133
absorption
  drift wave, 90
  wave, 91, 92
AC, *see* Alfven Cascade
accumulation of energy, 68
accumulation point, 232–234, 239
  continuum, 234
adiabatic, 235
advection
  algorithm, 14
  chaotic, 23
  flow, 125
  Hamiltonian, 93
  incompressible, 93
  mean flow, 36
  term, 47
  time, 47
advective, 12
  mixing, 37
  nonlinearity, 5, 13, 20
  time scale, 116
  transport, 3
AE, *see* Alfven Eigenmode
aerodynamic, 16
aerofoil, 118
Alfvén
  branch, 219
  continuous spectrum, 241
  fluctuation spectrum, 239
  frequency, 146
  mode, 220
    linear stability, 229
  perturbation, 232
  ratio, 59, 60
  spectrum, 220
  time, 48, 58, 71, 261
  velocity, 49, 146
  wave, 50, 57, 59, 61, 231
Alfvén Cascade, 240, 241
Alfvén Cascades, 240
Alfvén Eigenmode, 233
Alfvén Eigenmode, 233, 234, 239–241
  frequency, 234
Alfvén wave frequency, 228
Alfvén-acoustic
  fluctuation spectrum, 239
alignment, 52, 53, 66–68, 71, 72
  degree, 50, 58, 67, 68, 72
alignment, 66
almost fast dynamo, 127
almost fast dynamo, 126
alpha particle, 219, 253
ambipolar, 192
  automatic, 183, 185, 191, 198
  constraint, 180
  particle flux, 180
  radial current, 183
  radial diffusion, 183

radial particle flux, 181
tokamak, 185
ambipolar diffusion, 175, 176, 198
ambipolar diffusion, 175
ambipolar electric field, 175, 176, 183, 251
ambipolarity, 106, 181, 183
automatic, 177, 178, 183, 185
balance, 107
breaking, 105
condition, 177, 181–183, 197, 198
constraint, 109, 178, 279
electric field
automatic, 177
equation, 197
momentum balance, 197
quasi-symmetry, 197
ambipolarity, 180, 181
ambipolarity paradox, 183
ambipolarity paradox, 183
Ampère's law, 140, 274
low frequency, 225
perpendicular, 221
angular momentum diffusion, 193, 194
anisotropic
damping, 47
diagonal pressure tensor, 227
dispersion, 59
energetic ions, 227
flow, 65
Kolmogorov law, 60
spectrum, 67
anisotropy, 63, 67
anomalous resistivity, 81, 95, 96
anomalous scaling indices, 62
anomaly, axisymmetric, 10
anti-aligned, 66
anti-dynamo, 279
action, 279
anti-fast-dynamo, 126
anti-frictional, 2, 3, 12
jet, 17, 25, 35
anti-symmetry, 50
Arnold or Arnol'd, 27, 28, 33
aspect ratio, 48, 124, 139, 140, 145, 237, 239
astrophysical
body, 133, 134
dynamo, 115, 118, 133, 252
fast dynamo, 116
flow, 45, 62, 134
large-scale magnetic field, 118
magnetic Reynolds number, 133
MHD, 115, 134
turbulence, 50
astrophysics, 116
observation, 60
Reynolds number, 71
spectra, 48
asymptotic
analysis, 232
approximation, 20
eigenfunction, 130
expansion, 207, 236
limit, 223
matching, 232
regime, 131, 188
Asymptotic matching, 236
asymptotic Reynold number, 67
atmosphere, 4, 6, 9, 12, 13, 17, 21, 24, 31, 35
dynamics, 69
rotation, 31
stratification, 4
stratified, 15
atmosphere-ocean, 3, 7, 12, 15, 21, 23, 26, 29, 37
β, 23
dynamics, 1, 2, 5–8, 12, 36
jet, 2, 35
model, 23, 28, 29
atmospheric
jet, 3, 10
jet stream, 2
model
stably-stratified, 9
nuclear test, 4
PV field, 13
atmospheric wave–turbulence, 1
auto-correlation time, 98, 102

automatic ambipolarity, 177, 178, 183, 185, 198
condition, 183
automatic ambipolarity, 176
axial symmetry, 177, 178, 275
axis of symmetry, 139, 142
axisymmetric, 11, 198
anomaly, 10
boundary, 269
geometry, 178
model, 10
ohmic pinch, 274
perturbation, 189
tokamak, 178, 192
toroidal plasma, 226

BAAE, *see* Beta Alfven Acoustic Eigenmode
background state, 7
bad curvature, 142
region, 142, 143, 145
BAE, *see* Beta Alfven Eigenmode
baker's map, 120–123
balance
ambipolarity, 107
charge, 174
critical, 61, 64, 70, 72
cyclostrophic, 11
density, 174
energy, 51, 174
enstrophy, 92, 102
force, 22, 140, 173
geostrophic, 20, 22
heat, 208
hydrostatic, 11, 21, 22, 25
moment, 197
momentum, 81, 87, 92, 99, 103, 106, 173, 174, 177, 178, 181–184, 202, 203, 208
parallel momentum, 185, 187, 197
pressure, 225, 230
radial current, 189
radial momentum, 177
toroidal momentum, 181, 192–197
balance condition, 11, 20
Balescu, *see* Dupree-Leanrd-Balescu theory
ballerina effect, 21, 22
ballistic frequency, 99, 108
ballooning
angle, 162
approximation, 151
coordinate, 149
equation, 148–151, 154, 155, 159, 162
formalism, 236
instability, 137, 145
interchange pressure, 235
representation, 151, 235
space, 149, 162
symmetry, 147
theory, 137, 138, 147, 148, 154, 156
flows, 137
transform, 148, 149, 155, 156, 162, 235
transformation, 156
unstable, 240
variable, 162
Ballooning Instabilities, 137
Ballooning mode, 152
ballooning mode, 138, 140, 142, 144, 147, 152, 159
amplitude, 158
eigenfunction, 151
equation, 157
flow, 160
flow shear, 152, 154
Fourier mode, 148
growth rate, 154
MHD code, 151
MHD theory, 144
peeling mode, 145
physics, 137, 146, 147
plasma edge, 151
rational surface, 138, 151
rotation, 154, 158
stability, 148, 152
theory, 140, 152
toroidal flow, 138, 155
Ballooning mode physics, 140
Ballooning theory, 144

banana
  bounce frequency, 201
  collision time, 201
  collisionality regime, 201
  diffusion, 179, 194
  life-time, 179, 201
  orbit, 253
  particle, 179, 200, 201
  regime, 179, 181, 186, 192–194, 198, 201, 210, 211
  shape, 200
  trajectory, 201, 202
  width, 102, 179, 200, 229
banana-plateau boundary, 179
baroclinic
  instability, 3, 12, 17, 22
  vector product, 8
barotropic, 25
barrier, 3, 4, 14
  eddy transport, 3, 4, 13, 14, 17–19, 35
  external, 269
  internal, 247, 248, 267–270, 275, 281
  potential-vorticity, 3
  transport, 3, 251, 265, 270
Berk Breizman model, 85, 100
Berk Breizman nonlinear structure, 101
Bessel function, 24, 221, 223, 229
β, 12, 23, 29, 35, 36
  atmosphere-ocean, 23
beta (plasma), 12
β (plasma), 249, 253, 262
Beta Alfvén Acoustic Eigenmode, 239
Beta Alfvén Eigenmode, 229, 237, 239
beta effect, 12, 90
beta-turbulence, 32
biasing, 179, 192
bifurcation, 248, 261, 266, 267, 269, 281
  saddle-node, 261
  successive, 269
  tearing mode, 263
blob, 82, 83, 93, 95–97, 99, 104, 106, 109
  potential vorticity, 84, 87
  vorticity, 89
Boltzmann electron response, 102
Boltzmann kinetic equation, 202
bootstrap current, 145
bounce
  average, 101, 102
  frequency, 95, 227
    banana, 201
  period, 201
  time, 200, 201
boundary, 62, 72, 73
  banana-plateau, 179
  condition, 20, 22, 86, 148, 149, 157–160, 162, 196, 236
  plateau
    Pfirsch-Schlüter, 180
  trapped-passing, 188, 196
boundary layer, 48, 73, 118, 127
box counting dimension, 121, 122
box size, 65
Braginskii polynomial approximation, 207
Brownian motion, diffusion, 175
bump on tail, 100, 101
burning plasma, 219, 220, 226, 251
burning plasma, 226

cancellation
  gyroviscous, 208
  property, 123, 134
  stretching, 123
cancellation exponent, 122–125
Cantor ternary set, 121
cascade, 2, 19
  direct, 47
    energy, 69
  energy, 32, 68, 91
  enstrophy, 32, 91
  forward
    enstrophy, 91
    potential vorticity, 90
  inertial range, 47
  inverse, 47, 49, 51, 53, 69
    magnetic helicity, 53, 54, 69, 72

turbulent, 66
Casimir, 49
Cauchy problem, 131
Cauchy solution, 119, 131
causality condition, 234
causality constraint, 236, 239, 241
centrifugal force, 11, 20, 152
channel flox, 72
chaos, 2, 247, 261, 265–267, 280, 281
flow, 133
Lagrangian, 129
magnetic, 248, 254, 265, 266, 268
chaos-free, 281
chaotic
advection, 23
domain, 265
flow, 124–126, 134
motion, 132, 170
perturbation, 261
property, 127, 133
region, 129, 169, 265
transport, 265
charge
balance, 174
density, 9, 174
polarization, 102, 105, 106
charge separation, 173, 174
Charney, *see*
Rayleigh-Kuo-Charney-Stern, 8
Charney-Drazin, 27, 84, 85, 92, 94, 99, 100, 103, 104, 106
Charney-Obukhov-Hasegawa-Mima, 21
circular
cross-section, 139, 140, 145, 196
flux surface, 238
shifted, 239
circulating particle, 227, 241
closure, 67
EDQNM, 71
parallel viscosity, 185
poloidal viscosity, 186
second-order, 66
statistical, 110
two-point, 54
turbulence, 69
two-point, 70
Closure relations, 210
clump, 82
coherence time, 109
coherent structure, 13, 45, 49
nonlinear, 15
coherent time, 48
coherent vortex, 15, 46, 83
collective
mode, 226
excitation, 231
response function, 82
collective Alfvén mode excitation, 220
collective modes, 226
collision
binary, 183
electron-ion, 176
electronn-neutral, 175
frequency, 176, 186, 201, 202, 206, 207
expansion, 192
ion-ion, 195
ion-neutral, 175
time, 175, 198
banana, 201
collision operator, 183, 185, 190, 195, 196, 202, 204, 207
collisional
classical diffusion, 178
damping, 190
diffusion, 177, 178
dissipation, 110
frequency, 187, 210
momentum exchange, 96
operator, 100, 207
parallel viscosity, 207
perpendicular viscosity, 208
Pfirsch-Schlüter regime, 193
regime, 179, 186–188, 193, 207
resistivity, 96
transport, 179
collisional mean free path, 191
collisionality, 181, 187, 192, 198
regime, 180, 186, 192, 198, 199, 208
banana, 201
collisionless, 97, 102, 104, 179, 208

dissipation, 233
friction, 110
Landau damping, 191
plasma, 206, 241
regime, 187, 193
time scale, 198
collisionless gyrokinetic equation, 221
compressibility, 62, 242
compressible
MHD, 260, 261
compression, 224, 234
kinetic, 221, 230
magnetic field, 225
non-adiabatic
particle, 224
compressional Alfvén Eigenmode, 228
condensation of magnetic excitation, 53
configuration space, 228
connection length, 179, 194, 200, 202
conservation
charge, 175
cross-helicity, 63
density, 104
dipole moment, 106
energy, 51, 63
flux, 69
mass, 115, 119
moment, 183
momentum, 82, 89, 91, 96, 104, 106, 183, 185
phase space density, 100
potential vorticity, 85, 90, 93
property, 73
pseudomomentum, 27, 101
toroidal momentum, 192
continuous
spectrum, 232, 233, 239, 241
transition, 241
continuum, 233, 234
damping, 233, 234
eigenvalues, 161
stable, 163
ideal MHD, 161
continuum accumulation point, 234
convection, 33
layer, 3, 38
solar surface, 133
turbulent, 133
convective
derivative, 175
instability, 234
plume, 72
thermostat, 3
zone
stellar, 115
Coriolis
effect, 4, 9, 21
force, 11, 20, 28, 152
parameter, 12, 20–22
pressure anomaly, 20
coronal mass ejection, 45
correlation
2-point, 110
cross, 50, 55, 63, 64, 66, 72
flow, 66
function
two-point, 49
incoherent, 110
large scale, 54
local, 66
scale, 82
spectrum, 54, 55
time, 66, 82, 92, 99
correlator, 64, 65
third order, 67
corrugation, 73
cosmical magnetic field, 115
Coulomb
collision, 201
gauge, 220
Cowling theorem, 259, 260
critical
layer, 90
critical balance, 61, 64, 70, 72
critical opalescence, 266
cross-correlation, 50, 63, 64, 66, 72
cross-helicity, 49, 63, 68
conservation, 63
cumulant, 58
current

bootstrap, 145
density, 71
diamagnetic, 184
displacement, 46, 175
dissipative, 190
electric, 174
Hall, 61
helicity, 52
inertia
neoclassical, 184
ion inertial, 184
non-ambipolar, 189
parallel, 183
perpendicular, 183
plasma inertia, 189
polarization, 184, 224
radial, 190, 191
sheet, 46, 55, 61, 72, 73, 265
curvature, 142, 145, 146
average, 142, 143
bad, 142, 143, 145
drift, 225, 231
good, 142
good region, 142
magnetic drift, 220
magnetic field, 137, 142
terrestrial plasma, 137
vector, 142
cutoff, 11, 14, 35
diffusive, 125
cyclostrophic
balance, 11
cyclotron frequency, 176, 205, 220
cylindrical
geometry, 178, 196
symmetry, 256

damping
anisotropic, 47
collisional, 190
continuum, 233, 234
external, 100
frictional, 92
GAM, 190, 191
infrared radiative, 27
Landau, 95–97, 191
nonlinear Landau, 91, 96
poloidal flow, 152, 186, 189–191, 197
poloidal rotation, 183, 185, 188, 191, 198
time, 183, 185, 186
poloidal velocity, 182, 185, 186
rate, 190, 191
rotation, 198
wave, 100
damping of poloidal rotation, 183
Darmet model, 101–104
Darmet model, 101
Debye length, 174, 223
decay rate, 188
decaying flow, 65
density
balance, 174
conservation, 104
moment, 202
stratification, 1
density limit, 251
diagonal pressure tensor
anisotropic, 227
diamagnetic, 238
contribution, 184
current, 184
drift, 24, 176, 189, 210
effect, 208
frequency, 228, 238
sideband, 242
dielectric function, 82, 99
diffusion
ambipolar, 175, 176
angular momentum, 193, 194
banana, 179, 194
Brownian motion, 175
classical, 179, 180, 193
coefficient, 175
collisional, 177, 178
eddy, 18
irreversible, 194
magnetic, 117, 124
molecular, 116
momentum, 193, 194

neoclassical, 179, 180, 185
toroidal momentum, 198
non-ambipolar, 191
perpendicular, 193
Pfirsch-Schlüter, 180
plateau, 179
quasilinear, 98, 110
radial, 183, 198
random walk, 179
time, 195
toroidal momentum, 194, 196–198
transverse, 176
Diffusion in fully ionized magnetized plasma, 176
Diffusion of toroidal momentum, 193
diffusive
cut-off, 125
layer, 131
structure, 128
diffusivity
hyper, 61
magnetic, 49, 115
momentum, 196
negative
turbulent, 47
toroidal momentum, 198
dimensional analysis, 48, 57, 58, 70
Kolmogorov, 69, 70
dipole moment, 105, 106
direct cascade, 47
energy, 69
direct numerical simulation, 47, 48, 54, 55, 59, 66, 67, 69–71
discontinuity, rotational, 55
dispersion
anisotropic, 59
current, 190
property, 34–36
relation, 6, 7, 23, 33, 35
dispersion relation, 171, 219, 220, 232–234, 238
fishbone, 233
fishbone-like, 232, 235, 241
GAM, 189, 190
dispersion relation, 219
displacement, 25
current, 46, 175
field, 16
fluid element, 89–91
pattern, 16
random, 175
disruption, 254, 256
disruption-free, 249, 252, 253, 280
dissipation
collisional, 110
collisionless, 233
forcing, 6, 26, 32, 33, 48, 55
operator, 61
peak, 59, 60, 65
range, 59
rate, 59
scale, 68
viscosity, 90
viscous, 107
dissipative
current, 190
range, 68
scale, 51, 54
structure, 46, 48
weakly-dissipative, 33, 36, 38
DNS, *see* direct numerical simulation
Doppler shift, 152, 154
Drazin, *see* Charney-Drazin
drift
curvature, 220, 225
diamagnetic, 24, 189, 210
electric, 176, 224, 231, 248, 250, 280, 281
friction force, 176
kinetic, 195–197
magnetic, 179, 180, 194, 195, 200, 231
orbit, 229
precession, 102
velocity, 95
viscosity, 189
width, 229
zonal flow, 85
drift hole, 107
drift turbulence, 81, 85, 101
drift turbulence, 101

drift wave, 6, 7, 12, 15, 16, 23, 26–28, 90, 93, 107, 188
turbulence, 37, 82, 92, 93, 104, 106, 188
drift wave, 15
drift wave-zonal flow interaction, 82
drift kinetic theory, 194
Dupree mechanism, 81
Dupree-Lenard-Balescu theory, 85, 109
Dupree-Lenard-Balescu theory, 109
dynamics
atmosphere-ocean, 1, 2, 5–8, 12, 36
geophysical fluid, 59
Hamiltonian, 261
ideal, 53, 55–57, 68, 70, 72
Lagrangian, 65, 72
nonlinear, 51, 70, 229
passive tracer, 63
Dynamics of Structures, 81
dynamo, 115–118, 126, 127, 130, 132–134, 250, 252, 257, 259, 260, 263, 266, 279
ABC flow, 128, 133
action, 115, 116, 121, 125, 127, 128, 133, 134, 252, 279
almost fast, 127
astrophysical, 115, 118, 133, 252
driven, 132
effect, 71
evolution, 132
fast, 116–118, 120, 121, 123–127, 129–131
growth rate, 131
nonlinear, 132
flow, 126
growth rate, 116, 119, 126, 128, 131
incompressible fluid, 252
intermediate, 130
kinematic, 63, 116, 125, 130, 133
large-scale, 127, 134
map, 120, 122, 125
mean field, 118, 134
mechanism, 281
MHD, 116
mode, 128
problem, 68
process, 116
regime, 70
region, 124
RFP, 258, 279
Roberts, 126
rotating fluid, 71
saturated, 133
self-consistent, 134
slow, 116, 126, 130
small-scale, 134
solar, 132
successful, 124
tearing mode, 258, 280
velocity field, 248, 258, 264, 279, 280

eddy, 82, 109
cascading, 83
noise, 47
resistivity, 47
viscosity, 47, 51
Eddy Damped Quasi Normal Markovian, 71
eddy flux of momentum, 25
eddy flux of potential vorticity, 25, 28
eddy transport barrier, 3, 4, 13, 14, 17–19, 35
eddy transport barrier, 3
eddy turn over time, 48, 58, 71
EDQNM, *see* Eddy Damped Quasi Normal Markovian
eigenfunction, 130, 146, 162
ballooning mode, 151
eigenmode, 68, 85, 87, 89, 145, 152, 154–157, 162
equation, 148
Floquet, 157
GAM, 189
growth rate, 160
structure, 148
eigensolution, 119
eigenvalue, 124, 128, 148, 150, 159–161
condition, 150

continuum, 161
Floquet, 160, 162
eikonal, 155, 156, 160, 162
form, 146, 149, 150, 155
time-dependent, 155
Ekman friction, 47
elastic structure, 17
elasticity wave, 18
electromotive force, 69, 134
electron-neutral collision, 175, 176
Eliassen-Palm flux, 25
Elsässer variables, 50, 62, 63, 66, 67
Energetic Particle continuum Mode, 233, 234
Energetic Particle Mode, 224, 233, 234
Energetic Particle Modes, 233
energy
accumulation, 68
balance, 51, 174
cascade, 32, 51, 68
confinement time, 227, 255, 267, 281
conservation, 51, 63
flux, 67
free, 82, 87, 97–99, 101, 108, 109, 145
kinetic, 49, 53, 58, 60, 64, 69, 70, 145
magnetic, 53, 54, 60, 64, 65, 69, 70, 116, 119, 128
modal, 53
potential, 232, 234
scaling, 65
spectral index, 56, 64
spectrum, 51–57, 60–65, 68, 69, 71, 72
total, 56, 57, 59, 63, 64, 69
spectrum, 63
transport, 234
energy principle, 104, 223
energy scaling, 69
enstrophy, 92, 93, 103
cascade, 32, 91
forward cascade, 91
potential, 26
enstrophy balance, 102
enthalpy, 87
EPM, *see* Energtic Particle Mode
equilibrium
current, 223
distribution function, 221
flow, 227
force-free, 266
MHD, 225–227, 270
quasi single helicity, 270
single helicity, 264, 271
equilibrium poloidal rotation, 181
equipartition, 51, 53, 59, 69, 71
pinch, 199
spectrum, 51, 57
error field, 173
expansion
collision frequency, 192
WKB, 149
external biasing, 192
external resonant magnetic perturbations, 173

fast dynamo, 116, 118, 121, 126, 127, 130–132
action, 117, 118, 124–127, 129, 133, 134
flow, 118, 120
growth rate, 123, 127, 131
nonlinear, 132
fast dynamo growth rate, 123
Fast Dynamos, 115
Fast Dynamos Action, 125
Fast Dynamos Flows, 126
fast magneto-acoustic wave, 228
fast particle
MHD mode, 251
fast-dynamo, anti, 126
Fickian eddy, 18
field line bending, 223, 234, 235
field line bending, 223
field, mean field theory, 81, 90, 98
filament, vortex, 46, 48, 61, 69, 72
filtering methodology, 71
fishbone dispersion relation, 233

fishbone like dispersion relation, 219, 232, 235, 239
fishbone-like dispersion relation, 219, 220, 232, 233, 235, 241
Floquet
approach, 157, 162
eigenmode, 157
eigenvalue, 160, 162
form, 157, 162
growth rate, 161
mode, 154, 155, 157, 162
flow
ABC, 128, 133
forcing, 133
advection, 125
anisotropic, 65
astrophysical, 45, 62, 134
channel, 72
chaos, 133
chaotic, 124–126, 134
correlation, 66
decaying, 65
divergence free, 152
dynamo, 118, 126
fast dynamo, 120
forced, 48, 132, 170
geophysical, 45
growth rate, 127
helical, 126, 127
helicity, 134
ideal, 28, 118
ideal fluid, 5, 7, 8, 27, 28
ideal MHD, 54
incompressible, 124, 126
inflection point, 86, 89
inhomogeneous, 71
integrable, 126
inviscid, 118
jet, 25, 35
Lagrangian, 28, 29
magnetized, 46
mean, 84, 86, 87, 91, 94
MHD, 71
momentum, 89, 92, 101
non-chaotic, 126
parallel, 187, 188
pattern, 34
poloidal, 152, 186, 191
damping, 186, 189–191, 197
Roberts, 126, 127, 129
rotating, 7, 59, 61, 69
shear, 85, 86, 92, 152, 154, 155, 157, 158, 160–163
stochastization, 90
stratified, 7, 59
stretching, 133, 134
subsonic, 60
symmetrical, 128
Taylor-Green, 56, 64, 70
toroidal, 152, 155, 186, 191
turbulent, 46, 47, 51, 52, 61, 73, 115, 133, 252
zonal, 3, 28, 32, 37, 38, 81, 82, 84, 85, 88, 90–92, 101, 103, 104, 106–110
shear, 109
fluid
conducting, 117
geophysical, 59
ideal, 51, 52
incompressible, 49, 119
neutral, 46, 47, 62, 63
perfectly conducting, 118, 119
quasi-geostrophic, 87, 88
rotating, 47
turbulent, 47, 69
fluid turbulence, 49, 61, 63, 69, 72
Fluid, perfectly conducting, 118
flux
ambipolar, 191
conservation, 69, 256
Eliassen-Palm, 25
energy, 67
heat, 102, 174, 180, 193, 197–199, 204, 205, 207–210
ideal
growth rate, 131
mass, 28, 29
momentum, 25, 26, 28
polarization, 105, 106
potential enstrophy, 92
potential vorticity, 25, 28, 88, 90

reversal, 132
vorticity, 90
flux conjecture, 119, 123, 131
force
centrifugal, 152
Coriolis, 152
electromotive, 69, 134
friction, 176, 203
restoring, 89, 223
Reynolds, 88
thermal, 176, 203
viscosity, 183
viscous, 173, 185, 186
force balance, 22, 140, 173
force-free, 54, 63, 271, 274
condition, 272, 273
equilibrium, 266
forced flow, 48, 132, 156, 170
forcing, 64, 66, 68, 92
ABC flow, 133
artificial, 32
dissipation, 6, 26, 32, 33, 48, 55
external, 33
mechanism, 54
non-stochastic, 32
quasi-stochastic, 33
self-excitation, 32, 33
stochastic, 92
time-scale, 48
form stress, 25, 26
forward cascade
enstrophy, 91
potential vorticity, 90
Fourier
expansion, 149
harmonic, 142, 143, 262
mode, 45, 51, 55, 141, 143, 144, 146, 148, 151, 155, 264, 267
representation, 156, 157
series, 146, 149
space, 47, 69, 236
spectrum, 52, 56
transform, 49, 147–149, 156, 157, 162, 235
truncation, 49
Fourier shell, 51
Fractal dimension, 121
fractal dimension, 121
free energy, 82, 87, 97–99, 101, 108, 109, 145, 228, 233
free lunch, 101
friction, 49, 109, 110
Ekman, 47
force, 176, 203
drift, 176
frontogenesis, 12, 22

g-mode, 281
GAM, *see* geodesic acoustic mode
gap
frequency, 241
mode, 233, 239–241
toroidal, 232, 239
gauge, 271
Coulomb, 220
generic dynamics, *see* potential vorticity dynamics, 7–10
geo-magnetic reversal, 69
geodesic acoustic mode, 188–191, 254
damping, 190, 191
eigen mode frequency, 189
weak damping, 190
geomagnetic storm, 46
geophysical flow, 45
geophysical fluid
dynamics, 59
geophysics
Reynolds number, 71
geostrophic, *see* quasi-geostrophic, 20
balance, 20, 22
Grad-Shafranov
magnetic equilibrium, 178
Grad-Shafranov equation, 272, 276
granulation, 108–110
phase space, 85, 109
granule, solar surface, 133
gravity, 21, 28, 65, 137
self, 104
wave, 18, 21, 22, 24
Greenwald density, 251, 271
Greenwald density limit, 252
group velocity, 35, 236

growth rate, 95, 97–99, 122, 126, 129–131, 134, 146, 148, 150, 152, 154, 157–160, 162, 163
  asymptotic limit, 127
  ballooning mode, 154
  computation, 128
  dynamo, 116, 117, 119, 126, 128, 131
  eigenmode, 160, 162
  fast dynamo, 118, 123, 131
  Floquet, 161
  flow, 127
  flux, 119
  ideal flux, 131
  instability, 160
  local, 160, 161
  magnetic flux, 131
  maximum, 127
  time-averaged, 157
guiding center, 102, 105, 108, 221
gyro
  frequency, 22
  radius, 22
gyro-averaged, 231
gyrokinetic
  code, 237
    linear, 242
  description, 221
  distribution function, 221
  equation, 221–223, 230
  MHD
    hybrid, 230, 242
  response, 220
  simulation, 242
  Vlasov, 107
gyrokinetic Poisson equation, 102, 105, 109
gyrokinetic turbulence, 81, 85
gyroviscosity, 192, 193, 197, 199, 207–210
  tensor, 193, 199, 208
gyroviscosity, 205
gyroviscous cancellation, 208

Hall current, 61
Hamiltonian, 93, 100
  advection, 93
  dynamics, 261
Harris current sheet, 46
Hartmann number, 260, 262, 263
Hasegawa, *see* Charney-Obukhov-Hasegawa-Mima
Hasegawa-Mima, 4, 20, 21, 36, 87, 104
  turbulence, 104
Hasegawa-Wakatani, 4, 87, 106, 107
heat balance, 208
heat flux, 102, 173, 180, 193, 197–199, 204, 205, 207–210
  parallel, 206
heat transport, 264, 267, 281
helical
  axis, 264, 269, 277, 281
  boundary condition, 270
  dipolar structure, 263
  distortion, 258, 260, 264, 280
  equilibrium, 270, 277, 281
  flow, 126, 127
  Grad-Shafranov equation, 272, 276
  magnetic field, 252–254, 270, 272, 274, 275, 280
  MHD instability, 249
  path, 138
  perturbation, 274
  rotation, 134
  self-organization, 254
  self-organized configuration, 280
  self-organized state, 247
  state, 247, 269, 276, 281
  structure, 269, 270
  turbulence, 68
helical deformation, 250, 251, 259, 260, 262, 265, 266, 280, 281
helical deformation, 259
Helical Grad-Shafranov, 271
helicity, 50, 52, 70, 71, 126
  cross, 49, 63, 68
  current, 52
  dynamo, 127, 134
  kinetic, 50, 53, 54, 63, 69
  magnetic, 49, 50, 53–55, 58, 59, 63, 64, 69, 72
    spectrum, 69

modal, 53
parallel, 54
pulsed flows, 129
spectrum, 52, 54, 71
hole, 82, 83, 97, 99, 104, 106, 108, 109
drift, 107
phase space, 107
homogeneous turbulence, 73
horse-shoe vortex, 72
hydrostatic
balance, 11, 21, 22, 25
hyper-diffusivity, 61
hyperviscosity, 70

ideal
kink, 256
shell, 268, 274
ideal ballooning, 154
ideal constraints, 70
ideal dynamics, 53, 55–57, 68, 70, 72
ideal flow, 28, 118
ideal fluid, 8, 15, 25, 26, 51, 52
flow, 5, 7, 8, 27, 28
ideal flux, growth rate, 131
ideal inviscid flow, 118
ideal MHD, 53–57, 151, 161, 168, 226, 227, 230, 232, 240
ideal Ohm's law, 230
ideal plasma equilibrium, 226
IK, *see* Iroshnikov-Kraichnan
impermeability theorem, 13
improved confinement regime, 173, 247, 281
incompressibility, 47, 49, 62, 124
condition, 185
incompressible
advection, 93
flow, 124, 252
fluid, 49, 119, 252
low frequency, 189
perturbation, 144
reduced MHD, 68
steady flow, 126
inertia, 145, 146, 223, 224, 231, 232, 234, 236, 239
current, 184
neoclassical, 186
neoclassical enhancement, 186, 188
poloidal rotation, 188
radial current, 189
inertia charge uncovering, 224
inertia term, 262
inertia term, 223
inertial
layer, 232, 235
region, 237
term, 132, 133
inertial index, 55
inertial layer, 235
inertial range, 47, 57, 59, 60, 70, 71
inflection point, flow, 89
infrared radiative damping, 27
inhomogeneous, 13, 17, 18
flow, 71
magnetic field, 178, 199, 200
PV mixing, 12, 15, 36, 37
space, 4
wave-turbulence, 2, 15, 35
instability
ballooning, 137, 145
baroclinic, 3, 12, 17, 22
convective, 234
critical pressure gradient, 144
drive, 145
growth rate, 160
ideal MHD, 240
ion acoustic, 81, 95
Jeans, 104
kink, 280
kink mode, 145
modulational, 90
resistive, 12
resistive kink, 249
shear, 12, 27
tearing, 276
integrable flow, 126
interchange pressure
ballooning, 235
intermediate dynamo, 130
intermittency, 48, 61, 62, 260, 266
intermittency exponent, 62
intermittent, 14, 62

structure, 46
internal transport barrier, 247, 248, 267–270, 275, 281
interstellar
medium, 45, 72, 116
turbulence, 72
intrinsic rotation, 108, 110
inverse cascade, 47, 49, 51, 53, 69, 91
magnetic helicity, 53, 54, 69, 72
inviscid
displacement, 87
flow, 85, 118
instability, 86
invariants, 93
ion
heat flux, 180, 198, 199
inertial current, 184
momentum balance, 182
polarization current, 184
poloidal velocity, 182
precession, 229
ion acoustic instability, 81, 84, 95
ion cyclotron resonance heating, 237
ion-ion collision, 186, 195, 198
ion-neutral collision, 175
ionospheric, 46
Iroshnikov-Kraichnan, 61, 65
Iroshnikov-Kraichnan spectrum, 59, 61, 64, 67
irreversibility, 90, 102
irreversible, 17
diffusion, 194
transport, 90
island
chain, 265
magnetic, 248, 261, 265, 268, 269
O-point, 261, 269
rotation, 187
X-point, 261, 269
ITB, *see* internal transport barrier

Jacobian, 139
matrix, 119, 124
Jeans
equilibrium, 101
instability, 104
mode, 107
jet
anti-frictional, 17, 25, 35
atmosphere-ocean, 2, 35
atmospheric, 3, 10
flow, 35
zonal, 3, 30, 38
jet self-sharpening, 33
jet core, 3, 4, 13–15, 17, 34, 35
jet stream, 5, 12
atmospheric, 2
jet structure, 10, 11
jet structure, 9
Joule damping, 47

K41, *see* Kolmogorov
KBM, *see* Kinetic Ballooning Mode
Kelvin
cat's eye, 34
circulation, 7, 8, 28, 29
impulse, 29
sheared-disturbance, 35
theorem, 93, 94
kinematic, 116, 118, 134
phase, 64, 132
regime, 63, 131–134, 170
viscosity, 49, 261
kinematic dynamo, 63, 116, 125, 130, 133
kinetic, 85, 104
Boltzmann equation, 202
compression, 221, 230
energy, 49, 53, 58, 60, 63, 64, 69–71, 145
equation, 228
helicity, 50, 53, 54, 63, 69
modal energy, 65
modeling, 220
pseudomomentum, 103
singular layer, 239
turbulence spreading, 103
kinetic 'phasetrophy' density, 103
Kinetic Ballooning Mode, 226, 229, 239
kinetic Beta Alfvén Eigenmode, 242
kinetic effect, 46

kinetic non-adiabatic particle compression, 224
kinetic Reynolds number, 50
kink, 250, 256, 260, 280
  ideal, 256
  mode, 145, 256
  non-resonant, 261
  resistive, 249, 258, 262, 267, 280
kink-like magnetic surface, 248
Kolmogorov, 47, 64, 65, 73
  constant, 58, 67
  dimensional analysis, 69, 70
  range, 59
  spectral index, 61
  spectrum, 51, 59, 60, 63, 64
  turbulence, 48
Kolmogorov law, 47, 48, 52, 63
  anisotropic, 60
Kolmogorov law, 47
Kraichnan, *see* Iroshnikov-Kraichnan
Kruskal-Shafranov threshold, 249, 258
Kubo number, 81, 82, 85, 102, 109
Kuo, *see* Rayleigh-Kuo-Charney-Stern
kurtosis, 65

Lagrange multiplier, 52, 72
Lagrangian
  chaos, 129
  description, 46, 67
  dynamics, 65, 72
  information, 29
  pseudomomentum, 29
  trajectory, 71
Lagrangian flow, 28, 29
laminar
  phase, 266
  regime, 263
laminar streamline, 118
Landau
  damping, 95–97
    collisionless, 191
  nonlinear damping, 91, 96
  resonance, 90, 92
Langmuir wave, 46
Large Eddy Simulation, 70
large helical deformation, 265–267
large-scale
  dynamo, 134
  magnetic chaos, 265
  shear flow, 134
large-scale dynamo, 134
  action, 127
Larmor
  finite Larmor radius effect, 221, 231, 242
  frequency, 200
  radius, 176, 181, 209, 229, 235
    correction, 194
    expansion, 191
Lawson criterion, 255
Lawson criterion, 255
Lenard, *see* Dupree-Leanrd-Balescu theory
LES, *see* Large Eddy Simulation
life-time
  banana, 179, 201
  Sun, 132
linear
  bump on tail, 100
  eigenmode, 87
  neoclassical theory, 191
  response, 97
  scaling, 62
  stability, 229
  theory, 85, 108, 143
  wave, 82, 97, 102, 103
local
  alignment, 66
  correlation, 66
  growth rate, 160, 161
  interaction, 46
  magnetic shear, 156
locality in Fourier space, 69
localized
  radially, 147, 150
  rational surface, 141, 144, 149
  structure, 48, 61
locked mode, 173
long-time memory, 69
low-β limit, 226
low-β ordering, 227, 231

low-collisionality, regime, 179
Lundquist number, 262
Lyapunov
 exponent, 125, 129, 132, 133, 169, 170
 number, 124, 125
Lynden-Bell, 93

magnetic
 boundary, 277, 281
 buoyancy, 134
 diffusion, 117, 124
 diffusivity, 49, 115
 drift, 180, 194, 195, 200, 229, 231
  curvature, 220
 energy, 53, 54, 58, 60, 64, 65, 69–71, 116, 119, 128
 excitation, 53
 fluctuation, 57, 64, 72, 265–267
 fluctuationn, 265
 helicity, 49, 50, 53–55, 58, 59, 63, 64, 69, 72
  spectrum, 69
 island, 248
 modal energy, 65
 moment, 200
 potential, 49, 53, 69
 relaxation, 258
 root mean square, 64
 self-organization, 250, 254, 256, 280
 shear, 148, 156, 232, 235, 236, 239, 270
 Taylor scale, 67
 turbulence, 247, 266, 281
magnetic axis, 264, 269, 274, 277, 281
magnetic axis, 269
magnetic chaos, 248, 254, 261, 265, 266, 268
magnetic field
 compression, 225
 cosmical, 115
 curvature, 137
 reversal, 256, 262, 265
 spectrum, 265
magnetic field compression, 225
magnetic flux surface, 23, 37, 180, 182, 183, 194–196, 200, 201
magnetic helicity, 69
magnetic island, 188, 265, 268, 269
magnetic Prandtl number, 50, 54, 64, 71, 262
magnetic Reynolds number, 47, 50, 116, 133, 134
magnetic self-reversal, 256
magnetic surface, *see* magnetic flux surface
magnetized flow, 46
magneto-hydro-dynamics, *see* MHD
magneto-sheath, 46
magnetosphere, 45, 51, 59, 61, 72
magnetospheric, 46
map dynamo, 120, 122, 125
Map Dynamos, 120
marginal stability, 27, 31, 33, 144, 234
marginally stable, 107
mass
 flux, 28, 29
Maxwell stress, 192, 196
mean field, 134
 dynamo, 118, 134
 generation, 127
 mechanism, 134
 MHD, 126, 133
mean field theory, 81, 90, 98
mean flow, 84, 86, 87, 91
mean free path, 202, 206, 207, 209
memory effect, 52
 long-time, 69
meso-scale, 109, 241
MH, *see* multiple helicity
MHD, 144
 astrophysics, 115, 134
 compressible, 260, 261
 dynamo, 116
 energy principle, 223
 equilibrium, 225–227, 270
 flow, 54, 71
 ideal, 53–57, 151, 161
 instability, 249
 mean field, 126, 133
 model, 278

momentum equation, 230
reduced, 64, 67, 68
relaxation, 249
resistive, 271
MHD non-adiabatic particle compression, 224
MHD simulation, 247, 256, 258, 261, 264, 266, 267, 271, 277, 280
MHD simulation, 260
MHD turbulence, 50, 57–62, 64, 66–69, 71–73
large scale excitation, 53
model, 53
spectral exponent, 49
spectral index, 49
weak, 59
MHD turbulence, 45
micro-scale, 241
Mima, *see* Charney-Obukhov-Hasegawa-Mima, *see* Hasegawa-Mima
mixing, 4, 12–15, 18, 23, 31, 35–37
advective, 37
nonlinear, 15
potential vorticity, 12, 13, 17, 18, 23, 25, 28, 31–36, 90, 91
potential vorticityhomogeneous, 37
potential vorticityinhomogeneous, 12, 15, 36, 37
mixing length, 82
mobility, 175
mobility coefficient, 175, 176
mode conversion, 233, 234
mode locking, 265, 267
mode structure, 147–149, 228
model
atmosphere-ocean, 23, 28, 29
atmospheric
stably-stratified, 9
axisymmetric, 10
Berk-Breizman, 85, 100
Darmet, 101–104
Hasegawa-Mima, 104
MHD, 278
MHD turbulence, 53
quasi-geostrophic, 19–21, 23, 29
stratified, 5
Tagger-Pellat-Diamond-Biglari, 101
wire, 256, 258, 260, 262
modulational instability, 90
moment balance, 197
moment conservation, 183
momentum, 8, 20, 25, 27–29, 115
conservation, 82, 89, 91, 96, 104, 106, 183, 185
diffusion, 193, 194
diffusivity, 196
exchange, 203
flow, 89, 92, 101
flux, 25, 26, 28
injection, 152
transport, 17, 173, 178, 220
turbulent transport, 199
momentum balance, 81, 87, 92, 99, 103, 106, 173, 174, 177, 178, 181–184, 202, 203, 208
momentum damping, 194
momentum diffusion, 191
multiple helicity, 261, 264–266, 268, 270
bifurcation, 281
ohmic state, 264
plasma, 267
regime, 264, 265
state, 260, 261, 264–266, 270, 280
transient, 266
Multiple helicity, 264, 267

Navier-Stokes, 51
Navier-Stokes turbulence, 63
negative dissipation, 6, 12, 20, 23
negative turbulent diffusivity, 47
negative viscosity, 2, 12, 24, 25
neoclassical, 102
diffusion, 179, 180, 185
diffusivity, 198
equilibrium, 191
inertia, 184, 186–188
ion heat flux, 198
particle flux, 180, 198
particle transport, 190
rotation, 179

theory, 174, 179, 181, 182, 191, 198, 210
transport, 174, 179, 180, 191
viscosity, 184, 187
viscosity coefficient, 187
viscous current, 184
viscous force, 186
neoclassical diffusion, 178
Neoclassical plasma inertia, 186
neutral fluid, 46, 47, 62, 63
no-slip boundary condition, 85
noise
1/f, 52
eddy, 47
red, 48
white, 48
non-adiabatic
particle compression, 224
non-adiabatic electron, 108
non-adiabatic ion response, 230
non-adiabatic particle compression, 224
non-ambipolar
current, 189
diffusion, 191
flux, 191
particle flux, 198
radial current, 183, 191
radial electric field, 191
non-axisymmetric, 11, 173, 197
geometry, 197
magnetic field, 253
non-axisymmetric effects, 197
non-cancellation, 123
non-chaotic flow, 126
non-disruptive, 248
non-integral measure, 121
non-local interaction, 68–70
non-Maxwellian, 226
non-resonant, 97, 103
fast-ion response, 231, 234
particle response, 224
wave-particle interaction, 220
non-rotating, 18
non-stochastic forcing, 32
non-universal behavior in MHD, 64
non-universality, 65
non-zero helicity, 126
nondissipating wave, 27
nonlinear
advection, 47, 50
coherent structure, 15
coupling, 45, 47, 90
dissipation, 96
dynamic, 1, 263, 264
dynamical system, 118
dynamics, 51, 70
equilibration, 132, 134
evolution, 132
fast dynamo, 132
growth, 95, 101, 106, 109
growth rate, 99
instability, 84, 85, 110
interaction, 58, 68, 69
mean field, 134
MHD, 260, 281
mixing, 15
process, 91
quadratic, 47
regime, 118
region, 2
Reynolds stress, 191
saturated regime, 133
saturation, 11, 13, 82, 256, 280
scattering, 91
simulation, 247, 258, 261
stability, 89
structure, 100, 101
tearing mode, 248, 250, 280, 281
time, 47, 48, 51
transfer, 58, 61
trnsfer, 58
viscosity, 207
wave, 4, 82
Nonlinear Fast Dynamos, 131
nonlinear Landau damping, 91, 96
nonlinearity
advective, 5, 13, 20
Reynolds stress, 188
nonlinearity, 12

Obukhov, *see*
Charney-Obukhov-Hasegawa-Mima
ocean, *see* atmosphere-ocean
ordering, 99, 226, 227, 229, 230
frequency, 220
low β, 227
wavelength, 229, 231
ozone hole, 4, 13

parallel
conductivity, 231
current, 183
electric field, 223
flow, 187, 188
flux, 192
heat flux, 197
helicity, 54
ion viscosity, 211
momentum
balance, 187, 197
nonlinear viscosity, 207
plasma pressure, 226
viscosity, 182, 184–186, 189, 190, 192, 197, 205–207, 210
viscosity force, 183
viscosity gradient, 184
parallel momentum balance, 185
parallel Ohm's law, 264, 273, 275, 279
parallel Ohm's law, 272
parallel viscosity, 210
paramagnetic pinch, 256, 260, 262, 264
passing
particle, 179, 194, 196, 201
trapped-passing boundary, 188, 196
passive tracer, 63
pattern, 16
Roberts flow, 129
peeling mode, 145
perpendicular
diffusion, 193
plasma pressure, 226
pressure balance, 225
tensor, 193
viscosity, 192, 194, 205, 208–210
Perpendicular Ampère's law, 225
perpendicular Ampère's law, 221
Pfirsch-Schlüter coefficient, 180
Pfirsch-Schlüter diffusion, 180
Pfirsch-Schlüter regime, 179, 180, 192, 193, 202
Phase Space, 81
phase space, 1, 63, 83, 97, 100, 104, 110, 201
blob, 106
density, 93
density hole, 83
depression, 107
dynamic, 81
granulation, 85, 109
integration, 229
structure, 82, 83, 85, 93, 95, 97–102, 104, 106, 107
trapped cone, 200, 201
vortex, 98
Phase Space Structure Dynamics, 93, 101
phasetrophy density, 103
pinch
current driven, 252
equipartition, 199
ohmic, 274
paramagnetic, 256, 260, 262, 264, 280
parameter, 262, 263
ultimate, 274, 276–278
pinch, ultimate, 275
plasma
burning, 219, 220
collisionless, 206, 241
inertia, 186, 189, 224
polarization, 180
poloidal rotation, 182, 188, 189
potential, 189
rotation, 173, 174, 191, 198, 251
thermonuclear, 251, 252
turbulence, 174
plasma dispersion relation, 238
plasma poloidal rotation, 188
plasma sheet turbulence, 46
Plasma viscosity tensor, 205
plateau, 100

banana-plateau boundary, 179
coefficient, 180
diffusion, 179
regime, 179, 181, 185, 192, 202, 210, 211
plume, convective, 72
Poisson equation, 93, 102, 105, 109, 174
polarization, 105
charge, 102, 105, 106
current, 184, 224
neoclassical, 186
plasma, 180
polarization flux, 105
poloidal
electric field, 173
flow, 186
viscosity, 186
poloidal flow, 191
damping, 186, 189–191, 197
poloidal rotation, 155, 182, 183, 185, 188, 189, 191, 197, 198
damping, 183, 185, 186, 188, 191, 198
poloidal velocity
damping, 182, 185, 186
relaxation, 198
potential energy, 232, 234
potential enstrophy, 26, 91–93, 103
flux, 92
potential temperature, 7
potential vorticity, 5–8, 11, 12, 18, 20, 25, 29, 32, 37, 38, 85, 93
anomaly, 6, 10, 11, 15, 16, 19, 32, 33
barrier, 3
blob, 84, 87
conservation, 5, 85, 87, 88, 90, 93
contour, 15–18, 31, 34–36
diffusivity, 90
distribution, 12, 14
field, 5, 6, 9, 12, 13, 15, 37, 38
atmospheric, 13
flux, 25, 28, 88, 90
forward cascade, 90
gradient, 3, 11–14, 16–18, 23, 31, 33, 87, 90
inversion, 6, 9, 16, 19, 28, 31
invertibility principle, 6, 8
jump, 30, 36
map, 13, 14, 33, 167
mixing, 12, 13, 17, 18, 21, 23, 25, 28, 31–36, 90, 91
inhomogeneous, 12, 15, 36, 37
passive, 37
perturbation, 91
slug, 89
staircase, 7, 27, 29, 30
structure, 82
transport, 90, 91
Potential Vorticity, 85
Potential Vorticity Dynamics, 5, 87
potential vorticity dynamics, 7–10, 87, 93
Potential Vorticity Staircases, 29
power-law, 65
index, 52
scaling, 58
spectrum, 69
Prandtl number, magnetic, 50, 54, 64, 71, 262
precession
bounce
toroidal, 227
drift, 102
energetic ion, 229
frequency, 228
precession-bounce resonance, 229
pressure balance, 225, 230
pseudoenergy, 27
pseudomomentum, 24–28, 91–93, 97, 99, 100, 103, 106
conservation, 27, 92, 101
Eulerian, 29
kinetic, 103
Lagrangian, 29
pseudomomentum, 24
PV, *see* potential vorticity

QSH, *see* quasi single helicity

quasi single helicity, 260, 266–268
  equilibrium, 270
  phase, 260
  regime, 268
  state, 260, 268–270
Quasi single helicity, 266, 267
quasi-Alfvénization, 70–72
quasi-coherent phase space, 102
quasi-elasticity, 2, 16–18
quasi-equipartition, 46, 65, 70
quasi-geostrophic, 20, 88, 92, 93, 103
  fluid, 87
  model, 19–21, 23, 29
  theory, 35
  turbulence, 93, 102, 103
quasi-particle, 92, 93, 104
quasi-particle field, 92
quasi-singular structure, 46, 48
quasi-static, 47
quasi-stochastic forcing, 33
quasi-symmetric, 197
quasi-zonal, 2
quasilinear
  diffusion, 98, 110
  process, 91
  theory, 81, 82, 90, 109
quasimomentum, 26
quasineutrality, 22, 174, 175, 183, 185, 187, 190, 192, 221, 223–225, 229, 230
Quasineutrality condition, 174, 225

radial
  build up, 253
  current, 180, 183, 189
    balance, 189
  diffusion, 183
  electric current, 196, 280
  electric field, 173, 177, 178, 180–183, 185, 186, 189, 191, 194, 196–198, 220
  envelope, 148–150
  equilibrium current, 184
  ion current, 190, 191
  magnetic field, 249, 254, 267, 277
  mode structure, 147, 149, 228, 235
  momentum balance, 177
  particle diffusion, 198
  particle flux, 180, 181
  particle transport, 189
  pinch velocity, 264
  singular structure, 233
  structure, 235
  two-scale structure, 232
radiating layer, 251
radiation stress, 17
radiative transfer, 62
random
  displacement, 175
  motion, 194
  process, 180
  velocity, 203, 204
random walk, 18, 37, 175, 176, 179
  diffusion, 179
rate
  damping, 190, 191
  stretching, 125
rational surface, 138, 140, 141, 144, 146–149, 151–153, 162
Rayleigh
  inflection point, 83, 86, 87
  stability criterion, 81
Rayleigh Stability Criterion, 85
Rayleigh-Kuo-Charney-Stern
  marginal stability, 27
  marginally stable, 31
  shear stability, 27
  stability criterion, 33
  stability theorem, 28
  stable state, 29
reconnection, 51, 66, 68, 121
recurrence of events, 69
reduced MHD, 64, 67, 68
  vorticity equation, 223
reflection point, 200
reflectional symmetry, 126
regulation, drift wave turbulence, 188
relaxation, 82, 101, 109, 110
  event, 83, 265
  MHD, 249, 258
  poloidal velocity, 198

rate, 95
Taylor theory, 266, 271, 275
time, 99
violent, 93
resilient
chaos, 261
disruption, 256
resistive
g-mode, 281
instability, 12
kink, 249, 258, 262, 267, 280
MHD, 271
wall mode, 249, 254, 268, 281
resistivity
anomalous, 81, 95, 96
collisional, 96
resonance
broadening, 98
Landau, 90, 92
wave, 96
resonant magnetic perturbation
control, 173
resonant mode, 269
resonant mode absorption, 233
resonant particle, 98, 100, 104
resonant triad interaction, 12
restoring force, 89, 223
reversal
parameter, 262, 265, 274
RFP magnetic field, 256, 259, 262, 265, 275, 277, 280
Reversed Field Pinch, 247
reversed field pinch, 134, 247, 248, 251, 277
configuration, 253, 280
dynamo, 258, 279
Lawson criterion, 255
non-axisymmetric, 259
reactor, 252, 255
self-organization, 254, 260
Reynolds force, 88
Reynolds number, 46, 48, 50, 51, 53, 61, 62, 65, 68, 70, 71, 73, 118, 133, 134
asymptotic, 67
kinetic, 50
magnetic, 47, 50, 116, 133, 134
Reynolds stress, 27, 109, 188, 192, 199, 204, 208
nonlinear, 191
RFP, *see* Reversed Field Pinch
Rhines scale, 30, 34
ripple
tokamak, 198
ripple, tokamak, 197
RMP, *see* resonant magnetic perturbation
Roberts dynamo, 126
Roberts flow, 126, 127, 129
roll-up, 46, 61, 72
Rossby
length, 21
wave, 6, 7, 14–19, 26, 27, 36, 90, 91
wave breaking, 17, 19, 24, 33, 37
wave dispersion, 35
waveguide, 17, 33
Rossby number, 20, 23, 52
Rossby wave, 15
rotating, 4, 6, 7, 9
Earth, 5
flow, 7, 59, 61, 69
fluid, 47
system, 4
turbulence, 52, 59, 65
rotating fluid
dynamo, 71
rotation, 65
atmosphere, 31
ballooning mode, 158
damping, 183, 198
equilibrium, 183
GAM, 198
helical, 134
intrinsic, 108, 110
island, 187
liquid sodium, 252
neoclassical, 179
plasma, 173, 174, 198, 251
poloidal, 154, 155, 173, 182, 183, 185, 188, 189, 191, 197, 198
damping, 183, 185, 186, 188, 191, 198

rigid body, 152
time, 82
tokamak, 173, 191, 198
toroidal, 152, 183, 191, 193–195, 197, 198
transport, 220
uniform, 54
rotational discontinuity, 55

$s - \alpha$ model, 236, 239
saddle-node bifurcation, 261
safety factor, 138, 178, 248, 249, 260, 261, 264, 274, 277
saturated
dynamo, 133
turbulence, 82
saturation
nonlinear, 11, 13, 82, 133
SAW, *see* Shear Alfven Wave
scale separation, 2, 4
scaling
energy, 65
exponent, 62, 72
index, 62
power law, 58
self-similar, 48
scaling law, 46, 49, 61, 73
scattering, stochastic, 91
second order phase transition, 266
self-acceleration, 88
self-advected nonlinearity, 50
self-advection, 66
self-binding, 107
self-coherence, 98, 109
self-consistent dynamo, 116, 134
self-excitation, 6, 12, 20, 23, 27, 32, 33
self-gravitating, 104
self-gravity, 46, 104
self-organization, 91, 247–250, 252, 254, 256, 260, 264, 269, 280, 281
self-organized, 2
self-organized criticality, 62
self-reversal, 256
self-similar, 48, 62
self-sustained structure, 17
self-trapping, 98
self-wrapping-up, 14
sensitivity to initial conditions, 188, 191
separation, scale, 2, 4
SH, *see* single helicity
Shafranov shift, 142, 239
shallow-water, 20–23
shear
instability, 12, 27
magnetic, 148, 156, 232, 235, 236, 239, 256, 270
stability, 27
zonal, 108
Shear Alfvén
continuous spectrum, 232, 239
gap, 239
mode, 232, 233
resonance, 232
spectrum, 220, 228, 229, 231
Shear Alfvén Wave, 220, 231, 232, 241
accumulation point, 234
continuous spectrum, 233
continuum, 233
spectrum, 231
shear flow, 85, 86, 92, 152, 154, 155, 157, 158, 160–163
large-scale, 134
toroidal, 152, 162
shear layer, 73, 89
sheared-disturbance, Kelvin, 35
short mean free path regime, 191
simulation, 53, 54, 60–63, 65–67, 70
direct, 47, 55
direct numerical, 48
dynamo, 133
fast dynamo, 118
single helicity, 263–266, 271, 279
bifurcation, 281
equilibrium, 264, 271
field, 264
ohmic state, 274, 277, 278, 280
perturbation, 258
plasma, 280
state, 258, 260–267, 271, 280
transient, 266
single helicity, 262

Single helicity ohmic state, 275
single helicity RFP, 271
singular
  inertial region, 232
  layer, 239
  structure, 232, 233
singularities in a finite time, 55
skewness, 65
slab geometry, 140, 141
  periodic, 140
slow dynamo, 116, 126, 130
slowing down time, 227
small-scale dynamo, 134
solar
  dynamo, 132
  flare, 45, 49, 51, 62
    weak, 62
  photosphere, 62
  surface convection, 133
  wind, 45, 46, 49, 53, 54, 59, 60, 62, 66–68, 72
specific entropy, 7, 10
spectral exponent, MHD turbulence, 49
spectral index, 56, 61
  energy, 64
  MHD turbulence, 49
spectrum
  Alfvén, 220
  continuous, 232, 233, 239, 241
  correlation, 54
  energy, 51–57, 60–65, 68, 69, 71, 72, 168
  equipartition, 51, 57
  Fourier, 52, 57
  helicity, 52, 54, 71
  Iroshnikov-Kraichnan, 59, 61, 64, 67
  Kolmogorov, 51, 59, 60, 63, 64
  Shear Alfvén, 220, 228, 229, 231
  Shear Alfvén Wave, 231
spreading, 2, 92, 103
  turbulence, 103, 107
stable continuum, 163
  ideal MHD, 161
stably-stratified, 9
  atmospheric model, 9
Staircases, 29
staircase, 30–33, 36
  potential vorticity, 27, 29, 30
  structure, 31
statistical closure, 110
  two-point, 54
stellar convective zone, 115
stellarator, 2, 197, 250–255, 275, 280, 282
  pinch equation, 274, 275
Stern, *see*
  Rayleigh-Kuo-Charney-Stern
stochastic, 32
  forcing, 92
  scattering, 91
stochasticity of streamline, 91
stochastization of flow, 90
strange attractor, 47
stratification, 8, 18, 21, 23, 62
  atmosphere, 4
  density, 1
  field, 8
stratification surface, 7–15, 18, 25, 31
stratified, 18, 20
  atmosphere, 15
  flow, 7, 59
  model, 5
stress
  form, 25, 26
  Maxwell, 192, 196
  radiation, 17
  Reynolds, 27, 109, 192, 199, 204, 208
    nonlinear, 188, 191
stress tensor, 70, 206
stretch-twist fold, 117
stretching
  amplification, 122
  cancellation, 123
  field, 121
  flow, 125, 133, 134
  magnetic field, 125
  non-uniform, 122
  vortex, 7, 21
structure, 96

coherent, 13, 15, 45, 49
diffusive, 128
dissipative, 46, 48
elastic, 17
filamentary, 14, 19
fine-scale, 130
generic, 4
helical, 269, 270
intermittent, 46
jet, 5, 10, 11
localized, 48, 61, 81, 82, 93
mode, 228, 237
nonlinear, 101
quasi-singular, 46
real space, 81
single helical axis, 271
singular, 232, 233
staircase, 31
vortex, 82
wave, 236
wrapping-up, 14
structure function, 48, 61–64
subsonic flow, 60
supercell storm, 69
supergranule, solar surface, 133
supersonic, 46, 60
turbulence, 62
sweeping oscillation, 240
Sweet Parker current sheet, 265
symmetrical flow, 128
symmetry
axial, 177, 178, 275
axis, 138, 139, 142
ballooning, 147
breaking, 147, 148
cylindrical, 256
reflectional, 126
toroidal, 138, 143, 178, 181, 192, 197
translational, 28

TAE, *see* Toroidal Alfven Eigenmode
Taylor identity, 17, 24–26, 28, 88, 90, 94, 102, 106
Taylor identity, 24
Taylor relaxation theory, 266, 271, 275
Taylor scale
magnetic, 67
Taylor Stability Criterion, 87
Taylor-Green
flow, 56, 64, 70
vortex, 55, 64
Taylor-Proudman, 8
tearing
instability, 276
tearing mode, 267, 278
bifurcation, 263
nonlinear, 248, 250, 280, 281
saturation, 258, 278
tearing mode dynamo, 258
thermodynamic, 7, 8
time-dependent eikonal, 155
tokamak, 2–4, 6, 7, 9, 12, 15, 16, 20–24, 32, 37, 247–251, 253–255
aspect ratio, 239
axisymmetry, 192
collisional transport, 178
disruptive, 252
neoclassical theory, 174
performance, 251
ripple, 197, 198
transport, 174
trapped particles, 179
topological entropy, 125, 126
toroidal
flow, 138, 152, 155, 162, 186, 191
geometry, 178–180, 186, 191, 194, 197
rotation, 152
symmetry, 138, 143, 181, 192, 197
Toroidal Alfvén Eigenmode, 100, 229, 237, 239, 240
Toroidal Alfvén Eigenmodes, 239
toroidal momentum, 191, 192, 199
balance, 181, 192–197
conservation, 192
diffusion, 194, 196–198
diffusivity, 198, 199
transport, 181, 194, 199, 209
toroidal momentum, 191

toroidal momentum diffusion, 194
toroidal plasma, 180
toroidal precession frequency, 227
toroidal rotation, 183, 191, 193, 194, 197, 198
toroidally symmetric, 197
toy model, 256
toy model, 33
tracer, passive, 63
transient state, 247, 266
transit frequency, 227
transport
  advective, 3
  alpha particle, 254
  chaotic, 265
  collisional, 179
  fast ion, 220
  heat, 264, 267, 281
  ion, 190
  ion heat, 199
  momentum, 17, 173, 178, 194, 199, 209, 220
  neoclassical, 174, 191
  particle, 173, 178, 189–191
  potential vorticity, 90, 91
  rotation, 220
  toroidal momentum, 181
  turbulent, 199
  vorticity, 88
transport barrier, 3, 251, 265
  external, 269
trapped
  electron, 102, 106, 228
  energetic ion, 240
  ion, 98, 102, 106
  particle, 227, 228, 238, 242
  thermal particle, 241
trapped particle, 179, 194, 198, 200, 201, 211
trapped-passing boundary, 188, 196
trapping
  particle, 179
  poloidal, 198
  toroidal, 198
triadic interaction, 47
turbulence, 32, 33, 36
  astrophysical, 50
  beta-turbulence, 32
  closure, 54, 69
  drift, 85, 101
  drift wave, 15, 37, 82, 92, 93, 104, 106
  excitation, 92
  fluid, 49, 61, 63, 69, 72
  geomagnetic, 46
  gyrokinetic, 81, 85
  Hasegawa-Mima, 104
  helical, 68
  homogeneous, 2, 4, 13, 30, 35, 73
  inhomogeneous, 18
  interstellar, 72
  Kolmogorov, 48
  magnetic, 247, 266, 281
  MHD, 49, 50, 53, 57–62, 64, 66–69, 71–73
  Navier-Stokes, 63
  packet, 67
  plasma sheet, 46
  quasi-geostrophic, 93, 102, 103
  regulation, 188
  rotating, 52, 59, 65
  saturated, 82
  spreading, 92, 103, 107
  Vlasov, 93, 99, 103
  weak, 51, 58, 59, 65, 67, 91
turbulent
  cascade, 66
  convection, 133
  eddy, 46
  flow, 46, 47, 51, 52, 61, 73, 115, 133, 252
  fluid, 47, 69
  mixing, 12
  momentum, 199
  motion, 133
  negative
    diffusivity, 47
  pinch, 199
  state, 82
  transport, 3, 199
  wake, 118
  zone, 28

two-point
correlation function, 49
two-point closure, 70
two-point statistical closure, 54
two-scale structure, 235
mode, 232
radial, 232

ultimate pinch, 274–278
universality, 65
breaking, 63, 65
MHD, 50
unstratified, 8
unstratified fluid, 4

viscosity, 186, 192, 204, 205, 207, 208, 210
coefficient, 187, 188, 194, 211
component, 191
eddy, 47, 51
force, 183
gradient, 184
hyper, 70
kinematic, 49, 261
moment, 203
neoclassical, 187
nonlinear, 207
oblique, 208
parallel, 182, 184–186, 189, 190, 192, 197, 205, 207, 210, 211
perpendicular, 192, 194, 205, 208–210
plasma, 178, 202
poloidal, 186
tensor, 196, 202, 205, 206
viscous
current, 184
dissipation, 107
force, 173, 185, 186
VKS, *see* von Karman Sodium
Vlasov
equation, 99
gyrokinetic, 107
plasma structure, 94
structure, 95
turbulence, 93, 99, 103
void, 83
von Karman geometry, 252
von Karman Sodium experiment, 252
vortex, 82, 89, 98
coherent, 15, 46, 83
cyclonic, 11
dynamic, 13, 24
horse-shoe, 72
interaction, 19
interchange, 89
polar, 4
stretching, 7, 21
structure, 82
Taylor-Green, 55, 64
tilting, 7
wing tip, 16
vortex filament, 46, 48, 61, 69, 72
vorticity, 221, 229, 230
equation, 222, 223, 230, 235–237
reduced MHD, 223
flux, 90
potential, 90
transport, 88
Vorticity Equation, 222
Vorticity Equation, 223

Wakatani, *see* Hasegawa-Wakatania
wave
absorption, 91, 92
action, 103
Alfvén, 50, 57, 59–61, 231
breaking, 17, 19, 28
counter-propagating, 58
drift, 12, 15, 16, 23, 26–28, 37, 90, 92, 93, 188
elasticity, 18
energy, 27
fast magneto-acoustic, 228
gravity, 18, 21, 22, 24
helically polarized, 68
Langmuir, 46
linear, 15
mean interaction, 17
momentum, 92, 96, 97, 103
nondissipating, 27
nonlinear, 4, 82

packet, 46, 92, 98
pseudomomentum, 92
resonance, 96
Rossby, 6, 7, 14–19, 27, 36, 90, 91
Shear Alfvén Wave, 220, 231–234, 241
standing, 61
zonal flow
interaction, 82
Wave–Flow Interaction, 89
wave-mean flow interaction, 84, 91
wave-turbulence, 4, 17, 25, 30, 33, 34
inhomogeneous, 2, 15, 35
wave-turbulence jigsaw, 1
wavelet, 49
weak
collision, 186, 190, 195, 201
damping, 190
flares, 62
weak collision regime, 193
weak MHD theory, 59
weak MHD turbulence, 58, 59
weak MHD turbulence, 57
weak turbulence, 4, 51, 58, 59, 64, 65, 67, 91
break-down, 59
spectrum, 58, 61, 64
weakly turbulent regime, 58
width
banana, 179, 200, 229
drift, 229
WKB
expansion, 149
form, 162
wrapping-up structure, 14
WT, *see* weak turbulence

Zeman scale, 59
zonal, 2, 35
acceleration, 90
average, 21, 37
contour, 28, 29
jet, 3, 30, 38
mean, 25, 26, 29
mean flow, 22
mean kinematic, 29
mean state, 25
momentum, 92
region, 31
shear, 108
wave, 91
zonal avearage, 88
zonal flow, 3, 28, 32, 37, 38, 81, 82, 84, 85, 88, 90–92, 101, 103, 104, 106–110
drift, 85
shear, 109
wave interaction, 82
Zonal Flow Evolution, 89
Zonal Flow Momentum, 85
zonally symmetric, 30, 31, 38